AF449362

50 years Progress in Crystal Growth

A REPRINT COLLECTION

Key to illustrations on front cover[*]

<table>
<tr><td>A</td><td>B</td><td>C</td></tr>
<tr><td>D</td><td>E</td><td>F</td></tr>
<tr><td>G</td><td>H</td><td>I</td></tr>
</table>

Plate A: A 5″ diameter single crystal silicon boule being pulled from a melt by the Czochralski method. *From the front cover of the AACG Newsletter, Vol. 13, No. 2, July 1983. (Photo courtesy Siltec Corp.)*

Plate B: One of the entries from J.A. Burton's Bell laboratory's notebook (1951) leading to the development of the BPS theory. Plot shows solute concentration vs. distance in the solid and melt during unidirectional solidification. The $x = 0$ position is the growth interface at some time t, while $x = l$ defines the boundary layer. *From the front cover of the AACG Newsletter, Vol. 13, No. 3, November 1983. (Courtesy J.A. Burton and W.P. Slichter.)*

Plate C: A 50 mm diameter $\langle 100 \rangle$ dislocation-free silicon crystal growing by the float-zone method at NREL. *From the front cover of the AACG Newsletter, Vol. 23, No. 2, Autumn 1993. (Courtesy Ted Ciszek.)*

Plate D: A succinonitrile dendrite growing into an undercooled melt. The dendrite is about 7.5 mm long. *(Courtesy Ken Jackson.)*

Plate E: The bottom side of a Czochralski grown lithium niobate crystal which was rapidly heated to cause it to separate from the melt surface. The resulting structure reveals the internal 3-fold symmetry of this rhombohedral crystal. *From the front cover of the AACG Newsletter, Vol. 18, No. 3, November 1988 (Courtesy R. Feigelson, Stanford University.)*

Plate F: Crystals of copper indium diselenide, a solar cell material, grown by the chemical vapor transport method at SERI. The pyramid face is normal is the $\langle 112 \rangle$. *From the front cover of the AACG Newsletter, Vol. 19, No. 3, November 1989. (Courtesy Ted Cizsek.)*

Plate G: Convection patterns on the surface of a gadolinium gallium garnet (GGG) melt during Czochralski growth at the Hitachi Chemical Co. Ltd. *From the AACG Newsletter, Vol. 13, No. 3, November 1983. (Courtesy Mitsuru Ishii.)*

Plate H: Scanning electron microscope photograph of crystals of the high T_c superconductor $YBa_2Cu_3O_7$, and some $BaCuO_2$ crystals grown from solution at the IBM Research Laboratories, Yorktown Heights, New York. The large crystal the top right is $BaCuO_2$ as is the small cube near the center $(440\times)$. *From the front cover of the AACG Newsletter, Vol. 18, No. 2, July 1988. (Courtesy F. Gayle and D. Kaiser.)*

Plate I: Lithium niobate pyramidal crystallites lined up along terrace ledges on a heat-treated sapphire substrate. The crystallites were deposited at Stanford University by the solid source MOCVD method. *(Courtesy R. Feigelson and Sang Yun Lee.)*

---

[*] See p. 240 for key to illustrations on back cover.

50 years Progress in Crystal Growth

A REPRINT COLLECTION

Edited by

R.S. Feigelson

Geballe Laboratory for Advanced Materials
Stanford University, Stanford CA, USA

2004

ELSEVIER

Amsterdam – Boston – Heidelberg – London – New York – Oxford
Paris – San Diego – San Francisco – Singapore – Sydney – Tokyo

ELSEVIER B.V.
Sara Burgerhartstraat 25
P.O. Box 211, 1000 AE
Amsterdam, The Netherlands

ELSEVIER Inc.
525 B Street, Suite 1900
San Diego, CA 92101-4495
USA

ELSEVIER Ltd
The Boulevard, Langford Lane
Kidlington, Oxford OX5 1GB
UK

ELSEVIER Ltd
84 Theobalds Road
London WC1X 8RR
UK

First edition 2004

ISBN: 0 444 51650-6

∞ The paper used in this publication meets the requirements of ANSI/NISO Z39.48-1992 (Permanence of Paper).
Printed in The Netherlands.

1931–2001

This book is dedicated to the late Professor August (Gus) F. Witt of MIT. During the past 40 years his laboratory has produced many important contributions to the field of crystal growth and electronic materials processing. His insightful and meticulous research experiments have set high standards for crystal growers worldwide. Some of his major research results are reviewed in one of the papers presented in this publication. The crystal growth community will miss his leadership, scientific contributions and friendship.

Contents

Melt and Solution Growth

High Pressure Growth

Thin Film Growth: MBE

Thin Film Growth: MOCVD (OMVPE)

Preface

This book contains a collection of articles that review many of the major advances made in crystal growth science and technology since the end of World War II. It contains material from two sources: (1) lectures presented at a Symposium entitled "50 Years of Progress in Crystal Growth" [1], and (2) selected articles which were published over a number of years in the Newsletter of the American Association for Crystal Growth (AACG). Most of the newsletter articles were part of a series entitled "Milestones in Crystal Growth".

The Symposium was organized in conjunction with the American Association for Crystal Growth's Fourteenth American Conference on Crystal Growth & Epitaxy (ACCGE-14) held in Seattle, Washington on August 4–9, 2002. It was planned both as a celebration of the dramatic advances made in the fields of theoretical and experimental crystal growth during this time period, and also to provide an opportunity for young scientists and engineers to attend lectures given by a select group of still active crystal growth pioneers. It was a very popular Symposium, and, at the request of many attendees, the lectures were written-up and the Proceedings published in the Journal of Crystal Growth [1]. It occurred to me during the early stages of editing this journal publication that the subject matter might be of interest to a broader community. As a result, a separate book with wider distribution seemed appropriate. Since the published lectures did not cover the field as comprehensively as I would have liked (due principally to time constraints in the conference format), I decided to include a series of related articles from the AACG Newsletter written between 1983 and

1993. This series, begun by Dennis Elwell and myself during our newsletter co-editorship (1980–1985), contained some very interesting personal reflections by several prominent crystal growth pioneers. They revealed how their research results came about and some of the excitement generated by these discoveries. In addition, I decided to include, as background material, some concise, well-written newsletter articles covering crystal growth history prior to World War II.

While this book focuses mainly on the remarkable progress made in the field of crystal growth during the last 50 years, crystals have been of great interest to mankind for thousands of years. This has been largely due to the great beauty, symmetry and utility of naturally occurring minerals. It is not surprising, therefore, that attempts to grow single crystals also began back in ancient times. The first part of this book covers some of the early foundations of crystal growth. J. Bohm, author of the first article, gives us a concise history of the crystal growth field, starting with some of the earliest recorded experiments. An article by Kurt Nassau's on Verneuil and his method, and then Leon Merker's reminiscences on the flame fusion method follow this. To round off the historical section, I have included a newsletter article written by David Bliss on Kyropoulos and his growth method.

A useful summary of the most of the important contributions made in the field of crystal growth during the first half of the 20th century is given in H.E. Buckley's 1951 book entitled "Crystal Growth" [2]. This was probably the first book written on this subject (in English at least). Several years before, Buckley lectured at what may have been the first conference devoted exclusively to crystal growth. It was held in England in 1949 and was sponsored by the Faraday Society. The Principal topics included (1) Theory of Crystal Growth, (2) Nucleation and Normal Growth, (3) Abnormal and Modified Growth and (4) Mineral Synthesis and Technical Aspects. Sir N.F. Mott gave the introductory lecture. A short summary of this meeting was given by Elwell and Feigelson [3].

Despite the progress made prior to 1950 (and even during the next 15 years), the field of crystal growth (fundamental studies and experimental research, development and production) was fragmented. Part of the problem is related to the interdisciplinary nature of crystal growth, which involves, separately or in combination, various aspects of physics, chemistry, crystallography, mathematics, mechanics, fluid dynamics, chemical engineering and biology, etc. It was also complicated by the fact that crystals can be grown from most classes of materials, including elements and simple or complex compounds (metallic, semiconductor, inorganic, organic and biological). As a result crystal growth scientists and engineers were scattered amongst a variety of communities and, therefore, were essentially isolated from one another. They also had some difficulties in publishing papers on this topic in many journals, at least without reference to physical or chemical measurements.

During the 1950s, academic and industrial crystal growth research started to transform and undergo a rapid expansion. This can be closely linked to the invention of the transistor at AT&T's Bell Laboratories in Murray Hill, New Jersey in the late 1940s. More specific to crystal growth, it dates to the pioneering work of G. Teal and W. Little on the Czochralski growth of germanium and shortly afterward silicon single crystals. These technological achievements in turn, led to the electronics revolution and a rapidly expanding solid state electronics and photonics industry that were subsequently able to invest heavily into materials research, including strong efforts in crystal growth theory and practice.

The field continued to expand (worldwide) throughout the 1950s and 1960s, both in terms of the scope and the number of research workers and laboratories involved. As a result, crystal growers from all the specialties started to realize that they needed to communicate with one another to discuss problems of mutual interest. A committee chaired by D. Turnbull (General Electric), organized a second conference on crystal growth in 1958. Other committee members included B. Chalmers (Harvard University), N. Cabrera (University of Virginia), P.J. Flory (Mellon Institute), and D.A. Vermilyea (General Electric). This meeting was called the "International Conference on Crystal Growth: Growth and Perfection of Crystals" and was held in the small town of Cooperstown, New York. There were about 63 participants, including, Ken Jackson, one of the speakers in this symposium. Sir Charles Frank gave the introductory lecture.

Eventually all this activity led to the organization of the first in a series of comprehensive conferences devoted entirely to the subject of crystal growth. The so-called "First International Conference on Crystal Growth" (ICCG-1), actually the third, was held in Boston in 1966. Hurle [4], in his editorial for the 35th Anniversary of the Journal of Crystal Growth, states correctly that "It (the meeting and the formation of the international organization for crystal growth) came about principally as the result of the vision and determination of one man—Michael Schieber". The relationship between ICCG-1 and the formation of the International Organization for Crystal Growth (IOCG) and the Journal of Crystal Growth (JCG) was later described by Schieber [5] in an article in the AACG Newsletter. One of the motivating forces for starting the Journal of Crystal Growth (the first volume dated January 1967) was related to problems associated with publishing the proceedings of ICCG-1. The first JCG Editors were N. Cabrera, B. Chalmers and M. Schieber (Principal Editor).

The Organizing and Program Committees for ICCG-1 were made up of prominent members of the American crystal growth community. M. Schieber and B. Chalmers chaired them, respectively. These committee members formed the nucleus of an ad hoc American Committee for Crystal Growth, which a few years latter evolved into the American Association for Crystal Growth (AACG) under the co-chairmanship of R. Laudise and K. Jackson. Around this same time period, a number of other National Societies were formed. All of these societies joined the IOCG and participated in their activities. The AACG organized their first conference (ACCG-1) in 1969 at the National Bureau of Standards (now NIST) in Maryland. Professor Witt from MIT was the first speaker at

ACCG-1, and he lectured on the "Microdistribution of Impurities in Semiconductor Single Crystals".

It should be stated that although crystal growth covers many areas, including synthetic gems, biological materials and pharmaceuticals, the dominant focus of these conferences and organizations was essentially on crystals for new and improved devices, as well as improving our understanding of crystal growth as a science.

The ICCG-1 contained presentations on a wide variety of topics within three general areas. The first concerned the fundamental aspects of nucleation and growth, including kinetics, thermodynamics, morphological stability, surfaces, segregation and convection. The second dealt with the preparation of single crystals from melts, solutions (aqueous, flux & hydrothermal), and the vapor phase (both bulk crystals and epitaxial thin film deposition, a relatively new technology). This topic included discussions on the synthesis of new compounds and the development of new crystal growth methods. The third area was crystal characterization. The introductory speaker at the conference was Professor Bruce Chalmers who lectured on the topic "Theoretical Problems in Crystal Growth". While a lecture by Prof. Alex Chernov on the "Crystallization of Binary Systems as a Random Walk Problem" was scheduled to follow, he was, in the end, unable to attend. His paper, however, was included in the ICCG-1 Proceedings, and may represent the first computer simulation work applied to crystal growth.

While in 1950 crystal growth was more an art than science, today we have a profoundly improved understanding of the fundamental mechanisms involved in the some of the most complex crystallization processes. Concepts such as constitutional supercooling, morphological stability, component and impurity segregation, convective effects, etc. have helped us to better interpret the results of growth experiments and have led to significant improvements in crystal size and quality. Well known crystal growth methods such as Czochralski, Bridgman, Stockbarger, vapor transport, flux and aqueous solution growth, etc., have been refined and/or adapted to new materials problems. In addition, new methods have been developed to solve important growth problems that conventional techniques cannot handle. These methods include zone melting, the heat exchanger method (HEM), edge-defined film-fed growth (EFG), skull melting, accelerated crucible rotation (ACRT), molecular beam and metalorganic vapor phase epitaxy (MBE and MOVPE, respectively), liquid encapsulated Czochralski (LEC) growth, laser heated pedestal fiber growth (LHPG), high flow rate solution growth and growth under extreme pressure. One should also mention the use of magnetic fields and growth in microgravity to suppress buoyancy-driven convection.

Mathematical and experimental modeling of crystal growth processes became very important new tools in understanding and predicting the results of many bulk and thin film crystal growth processes. We learned to appreciate the role of fluid dynamics in mass and heat transport in melt and vapor growth processes, its influence on crystal perfection, and how to control it to achieve specific results. We learned new ways to purify materials and much more about how impurities and dopants affect the growth process, as well as the properties of the materials of interest. Many studies have been supported both by NASA and space programs in other countries. These have been designed to elucidate the influence of gravity on crystal growth processes, and in particular, to suppress buoyancy-driven melt convection. Although most experiments have been carried out by mission specialists, several members of the crystal growth community (Drs. Ludwig Van den Berg, Roger Crouch and Jean-Jacques Favier) have flown on Space Shuttles and performed crystal growth experiments in this low gravity environment.

Materials characterization has always been an integral part of the crystal growth field. Over the last 50 years numerous advances have been made in the (1) design and construction of electronic and chemical analytical equipment, (2) effective use of computer technology in analysis as well as process control, and (3) analysis theory. This progress has made it possible to study materials down to smaller and smaller length scales. Surface and bulk techniques now allow us to measure impurity and dopant concentrations down to the ppb levels and high resolution TEM allows us to visualize actual atomic and molecular structure and lattice perfection. One of the newer tools in our arsenal is atomic force microscopy. This technique can be used to both reveal surface topography down to sub-unit cell dimensions, and also to perform in situ studies on a growing crystal interface (particularly in aqueous solution growth systems near room temperature, for example KDP and many biological materials).

As mentioned above, all types of elements and compounds can be prepared in single crystal form-in principle at least. The number of single crystal materials grown during the last 50 years is truly remarkable. New materials have been discovered in a number of ways, including the popular technique of deriving them from known compounds (ex. $KD_2PO_4 \rightarrow KD_2AsO_4$ and $KTiOPO_4 \rightarrow RbTiOPO_4$, $LiNbO_3 \rightarrow LiTaO_3$, etc.) and from phase equilbria studies of various material systems. While these crystals are most often simple binary and ternary compounds, more complex chemical entities such as the high temperature superconductors (YBCO) and biological materials have also been successfully prepared in single crystal form, as well as a wide array of solid solutions.

The oxides, the largest class of compounds, have held the attention of crystal growers throughout this period. These include the binary oxides such as sapphire, quartz, cubic zirconia and paratellurite, and numerous ternary and higher order compounds such as rare earth garnets ($Y_3Al_5O_{12}$-YAG, $Y_3Fe_3O_{12}$-YIG, $Gd_3Ga_5O_{12}$-GGG), various niobates, tantalates, borates, cuprates, silicates, molybdates and tunstates, and the KH_2PO_4 group to name just a few. Other very important classes of compounds include the chalcogenides (ex. CdS, CdTe, ZnSe, and HgCdTe), pnictides (ex. GaAs, GaInAs, InP, GaSb, $CdGeAs_2$, etc.), nitrides and carbides (GaN, AlN, SiC, etc.), halides (CsI, HgI_2, MnF_2, etc.), biological macromolecules (canavalin, insulin, isocitrate lyase, etc.) and a large number of organic compounds (urea, l-arginine, etc.). Many of these man-made crystals made new technologies possible, enabled well known substances to perform better and/or facilitated the measurement of important property data for a wide range of materials. Since 1950, single crystals of one type or another have found their way into almost every home or village around the globe.

The methods used to grow the above materials are varied and depend largely on the thermodynamic properties of the system of interest. Knowledge of these properties, such as melting temperature, thermal decomposition behavior, vapor pressure, reactivity with the growth atmosphere, container materials or ampoules, etc., allow the crystal grower to choose the most appropriate method(s) for the compound of interest. Stoichiometry, dopants and impurities also play an important role as well as the need to control the segregation of various species, which is both a thermodynamic and kinetic problem. Since crystal growth is essentially a phase transition, the simplest way to categorize a growth method is whether the crystal is growing from a fluid medium (melt or solution), a vapor phase or within a solid. In this context then, probably all these techniques were available prior to 1950. However, a study of the development of crystal growth methodologies during the last half century would show that a variety of sophisticated refinements really account for the major progress in crystal production—look at Si for example.

Many corporate and government laboratories had large efforts in crystal growth research during this time period. In addition to AT&T's Bell laboratories, which not only pioneered materials for semiconductor electronics, but also optical and magnetic materials and devices, there were many other important crystal growth laboratories in the United States and elsewhere. Significant industrial and government research activities were present all over Europe, including Bulgaria, England, France, Germany, Hungary, Italy, the Netherlands, the countries within the former Soviet Union, and Switzerland. The expansion of crystal growth activities in Asia has been truly impressive. There are major (and well-known) industrial and government laboratories in Japan, China and Korea and India. Many new materials and commercial developments have come from these countries. Finally, I have to call attention to the many Universities around the World, which have, through more modest individual efforts, contributed significantly to the advances in crystal growth science and technology.

Several crystal growth societies initiated special awards to recognize important individual contributions to the field. At ICCG-8 (1986) in York, England, the IOCG committee created two awards. These Prizes were to be given out at all subsequent Triennial Conferences. The Prizes were named after the first two IOCG Presidents: Sir Charles Frank and Dr. Robert Laudise. The Frank Prize is given for seminal contributions to the theory of crystal growth, while the Laudise Prize is given for contributions to experimental crystal growth. The first Frank Prize was given in 1989 in Sendi, Japan to Alex Chernov (Russia). This was followed by Prizes to Robert Sekerka (USA) in 1992, Pieter Bennema (The Netherlands) in 1995, Kenneth Jackson (USA) in 1998 and jointly to Donald T.J. Hurle (UK) and Sam R. Coriell (USA) in 2001. The first Laudise Prize went to Jun-Ichi Nishizawa (Japan) in 1989 followed by a joint Award to Viacheslav V. Osiko (Russia) and Joseph Wenckus (USA) in 1992. In 1995 the Prize was presented to Robert S. Feigelson (USA) and Iaam Akasaki (Japan) and Georg Mueller received the Prizes in 1998 and 2001 respectively.

Almost a decade earlier, the American Association for Crystal Growth initiated two Crystal Growth Prizes, an International Crystal Growth Award and a Young Author Award. The first recipient of the International Award (1978) was Sir Charles Frank. This was followed by Robert A. Laudise (1981), Bruce Chalmers (1984), Donald Hurle (1987), Mort Panish and Alfred Cho (1990), Kenneth Jackson (1993), Martin Glicksman (1996), Gerald Stringfellow (1999), David Brandle (2000) and Lynn Boatner (2003). Among the notables given the young author awards were Robert Brown from MIT (1984), and Thomas Keuch, then at the IBM Yorktown Heights Research Center (1987)

The initial idea for organizing the special symposium on 50 Years of Progress in Crystal Growth originated during a mid-morning coffee break at the June 2000 NASA Microgravity Materials Science Conference held in Huntsville Alabama. Rohit Trevedi and I were discussing the unfortunate loss of a number of pioneering crystal growers to retirement over the past few years. About six months later I remembered this conversation and was curious whether a symposium to honor some of our still active pioneers and their contributions to the field of crystal growth might be of interest to the general community. After I formulated a preliminary plan, I sounded out various prominent members of the crystal growth community to gauge their enthusiasm for such a project and to help fine-tune the concept. The response was uniformly positive. The ultimate plan was to invite still active pioneering researchers to present invited lectures on topics which would cover all facets of crystal growth i.e. theoretical developments, experimental and mathematical modeling, and new and refined experimental procedures. Each invited speaker would be asked to discuss the history of their respective fields, their own personal contributions, current activities and where future research opportunities were headed.

I approached the American Association for Crystal Growth with this idea. After a brief discussion at the AACG executive committee meeting, there was a consensus that, rather than holding a separate meeting, it would be helpful to combine this Symposium with their upcoming American Conference on Crystal Growth and Epitaxy (ACCGE-14) being planned for Seattle, Washington in August of 2001. While there would be a number of constraints associated with holding this symposium together with a general meeting, it was felt that this plan would be beneficial to both. Of the several possible venues explored, such as having it before or after AACGE-14, or in parallel with it, the latter was chosen.

Invited speakers for this special symposium were chosen only after extensive discussions with prominent members of the crystal growth community, both in the United States and abroad. Speakers were selected on the basis of their contribution to a specific area of crystal growth research. The speakers had to be both crystal growth pioneers (making significant contributions within the 1950–1980 time period) and still active in the field. We needed to cover a range of topics representative of crystal growth research and the resulting technology. Geographical considerations had taken into account as well. As mentioned above major contributions to the crystal growth field during the past half century have come from many parts of the world. Because of the limited number of speaker slots, however, it was clear that not all active pioneers could be included, nor all the principal countries represented. The 12 speakers finally chosen were A.A. Chernov, C.D. Brandle, M.E. Glicksman, D.T.J. Hurle, B.A. Joyce, J.B. Mullin, K.A. Jackson, R. Brown, R.F. Sekerka, G. Stringfellow, I. Sunagawa and A.F. Witt. All but Professors Brown, Stringfellow and Witt gave presentations at ICCG-1 in 1966. However, Prof. Witt gave the first lecture at ACCG-1 in 1969 and the fields represented by Prof. Brown and Stringfellow (fluid dynamics and MOCVD) were not yet an important part of crystal growth research. Four of the speakers were recipients of the IOCG Frank Prize, four received the AACG International Crystal Growth Award and one the AACG Young Author Award.

The Symposium program encountered some serious problems early on. In the end, a third of my speakers developed some form of health problem, making their attendance doubtful and putting the Symposium in serious jeopardy. By conference time Gus Witt, Don Hurle and Bruce Joyce were not able to attend. The problem was not just a matter of finding new speakers. It was important to the concept of this Symposium that their individual and very important contributions to the field be included in this event. In the case of Witt and Hurle, I managed to come up with the idea of having them collaborate on their presentations with a knowledgeable colleague. In the case of Gus Witt, I asked Michael Wargo, a former student, to help put together what turned out to be an excellent retrospective of Prof. Witt's work. For Don Hurle I arranged for Peter Rudolph from Berlin to help write and deliver the lecture, which also turned into an excellent collaboration. In the case of Bruce Joyce I asked Tim Joyce, who also works in that general area to collaborate with his father. In the end, however, this latter presentation was doomed due to a last minute cancellation of a critical flight leg from England. In an attempt to put things right, however, I thought it would be appropriate to include their intended lecture into this proceedings anyway. Of the 12 invited talks planned for this Symposium I am only able to include 11 papers in this publication. Unfortunately the excellent lecture on "Macroscale Modeling and Simulation in Semiconductor Crystal Growth" by Prof. Robert Brown from MIT could not be included.

The actual program of the Symposium is given in the Table below. Since the order of the talks in Seattle was determined partly by a number of logistical issues rather than on a purely topical basis, I thought it would be better to reorganize the resulting papers in this book under the following general headings; "Early Foundations", "A National Perspective" and "Progress Since 1952". Under the latter, the subheadings are (1) "Crystal Growth Fundamentals", (2) "Melt and Solution Growth" and (3) "Thin Film Epitaxy". I also interspersed the Newsletter articles near their most appropriate subject matter, and as mentioned earlier, clustered the pre 1950s topics at the beginning of the book under the heading "Early Foundations".

I am very pleased to be able to dedicate this volume to the memory of Professor August F. Witt and his pioneering contributions to Crystal Growth. While he participated in the preparation of his invited lecture up to the time of the Symposium, he was too ill to attend and make the presentation himself. He died several months afterward. The Proceedings paper is based loosely on the lecture, and discusses the work in his laboratory over many years. It was co-written by a number of enthusiastic former students and colleagues, under the leadership of Chris Wang.

I would like to take this opportunity to acknowledge the help and advice of my many colleagues who helped contribute to the planning of this Symposium and Proceedings, and to the AACG for supporting this effort. I would like to especially thank Don Hurle for his advice and encouragement with this volume. Lastly, I would like to thank the speakers for their excellent presentations and for agreeing to contribute papers to this Proceeding.

Robert Feigelson
Editor

References

[1] R.S. Feigelson (Ed.), 50 Years of Progress in Crystal Growth, The Proceedings of a Symposium held during the Fourteenth American Conference on Crystal growth and Epitaxy, Seattle, August 4–9, 2002, J. Crystal Growth 264 (1) (2004).

[2] H.E. Buckley, Crystal Growth, John Wiley & Sons, New York, 1951.

[3] D.F. Elwell, R.S. Feigelson, Am. Assoc. Crystal Growth (AACG) Newslett. 14 (1) (1984) 5.

[4] D.T.J. Hurle, J. Crystal Growth 243 (2002) 1.

[5] M. Schieber, Am. Assoc. Crystal Growth (AACG) Newslett. 15 (3) (1985) 4.

Actual Program Scheduled for the Symposium "50 Years of Progress in Crystal Growth"

Monday AM, August 5, 2002

8:50   **MBE-From Small Beginnings to Nanostructures to ?**[**]

Tim Joyce[*] (a), Bruce Joyce (b); (a) University of Liverpool, Liverpool, UK, (b) Physics, Imperial College, London, UK

9:30   **Development and Current Status of Organometallic Vapor Phase Epitaxy**

Gerald Stringfellow; College of Engineering, University of Utah, Salt Lake City, Utah, USA

10:10   Break

10:30   **A History of Defect Formation, Segregation, Faceting and Twinning in Melt-Grown Semiconductors**

Peter Rudolph[*] (a), D.T.J. Hurle (b); (a) Czochralski Semiconductor Compounds, Institute of Crystal Growth, Berlin, Germany, (b) Bristol University, UK

Tuesday AM August 6, 2002

8:00   **Morphology: From Sharp Interface to Phase Field Models**

Robert Sekerka; Department of Physics, Carnegie Mellon University, Pittsburgh, PA, USA

8:40   **Dendritic Crystal Growth**

Martin Glicksman; Materials Science & Eng., Rensselaer Polytechnic Institute, Troy, NY, USA

9:20   **Fundamentals and Applications, Fifty Years Retrospective of Japanese Crystal Growth Community**

Ichiro Sunagawa; Yamanashi Institute of Gemmology and Jewellery Arts, Yamanashi, Japan

Wednesday AM August 7, 2002

8:00   **Surface Processes of Faceted Growth**

Alexander Chernov; USRA, Huntsville, Al, USA

8:40   **Recent Progress in the Melt Growth of III–V Compound Semiconductors**

Brian Mullin; Consultant Editor, EMC/University of Durham, UK

Thursday AM August 8, 2002

8:00   **Quantitative Microsegregation/Bridgman Growth**

Michael Wargo[*] (a), August Witt (b); (a) Physical Sciences Division, NASA, Wash, DC, USA, (b) Massachusetts Institute of Technology, Cambridge, MA, USA

8:40   **Macroscale Modeling and Simulation in Semiconductor Crystal Growth**

Robert Brown; Chemical Engineering, Massachusetts Institute of Technology, Cambridge, MA, USA

Friday AM August 9, 2002

8:00   **Czochralski Growth of Oxides**

C.D. Brandle; Agere Systems, Murray Hill, NJ, USA

8:40   **Constitutional Supercooling/Surface Roughening**

K. Jackson; University of Arizona, Tucson, AZ, USA

---

[*] Presenting Lecturer.

[**] T. Joyce was unable to attend to present this lecture.

About the Editor

Robert S. Feigelson is currently a Professor Emeritus in Materials Science and Engineering at Stanford University, where he has spent most of his professional career, as well as a member of the Geballe Laboratory for Advanced Materials. For over 30 years, starting in 1963, he was involved with the technical and administrative direction of the Crystal Growth Laboratory at Stanford's Center for Materials Research. He received his B.Cer.E. (1957) from the Georgia Institute of Technology, S.M. in Ceramics (1961) from the Massachusetts of Technology under Professor David Kingery, and his Ph.D. in Materials Science and Engineering (1974) from Stanford University under Professor Richard Bube. He has been actively involved in single crystal growth and materials preparation for over 40 years including bulk and epitaxial growth processes. He has worked on a variety of different classes of materials including nonlinear optics, ferroelectric and photorefractive oxides, thermoelectrics, semiconductors, superconductors, magnetic and biological materials. He is particularly well known for his work on single crystal fiber growth and chalcopyrite compounds. He was the recipient of the 1995 IOCG's Triennial Laudise Prize, awarded for outstanding achievements in experimental crystal growth. He is a current Editor of the Journal of Crystal Growth and was President of the American Association of Crystal Growth (1981–84) and other regional societies. He has authored or co-authored over 250 papers, organized many conferences and workshops, and was instrumental in the formation of the Western Section of the AACG and the international conference series on the growth of biological macromolecules.

Contributors

Symposium Authors

Professor **Ichiro Sunagawa**, throughout his career as a scientist, has been (and even now at the age close to 80) actively interested in understanding how crystals grow, and why they exhibit elaborately varied morphology, perfection and homogeneity. He graduated from Tohoku University in 1947 and received a D.Sc. degree from Hokkaido University in 1957. Afterward he served for 23 years as a mineralogist in the Geological Survey of Japan and for 17 years as Professor of Mineralogy at Tohoku University where he now holds an emeritus professorship. After retiring in 1988 and up to the present time, he has been acting as Principal of the Yamanashi Institute of Gemology and Jewelry Arts. His initial research work was on understanding the origins of habit variations in natural minerals, such as pyrite, chalcopyrite and calcite. This motivated him to investigate the surface microtopographs of the crystal faces of minerals like hematite, diamond, and clays formed under a variety of geological conditions. This led to a deeper understanding of the morphologies of complex systems based on crystal growth mechanisms at the atomic level. This approach opened new windows in Earth and Planetary Sciences and also in gemology, where he provided the basic concepts for identifying natural from synthetic gemstones. He has published many scientific papers and 10 books on crystal growth, mineralogy, gemology and diamonds.

A.A. Chernov became a Professor of Physics in 1970. In 1987, he was elected a Member of the USSR (in 1992, renamed the Russian) Academy of Sciences, Department of General Physics and Astronomy. His career consists of time with the Institute of Crystallography at the Russian Academy of Sciences in Moscow as a Technician, Junior Scientist, Senior Scientist, Head of the Elementary Processes of Crystal Growth Laboratory and the Crystallization and the Crystal Perfection Physics Department. He has ~ 290 publications on crystal growth and surface physics. Since 1996 he has held the positions of Program Director with the Universities Space Research Association and Chief Scientist with BAE Systems Analytical Solutions, both at the Biological and Materials Science Research Laboratory at NASA Marshall Space Flight Center in Huntsville, AL, USA. Prof. Chernov was awarded the first IOCG Frank Prize in Sendai, Japan in 1989.

Kenneth A. Jackson is Professor in the Department of Materials Science and Engineering at the University of Arizona in Tucson, where he has been since 1989. He received his Ph.D. degree from Harvard University in 1956, and was Assistant Professor there until 1962, when he joined AT&T Bell Laboratories. At Bell Labs he was head of Materials Physics Research for many years. His major scientific interests are in the kinetic processes of crystal growth, and his scientific contributions include constitutional supercooling, the surface roughening transition, defect formation in crystals, and studies of alloy crystallization. He pioneered computer simulation studies of the atomic scale processes during crystal growth. He has written and edited several books. He has served as President for both the American Association for Crystal Growth and the Materials Research Society. He has received awards for his scientific contributions from both the American and the International Crystal Growth Societies, and from the Materials Society of AIME.

Robert F. Sekerka, is currently a University Professor in Physics and Mathematics at Carnegie Mellon University. He received his B.S. in Physics from the University of Pittsburgh in 1960 and his A.M. and Ph.D. in physics from Harvard in 1961 and 1965 respectively. After graduation he worked as a Senior Scientist in the Department of Theoretical Physics at the Westinghouse Research Laboratories and later was Manager of the Materials Growth and Properties Department. In 1969 he joined the Metallurgy and Materials Science Department faculty at Carnegie Mellon University and later was Dean of the Mellon College of Science. He was an Associate Editor of the Journal of Crystal Growth and Metallurgical Transactions and since 2001 has been the President of the IOCG. He has won numerous awards including the IOCG Frank Prize in 1992 for "Seminal contributions to the theory of crystal growth, including the theory of morphological stability of growing crystals", the TMS Bruce Chalmers Award, and the Philip M. McKenna Memorial Award for outstanding research contributions to Metallurgy.

Martin Glicksman is Professor of Materials Science and Engineering and Chemical Engineering at Rensselaer Polytechnic Institute. He is also currently the Director of Microgravity Science and Applications for the Universities Space Research Association. He received his B.Met.E. (1957) and Ph.D. (1961) from Rensselaer. In 1963 he joined the Metallurgy Division of the Naval Research Laboratory in Washington D.C. where in 1967 he established NRL's Transformation and Kinetics Branch. His research concentrated on kinetic studies of solid–liquid and solid state transformations and the processing and properties of A15 superconductors. In 1975 he was appointed to the chairmanship of the Materials Engineering Department at Rensselaer and in 1986 was appointed to the John Tod Horton Distinguished Chair in the Department. He has co-authored over 250 technical publications, and is the author of a recent text entitled "Diffusion in Solids: Field Theory, Solid State Principles, Applications. Professor Glicksman has received numerous awards including the Stanley P. Rockwell Medal, the Kent Van Horn Award and, in 1996, the AACG International Crystal Growth Award.

Afina Lupulescu received her B.S. and M.S. degrees in technical Mineralogy at the University of Bucharest, Romania, and her Ph.D. in from Department of Earth and Environmental Sciences at Rensselaer Polytechnic Institute in 1996. Her Ph.D. thesis focused on melting connectivity & segregation in silicate systems of high anisotropy. She joined Professor Glicksman's group in June 1996, first as a Postdoctoral Research Associate, Research Assistant Professor, and now as a Research Associate Professor. Her research interests center on diffusion, melting and solidification, and crystal growth. Studies include dendritic growth, diffusion, and the kinetics of zero flux planes in multicomponent diffusion.

Donald Hurle is semi-retired after a career in electronic materials research at the Royal Signals and Radar Establishment, Malvern, UK where he rose to be a Deputy Chief Scientific Officer. The underlying theme of his work has been improvement in understanding and control of melt growth processes. He is credited with basic discoveries in (1) morphological stability of semiconductor growth, (2) the role of convection and its control using magnetic fields, (3) the dynamics and control of crystal pulling, (4) the mechanism of growth twin formation and (5) semiconductor point defect equilibria. Currently he is a Visiting Professor of Industrial Physics at the University of Bristol, is a Fellow of the Institute of Physics, a past President of BACG, the recipient of the 4th International Award of the AACG (1987) and joint recipient of the IOCG Frank Prize (2001). He edited the Handbook of Crystal Growth (Elsevier 1993/4) and has been on the editorial board of the J. Crystal Growth since 1980. For over a decade he has assisted ESA in the defining its microgravity program in the physical sciences.

Peter Rudolph was born in Germany in 1945. In 1969, he received a Diploma in Electronic Technology, and then, in 1972, a Ph.D. (Dr. Engineer) in Solid State Physics and Technology (on crystallization of CdSb layers), both at the Technical University of Lvov (Ukraine). From 1973 to 1993 he was employed at the Institute of Crystallography and Material Science of the Humboldt University in Berlin as lecturer of Kinetics of Phase Transitions, Crystal Growth and Technical Crystallography. In 1979, he obtained the D.Sc. (Dr. habil.) in Crystallography at Humboldt University. From 1980 he headed their Laboratory of Crystal Growth of IV–VI (PbTe) and II–VI materials (CdTe), and was also involved in the industrial growth of $LiNbO_3$ and $PbMoO_3$. Since 1994 he has been employed at the Institute of Crystal Growth in Berlin, dealing with the growth and characterization of III–V compounds (GaAs, InP) by a modified Czochralski method. He carried out space crystal growth experiments in 1978 & 1992, and was a member of ESA's expert group on melt growth experiments. From 1993 to 1994, and in 1998 he was a Guest Professor in Prof. T. Fukuda's Crystal Growth Laboratory at Tohoku University in Sendai (Japan). His publications include 14 monographs, 4 editions, 134 original publications and 17 patent descriptions.

Christine A. Wang is a senior staff scientist in the Electro-Optical Materials and Devices Group at Lincoln Laboratory, Massachusetts Institute of Technology. She received the S.B., S.M., and Ph.D. degrees from the Materials Science and Engineering Department at the Massachusetts Institute of Technology. Dr. Wang's doctoral thesis work in Prof. August F. Witt's laboratory (between 1979 and 1984), identified and characterized critical elements for optimized semiconductor crystal growth in vertical Bridgman configuration. She joined Lincoln Laboratory in 1984 where her principle research activities have included hydrodynamics studies for design of reactors for organometallic vapor phase epitaxy (OMVPE), alternative chemistries for OMVPE growth, in-situ growth monitoring, and ex-situ materials characterization of III–V semiconductors. She has developed GaAs, AlGaAs, InGaAs, and AlInGaAs OMVPE-grown materials and device structures for high-performance semiconductor quantum-well lasers used for optical pumping solid-state lasers and optical communications. Her current research is focused on the materials and device development of GaSb, InAs, GaInAsSb, and AlGaAsSb III–V alloys for a variety of optoelectronic devices, including state-of-the-art thermophotovoltaics and quantum-well lasers. She is an author or co-author of over 120 publications and 5 patents.

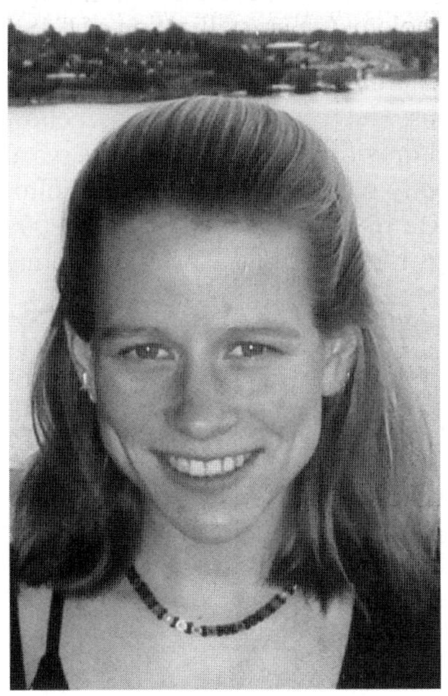

Michaela Wiegel was born in Germany. She came with her family to the United States in 1982 and studied Mechanical Engineering at Case Western Reserve University (Cleveland, OH) where she received her B.S. in 1995 and M.S. in 1997 under Dr David Matthiesen, a former student of Prof. August F. Witt. She received her Ph.D. in 2002 from the Massachusetts Institute of Technology under Prof. Witt, and was his last student. Her Masters and Doctoral work involved experimental crystal growth research on germanium, gallium arsenide, and bismuth silicate, using both Czochralski and Bridgman–Stockbarger techniques. She is currently working at MIT's Lincoln Laboratory (Lexington, MA) as a technical staff member in the optical systems engineering group.

Biographies and photographs of co-authors D. Carlson, S. Motakef and M.J. Wargo were not available at press time.

Professor **John Brian Mullin** retired in 1990 from a special merit DCSO post at The Royal Signals and Radar Establishment, now Qinetiq, Malvern, UK where he led a specialist research team working on infrared materials for detector applications. The main theme of his work has been on the scientific investigation of new methods of crystal growth as well as on the role of crystallization phenomena on defect formation during crystal growth, essentially the materials science of III–V and the II–VI compound semiconductors. He was trained as a physical chemist at the University of Liverpool where he took a Ph.D. in Chemical Oceanography. In 1983 Liverpool awarded him a D.Sc. for his work on II–VI and III–V compounds. He is a Visiting Professor at the University of Durham, Coventry University, and the University of Shanghai. He is an Associate Editor of the Journal of Crystal Growth and Editor-in-Chief of Progress in Crystal Growth and Characterization of Materials.

Charles D. Brandle, Jr. (Dave) received his BS (1963) and MS (1964) in Chemical Engineering from the University of Texas and his Ph.D. from the University of Texas at Austin in 1970. Throughout his career, he has been involved with oxide materials and their use in single crystal form for various optical device applications. He became a Bell Labs Fellow in 1998 and received the AACG (American Association for Crystal Growth) Award in 2000. He is currently serving as an Associate Editor for the Journal of Crystal Growth. He retired from Bell Labs in 2002 and currently is President of CrysTex, LLC, a crystal growth consulting company.

Bruce Joyce is presently an Emeritus Professor of Physics and Senior Research Investigator at Imperial College London, where he was Director of the Interdisciplinary Research Centre for Semiconductor Materials before his retirement. He studied chemistry at the University of Birmingham and joined the Allen Clark Research Centre of the Plessey Company in 1958, before moving to Philips Research Laboratories, where he worked from 1969 to 1988. He was awarded a D.Sc. by Birmingham University in 1973 and in the same year was elected a Fellow of the Institute of Physics. He was awarded the Duddell Medal and Prize of the Institute of Physics in 1981, the IBM (Europe) Science and Technology Prize in 1986, was elected a Fellow of the Royal Society (FRS) in 2000 and became President of the British Association for Crystal Growth (BACG) in 2003. His research interests are the physics and chemistry of semiconductor surfaces and reduced dimensionality structures and in the course of this work he has published some 350 papers.

Tim Joyce is currently Principal Experimental Officer for the nanotechnology research group in the Department of Materials Science & Engineering at the University of Liverpool. He was awarded a B.Sc. in Physics at the University of Bristol in 1981 and undertook research in epitaxy & CVD at the GEC Hirst Research Centre and later at the Allen Clark Research Centre of the Plessey Company. In 1988 he moved to the University of Liverpool to start a new research activity in MOMBE/CBE, which also led to a Ph.D. in 1991. His current research interests include CBE of dilute nitrides, MOCVD of novel oxides and GSMBE of III-nitrides for MEMS applications. He has been a member of the British Association for Crystal Growth for many years and is currently honorary treasurer, also acting as treasurer for the forthcoming IC-MBE 2004 in Edinburgh.

G.B. Stringfellow is a Distinguished Professor of Materials Science and Engineering and Electrical and Computer Engineering at the University of Utah. His research interests have focused on two main areas: the organometallic vapor phase epitaxial growth (OMVPE) process and the epitaxial growth and characterization of III/V semiconductors. His work includes more than 360 papers in the archival literature and 15 books, including "Organometallic Vapor Phase Epitaxial Growth: Theory and Practice" which is now in its' second edition. He was the editor of the Journal of Crystal Growth from 1998 until 2003 and is a member of the National Academy of Engineering and has, in recent years, won the AACG Crystal Growth Award and the TMS Bardeen Award.

Professor **August F. Witt** was born in Innsbruck, Austria. In 1953–54, he did graduate research in nuclear chemistry with Nobel Laureate Mme. Joliot-Curie in Paris, and afterward worked on radiation chemistry at the Atomic Energy Research Establishment in England. In 1959 he received his Ph.D. in Physical Chemistry from the University of Innsbruck, and in the following year became a Research Associate in MIT's Department of Metallurgy. In 1962, he was appointed to the faculty as an Assistant Professor and his primary research focus changed from surface chemistry to the processing and characterization of electronic materials. In 1972, he was promoted to Professor in Materials Science. He was widely published in the scientific literature and his passion and devotion to teaching undergraduates at MIT was widely recognized. Professor Witt received NASA's Outstanding Scientific Achievement Award in 1974 for his work in advancing materials processing in space, and in 1976 received Austria's Exner Medal for Outstanding Contributions to Science and Technology. Professor Witt was a member of the American Association of Crystal Growth (President 1975–1981). He chaired NASA's Electronic Materials Working Group from 1982–89. In 1990, he was the first recipient of the Amar Bose Award for Sustained Efforts in Undergraduate teaching as well as being named TDK Professor in Materials Science and Engineering. In 1992, he received the Space Processing Award from the American Institute of Aeronautics and Astronautics. In 1993, he was appointed Ford Professor of Engineering.

AACG Newsletter Authors

Joachim Bohm author of "The history of crystal growth" was born in Brandenburg, Germany. He received a diploma in Mineralogy from Humbolt University-Berlin in 1958 and his doctorate there in 1962. He was a research assistant from 1958–1964 in the Central Institute for Materials Research at the Reinststoffe Institute for Applied Physics in Dresden. From 1964 to 1991 he worked at the Central Institute of Optics and Spectroscopy of the GDR

Academy of Sciences in Aldershof. His main fields of activities have been crystal growth, structure, symmetry, crystallographic groups, teaching and writing books. He is currently retired and resides in Berlin.

Kurt Nassau author of the articles "Dr. A.V.L. Verneuil and the synthesis of ruby and sapphire" and "Early history of lithium niobate: personal reminiscences" was born in Austria. He received his B.Sc. Honors at the University of Bristol in England in 1948, and his Ph.D. at the University of Pittsburgh in 1959. He spent the next 30 years at AT&T Bell Laboratories in Murray Hill, NJ, in the field of crystal chemistry and physics, crystal growth, lasers, glass, etc. He retired in 1989 as Distinguished Research Scientist. He is the author of over 440 publications including five books (on gemstones and the physics and chemistry of color), and 17 patents. He was on the Board of Governors of the Gemological Institute of America for 20 years. He is currently retired and living in New Jersey.

Leon Merker author of "Remembrances of flame fusion" was born in Europe and studied chemistry at the University of Vienna before World War II. Forced to leave Europe for the United States due to anti-Semitism, he obtained his M.S. in chemistry from the University of Michigan. After graduation he helped start General Synthetics Corp in Newark, NJ to grow Verneuil ruby crystals. He is currently retired and living in New Jersey. Other details of his career are given in his article.

David Bliss, author of "Evolution and application of the Kyropoulos crystal growth method" received his B.A. in economics from Case Western Reserve University, and his S.M. and Ph.D. in materials science and engineering from the Massachusetts Institute of Technology and SUNY at Stony Brook, respectively. In 1981 he joined M/A-COM Inc. in Burlington, MA and since 1989 he has been program manager for bulk crystal growth at the AFRL Sensors Directorate's Hanscom Air Force Base in Massachusetts. His expertise is in the growth of III–V semiconductor materials.

Joseph A. Burton and William P. Slichter, authors of the article "Reminiscences about the early background of the papers on the distribution of solute in crystals grown from the melt" were both at the AT&T's Bell Laboratories (BL) in Murray Hill, NJ for most of their careers. **Dr. Burton** received his B.S. in chemistry at Washington and Lee University in 1934 and his Ph.D. in chemistry in 1938 from the Johns Hopkins University. After graduation he joined BL and worked on phosphors and thermionic materials before the invention of the transistor, and then became very interested in the growth of semiconductor single crystals. He led an important study on thermally driven convective mass transport during melt growth. In 1954 he became head of the Semiconductor Physics Research Department. He retired from BL as Director of Physics Research in 1976. From 1970–1985 he was treasurer of the American Physical Society and was a member of the governing board of the American Institute of Physics (1969–1984). **Dr. Slichter** received his BA (cum laude), MA and Ph.D. in chemical physics from Harvard University in 1947, 1949, and 1950 respectively. He joined BL in 1950, and shortly afterward joined a team working on the processes of diffusion in semiconductor crystals. During his 37-year career at BL, he served as Executive Director of Research of the Materials Science and Engineering Division from 1973–1987. Through his efforts, the diverse branches of materials science were developed into a coherent, unified, and effective organization. Examples include optical fiber technology (including glass compositions and processing as well as the plastic coating), resist chemistry for electron beam production of integrated circuit masks, wire and cable insulation and sheathing, magnetic telephone components, and novel alloys for connectors, etc. Dr. Slichter was awarded the American Physical Society High-Polymer Physics Prize in 1970 and his managerial accomplishments were recognized by two awards in 1988: the ACS's Earle B. Barnes Award for Leadership in Chemical Research Management and the Application to Practice Award of the Minerals, Metals and Materials Society. He was elected to the National Academy of Engineering in 1976 and was appointed to numerous committees including the NAS's Committee on Human Rights.

The work of Burton and Slichter led in 1953 to the seminal paper with R.C. Prim on solute segregation. Their findings were immediately applied to the preparation of silicon single crystals and were key to obtaining crystals satisfactory for early transistors. This became one of the building blocks of the field of semiconductor crystal growth and solid-state processing that continues until this day.

William A. Tiller, author of "How the constitutional supercooling formula was developed" received his B.A.Sc. in engineering physics in 1952 and his M.A.Sc. and Ph.D. in physical metallurgy in 1953 and 1955, respectively, all from the University of Toronto. He then joined the Westinghouse Research Laboratory as an advisory physicist where he conducted basic research on the science of crystallization. In 1964 he left for a full professorship in the Materials Science and Engineering Department at Stanford University. He was department chairman from 1966–1971, and became Professor Emeritus in 1992, shortly after completing his two books "The Science of Crystallization: Microscopic Interfacial Phenomena" and The Science of Crystallization: Macroscopic Phenomena and Defect Formation" both from Cambridge University Press (1991). In addition to studying crystallization problems for much of his career, he also spent many years conducting research in the fields of computer modeling, semiconductor processing and psychoenergetics.

Ernest Buehler, author of "The first Czochralski silicon", graduated from Brooklyn College in 1930 and joined AT&T's Bell Laboratories at Murray Hill, NJ shortly afterward as a shop apprentice. He became an instrument maker in 1934 and a group supervisor in 1940. In 1945 he switched to research and became a technical assistant. This started his long career as a crystal grower. During this time he also completed his studies at the Newark College of Engineering and in 1954 he became a member of the technical staff at Bell Laboratories. After retiring from Bell Labs in 1978 after nearly 49 years of service he began a second career at Allied Chemical Corp. He worked there until his death (in the laboratory) in 1988. During his long and distinguished career he worked at the forefront of crystal growth technology, first during WW II on ADP and EDT for sonar submarine detection, followed next by the hydrothermal growth of quartz crystals (devising a vital autoclave sealing technique). Together with A.C. Walker, he was one of the first persons to successfully grow single crystal quartz of a size and quality useful for devices. After this, he worked with Gordon Teal on the first growth of germanium and silicon single crystals followed by work with J. Kunzler and F. Hsu on superconducting wires of $SnMo_3$ and $SnNb_3$. His contributions led to the production of 100 kOe superconducting magnets.

William G. Pfann, author of "How zone melting was invented", was a member of the technical staff of AT&T's Bell Laboratories in Murray Hill, NJ from 1935 until he retired in 1982. He was born in New York City and received a B.S. in chemical engineering from the Cooper Union School of Engineering in 1940. He conceived and developed the prototype of the first transistor to be manufactured. In 1951 he developed a process, known as zone melting, which produced semiconductors of extreme purity and made possible the large-scale manufacture of transistors and other integrated circuits. He held 65 patents on zone melting, semiconducting devices and crystal growth techniques. He was elected to the National Academy of Sciences and the American Society for Metals and won the first Award for Creative Invention given by the American Chemical Society. Bill died in 1982 not long after this article was written.

Robert Mazelsky, author of "The origin of Czochralski growth through B_2O_3 glass: a step in the evolution of LEC growth" received his B.S. and Ph.D. degrees in solid state chemistry from Hofstra College in 1954 and 1959 respectively. He joined the Westinghouse Research in 1958 and remained there until he retired in 1994. He joined the laboratory to work on thermoelectric materials a project, which led him to his work on using boron oxide encapsulants. Before he retired from Westinghouse he was Department Manager and Technical Area Manager in Crystal Growth and Optical Materials and Cryogenic Electronics. In 1993 he was elected a Fellow of the ASM "for the development of new and innovative methods for the growth of crystals and the processing of advanced materials".

Hans J. Scheel, author of "The accelerated crucible rotation technique (ACRT)" was born in Bremen, Germany. From 1953–1957 he received his first chemical education at the ASTA Chemical Factory in Germany and in the period 1958–1968 took graduate courses at the University of Zurich and at ETH Zurich. In 1995 he received the Doctor of Engineering from Tohoku University in Sendai, Japan. He worked with Prof. Laves at the Institute of Crystallography at ETH Zurich from 1959 to 1968 and then joined the IBM Zurich Research Laboratory as research staff member. In 1983 he left to become Professor at the University of Sao Paulo in Sao Carlos, Brazil. After 1985 he worked in industry for a few years before joining, in 1988, the Institute of Micro- and Optoelectronics of the Swiss Federal Institute of Technology EPFL in Lausanne as head of the Cristallogenese group until his retirement in 2001. He is the author of more than 100 scientific publications and is the co-author (with Dennis Elwell) of a well-regarded book "Crystal Growth from High-Temperature Solutions", Academic Press 1975. His major contributions include ACRT; then the slider-free LPE method for multilayers, superlattices and atomically flat surfaces of GaAs, and solving experimentally and theoretically the long-standing striation problem. In addition, he was pioneer of LPE of high-Tc superconductors and of GaN. In 1972 he received the first award in crystal growth from the Swiss Crystallographic Society SCS. From 1979–1982 and 1999 to 2005 he was and is Vice President of SCS and Chairman of the Swiss Crystal Growth Section. He received awards from IBM, from the British and the Korean Associations of Crystal Growth, and became foreign member of the Russian Academy of Engineering Sciences. Recently Dr. Scheel started a consulting company and has been visiting professor at Osaka University 2002 and 2003.

A.I. (Ed) Mlavsky, author of "Shaped crystals from the melt by EFG" received his B.Sc. in chemistry with Honors in 1950 and his Ph.D. in physical chemistry in 1953 from Queen Mary College at the University of London. After graduation he worked at General Electric's Wembley Research Labs on the growth of silicon and germanium crystals. In 1956 he emigrated to the U.S. where he worked for Transitron in Wakefield, MA on silicon growth. In 1960 he joined the Materials Research Division of Tyco, Inc. where the EFG process was developed, and was appointed President of Tyco in the early 1970s, 18 months before leaving to form Mobil Tyco Solar Energy Corporation. At the time the article was written Dr. Mlavsky served as the Executive Director of the Israel-US Binational Industrial R & D Foundation. From 1979–1992 he was responsible for providing grants of over $100 million to some 300 projects between U.S. and Israel companies for the commercialization of (non-defense) high tech products. In 1993 he founded the Gemini Israel Funds where he is currently Chairman and Founding Partner. In 1989 he was elected a member of the Cosmos Club in Washington DC. He has 53 publications and 24 patents.

Harold E. LaBelle, author of "Experimental work leading to EFG" attended New Mexico State University. He joined Tyco Laboratories as a technician in 1962, focusing on solution growth techniques for semi-conductors. In 1965 he became project leader investigating novel techniques for the melt growth of sapphire filaments. In 1966 he invented the EFG technique. As staff scientist his work focused on the research and development of the process and in 1970 he became President of the newly founded Saphikon Division of Tyco. By 1971 his achievements ranged from the simultaneous growth of 25 sapphire filaments to the growth of large sapphire plates, 25 cm square. The EFG method was also applied to other crystalline materials at that time. Between 1971–1984 as President of the Saphikon Division he led the commercialization of EFG and ongoing research and development. This included the transfer of technology to several major international corporations. He has 17 US patents and published 16 scientific articles. In 1985 he became co-owner of Saphikon, Inc. with his wife, Judith J. LaBelle. They retired in 1987 and live in Naples, Florida.

Robert A. Laudise, author of "Hydrothermal synthesis of crystals" was Director of the Materials Chemistry Research Laboratory at AT&T Bell Laboratories in Murray Hill, NJ from 1987 until his death in 1998. He was also concurrently an Adjunct Professor at MIT. He received his B.S. in chemistry in 1952 from Union College in Schenectady, NY, and a Ph.D. in inorganic chemistry from MIT in 1956. He joined AT&T's Bell Laboratories as a member of the technical staff after graduation. He was a leader in research on the chemistry of materials, particularly the growth and properties of single crystals of electronic materials. He held many important managerial positions at Bell and served as a consultant and adviser to industry and the government. Laudise's major scientific contribution was the systematic study of the chemistry of hydrothermal syntheses. He and colleagues helped understand the hydrothermal method used to grow quartz and other materials such as ruby, KTP and ZnO. He published over 100 papers, held 14 patents and authored a book entitled "The Growth of Single Crystals". In 1970 he was awarded the Sawyer Price for contributions to piezoelectricity and in 1980 was elected to the National Academy of Engineering. In 1984 he received the AACG's International Crystal Growth Award and in 1989 the International Organization for Crystal Growth established a Laudise Prize for outstanding accomplishments in experimental crystal growth. He was a cofounder and first president with Ken Jackson of the AACG and was also president of the IOCG.

Howard Tracy Hall, author of "The transformation of graphite into diamond" was born in Ogden, Utah and received his B.S. from Weber State and his M.S., and Ph.D. degrees in physical chemistry from the University of Utah in 1942, 1943 and 1948, respectively. He also received honorary D.Sc.'s from Brigham Young University in 1971 and also Weber State and the Universities of Utah. His specialty has been ultra high pressure, high temperature techniques and phenomena, in particular the synthesis of diamonds. From 1948 to 1955 he was a member of the technical staff of General Electric where he helped develop the first process for making bulk synthetic diamonds. From 1967 he has been a Distinguished Professor of Chemistry and Director of Research at Brigham Young University. At the time the article was written he was Chairman of Novatek, Provo, Utah, a company producing diamonds based on his process. He is a Fellow of the American Academy of Arts and Sciences, the American Chemical Society (ACS), and the American Institute of Chemists. In 1972 he received the ACS Award for Creative Invention and in 1977 the American Physical Society's IBM International Prize for New Materials. He has over 110 publications and holds 19 patents.

Alfred Y. Cho, author of "Recollections about the early development of molecular beam epitaxy (MBE)" was born in Beijing, China. He received his B.S. and M.S. degrees in electrical engineering from the University of Illinois in 1960, and 1961, respectively. From 1961 to 1965 he worked first for the Ion Physics Corporation in Burlington, MA and then TRW-Space Technology Laboratories in Redondo Beach, CA. In 1965 he returned to the University of Illinois and received his Ph.D. degree in electrical engineering in 1968. Upon graduation he joined AT&T's Bell Laboratories in Murray Hill, NJ as a member of the technical staff. He was promoted to Department Head in 1984, named Director of the Materials Processing Research Laboratory in 1987 and Semiconductor Research Vice President in 1990. He has made seminal contributions to materials science and physical electronics through his pioneering development of the molecular beam epitaxy (MBE) crystal growth process, whose foundations he laid in the early 1970s. This led to the first MBE superlattice with AlGaAs/GaAs, and the precise fabrication of quantum wells structures which revolutionized electronic and optical devices for the consumer electronics, computer and communications industries. He has authored over 590 papers in surface physics, crystal growth and device physics and performance, and holds 75 patents. In addition to numerous awards he is also a Fellow of the Institute of Electrical and Electronics Engineering, the American Physical Society and the American Academy of Arts and Sciences.

Harold M. Manasevit, author of "The beginnings of metalorganic chemical vapor deposition (MOCVD)" received his B.S. and M.S. degrees in chemistry from Ohio University in 1950, and Pennsylvania State University in 1951 respectively. He received his Ph.D. in physical organic chemistry from the Illinois Institute of Technology in 1959. After graduation he first joined the U.S. Borax Research Corp. in Anaheim, CA, but in 1960 left to join the North American Aviation Co. In 1983 he joined TRW, Inc where he was a Senior Scientist. Dr. Manasevit has been involved in chemical vapor deposition of materials for 32 years. He was to first report on the epitaxial growth of silicon on sapphire (1963) and subsequently other insulators such as spinel, BeO and cubic zirconia. In 1968 he reported the first use of MOCVD for the epitaxial growth of GaAs and in the next few years extended this technology to other binary and ternary III–V (e.g., AlN, GaN) and II–VI semiconducting compounds, and later superconducting materials. He has authored over 50 publications and holds 16 patents. He has received a number of awards for his research work including from the Electrochemical Society in 1975, and in 1985 the IEEE Morris N. Liebmann Memorial award for his accomplishments in MOCVD.

The history of crystal growth

J. Bohm

Central Institute of Optics and Spectroscopy Academy of Sciences of the GDR, Berlin Adlershof, Federal Republic of Germany

A survey is given of the historical development of both theoretical and experimental knowledge from the early beginning to the fifties of our century. The survey is completed by a full bibliography of the most relevant papers, a timetable, and a list of conferences devoted to crystal growth.

The art of crystallization extends far back in the past and antedates considerably the written history of man. The crystallization of salt from sea water by evaporation was already practiced at many places in prehistoric time and can be considered one of the oldest technical methods of transforming materials—perhaps as well as the burning of earthenware. Crystallization procedures were recorded in written documents well before the Christian era. The Roman Plinius in his "Naturalis historia" mentioned the crystallization of a number of salts, for instance of vitriols. The medieval alchemists, European as well as Arabian, had arrived at a stage of detailed knowledge about many crystallization processes and phenomena. The alchemist Geber, whose papers are dated in the 12th or 13th century (cf. Darmstaedter 1922), described the preparation and purification of various materials by recrystallization as well as by sublimation and distillation.

Towards the end of the Middle Ages, the general technical progress led to corresponding progress in the techniques of material production and transformation, too. In the middle of the 16th century, Birringuccio (1540) recorded in detail the leaching of saltpeter and its purification by recrystallization, and the Saxonion

Previously published in AACG Newsletter Vol. 17(2) July 1987.

Fig. 1. Crystallization of vitriol. Woodcut from Agricola's "De re metallica" Basel: Froben 1556 (strings are used for seeding).

scientist Agricola (1556) in his famous, extensive work "De re metallica" gave instructions on how to produce various salts, alums and vitriols. (Fig. 1)

In the following century, the word "crystal" came into use more and more in the modern, general sense. Originally, Homer had used the expression "crystallos" for ice crystals only, antiquity had extended it to quartz crystals (rock crystals). Also in the 17th century the denotation "crystallization" came into use, replacing earlier expressions like "condensation" or "coagulation."

On New Year's night of 1611, a snowflake landing on Johannes Kepler's sleeve was the point of departure for his charming essay: "A New Year's Gift, or On a

Hexagonal Snowflake." Kepler (1611) concluded that snow crystals are built by closed packed spherical particles and posed, in such a way, the correct principle of crystallographic form and structure. Fifty years later Hooke (1665) claimed—in his "Micrographia" based on microscopic observations of many crystals—that every crystal form can be realized by arrangements of spherical particles. Looking at Kepler's drawings and remembering the work of Agricola and his precursors formerly cited, one can say that the very basic knowledge on both crystal growing and structure had already become available at this early time.

Indeed, a historical review reveals that the "modern" scientific development of crystallography started in the 17th century. In about 1600 Caesalpinus (1602) had already observed that crystals of specific materials, like sugar, saltpeter, alum, vitriols and so on, grown from solutions, exhibit typical forms, characteristic of each material. But it was not until as late as 1669 that Nicolaus Steno discovered the law of constancy of crystal angles—the fundamental law underlying the growth of crystals. Steno's work was extended and generalized by Guglielmini (1668; 1705) and finally confirmed about 100 years later by Romé de l'Isle (1772; 1783). According to them, every chemical species has its specific crystal form.

In contrast to the extensive experimental experience and the accuracy in describing crystals, the generation and the growth process of crystals long remained subject to the speculations and mystifications deriving from the Middle Ages. The growth of crystals was often considered to be similar to that of plants or animals, viz. connected with mystic powers and virtues. At that time, of course, it seemed difficult to understand how crystals can form from fluid, transparent and even microscopically clear solutions. It was the great experimentalist Boyle (1666; 1672) who observed that the nature of solution-grown crystals depends both on impurities and the rate of deposition, i.e. the growth rate. He also concluded from crystal forms and inclusions, partly fluid, that gem stones and other minerals are generated from solutions, too. Nevertheless, Boyle believed the growing process to be driven by non-materialistic and imponderable powers, and he also gave credence to the medical virtues of crystals, especially of gem stones. But Steno (1669) had already concluded that crystals grow by the attachment of material from outside and not by any veg-

etative mode of growth. However, Steno's statement became accepted only gradually, for instance by Hottinger (1698) and by Homberg (1692), who wrote that dendrites also grow in a simple way from outside. But even in the 18th century some notable scientists, for instance Leeuwenhoek (1685; 1703; 1705), still gave credence to vegetative growth modes.

During the 18th century significant progress was achieved both in the systematic description of crystals, especially minerals (cf. Capeller 1723; Linnaeus 1768), as well as in crystal growth experiments. Fahrenheit (1724) discovered the supercooling of water and noted the release of heat when ice formation occurred. Towards the end of the century Lowitz (1795) in his extensive work reaffirmed the earlier implied requirement of supersaturation or supercooling for the initiation of crystal growth and described the now well-known features of supersaturated solutions. The supersaturation of a solution can be achieved both by evaporation or supercooling; the degree of supersaturation that can be attained depends on the particular salt and on the pretreatment of the solution. He also used seeding and recognized a specificity of different nucleating agents. From a mixed supersaturated solution, the separate salt that is used for seeding will be deposited. The identity of the crystallizing salt and the nucleating agent is not required in all cases.

Lowitz's work as well as the extended investigations of Leblanc (1802), Beudant (1817; 1818), Gay-Lussac (1813; 1819), Fuchs (1815; 1816; 1817) and others prepared the way for Mitscherlich's (1819) general formulations regarding isomorphism and epitaxy. Somewhat later (1822) Mitscherlich also discovered the phenomenon of polymorphism (originally of dimorphism), that all together led to a revision of the basic crystallo-chemical principles. At the same time Schweigger (1813) made the significant observation that a seed or nucleus must be of a certain size in order to initiate crystallization—the point of departure for the subsequent concepts of critical-sized nuclei. In the meantime the outstanding work of Haüy (1782; 1784) had appeared; after some precursors (Guglielmini 1688; 1705, Westfeld 1767, Bergman 1773; 1779), he propounded the view that continued cleavage of a crystal should ultimately result in the smallest possible unit, a *"molécule intégrante,"* by a repetition of which the whole crystal is built up. The concept of a periodic crystal structure was now well

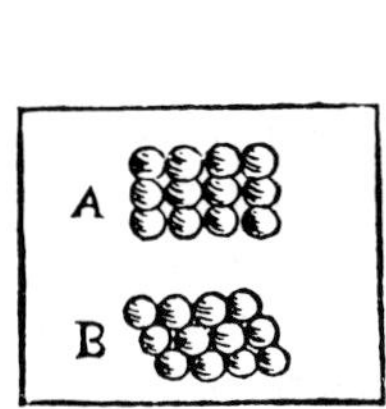 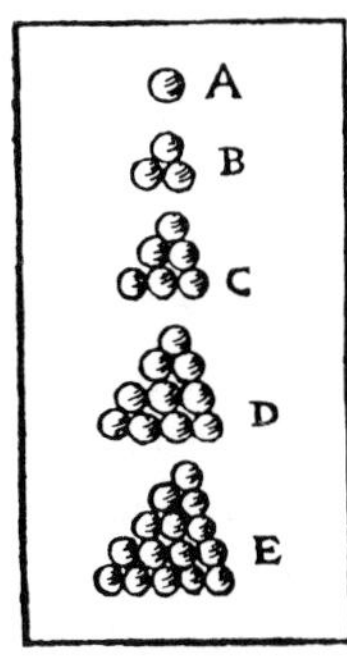

Fig. 2. Planar arrangements of dose packed spheres. Reproduction from Johannes Kepler's "Strena seu de Nive Sexangula." Frankfurt: Tampach 1611.

established and, furthermore, the idea of molecular growth units was introduced to the crystallographic community, too. Despite Haüy's work, Weiss (1804) considered crystals to be anisotropic continua; he derived the crystal systems (1815) and discovered the law of rational intercepts (1816) and the zone law (1820). On the other hand, Seeber (1824) discarded the concept of polyhedral cleavage nuclei: he concluded from the compressibility of crystals that they are built by a parallelepipedic arrangement of spheres—that means a lattice. In the middle of the century Bravais (1849) derived the 14 lattice types, setting the periodicity of crystals on a sound footing. He hypothesized a correspondence of lattice and morphology according to which crystal faces are planes with a high density of lattice points.

Referring to experimental investigations, mention must be made of the excellent work of Löwell (1857). Following many early papers on the $Na_2SO_4-H_2O$ system (e.g. Ziz 1815), he determined accurately the solubilities not only of the stable anhydrous Na_2SO_4 and its decahydrate (Glauber's salt, 1658), but of the metastable heptahydrate, too. His observation that from a supersaturated solution a crystal of the metastable heptahydrate crystallizes rather than the stable decahydrate afterwards prompted Ostwald (1897) to formulate his Law of Stages. Furthermore he established the expression of metastability (1893) and of a metastable region of supercooling or supersolubility (1897), nowadays called the Ostwald–Miers region. Ostwald was also concerned with the critical size of nuclei, he gave an interpretation of Liesegang's rings as a supersaturation phenomenon (1897) and derived the thermodynamic formula of the enhanced solubility of small particles (1900). Liesegang (1896) generated his rings by placing a droplet of a solution of silver nitrate onto a layer of gelatine containing potassium chromate, and we may credit this as the introduction of the technique of crystal growth in a gel. Intensive investigations and speculations on the aspects of nucleation were stimulated further by the work of de Coppet (1872; 1875). According to his observations supersaturated solutions or supercooled melts remain stable for a limited interval of time depending on the size of the sample. De Coppet extended his experiments over exceptionally long periods up to several years until crystallization took place; he did in fact maintain solutions of Glauber's salt in the supersaturated condition for nearly 35 years, and these were still intact at the time of his last report (1907). De Coppet explained his results by the formation of crystal embryos via ordinary collisions—a first theory of homogeneous nucleation. This found considerable criticism, and there was a lengthy controversy between the followers of homogeneous and of heterogeneous nucleation theories. The first quantitative measurements of the linear growth rates were performed by Gernez (1882): he crystallized sulphur and phosphorus from their respective melts, using long glass capillary tubes. Later, in the early 20th century, Tammann (1898; 1903) became the leading exponent for quantitative measurements, both of nucleation and crystal growth rates.

But foremost, there is the masterly theoretical work of Gibbs (1878) on heterogeneous equilibria, but the value of this work was generally recognized only with great delay. Gibbs determined the energy needed to generate a nucleus and derived the equilibrium form of a crystal that fulfills the condition of minimum total free surface energy. But in a footnote he pointed out that the equilibrium form may determine the nature of small crystals only whereas the larger ones will be confined finally by such faces onto which the attachment of material proceeds most slowly. Curie (1885) independently concluded in a short meaningful paper that the stable form, as he said, of a crystal is given by the minimum of the sum of the products of surface tension times surface area. This led to the well-known construction of a crystal form by Wulff (1895; 1901). While the kinematic theory of crystal growth, based on the velocities of advance of the individual crystal faces, was developed by Becke (1894), by

Johnsen (1910), and by Gross (1918) another alternative approach was introduced by Noyes and Whitney (1897) and by Nernst (1904) and Brunner (1904), who treated crystal growth as a diffusion-limited phenomenon. Early in this century Laue's invention of X-ray diffraction and the determination of crystal structures based thereon (first performed by Bragg) gave definite knowledge of the inner construction of crystals.

In the meantime, significant progress was achieved in crystal growth technology. A great deal of effort was made in the field of experimental mineralogy, stimulated not least by the search for recipes to make synthetic gem-stones. As early as 1837 Gaudin, and then Böttger (1839) and Elsner (1839) prepared small ruby crystals by melting a mixture of potassium alum and potassium chromate. Fremy (1891), cooperating with Verneuil in his attempts to grow gem-stones and other crystals from high temperature solutions, used large crucibles containing up to 50 liters of melt but they got mm-sized crystals only, mainly because of the poor temperature stability of their furnaces. Among the numerous efforts to synthesize diamonds, only those of Hannay (1880) and Moissan (1894) are mentioned here. Hannay accidentally found diamonds when he heated a mixture of lithium, paraffin and bone oil in thick-walled iron tubes. Moissan claimed to have obtained diamonds by quenching a melt of iron saturated with carbon. At the beginning of the present century Verneuil (1902; 1904) published his well-known flame fusion method by which he succeeded in growing large ruby crystals. Soon afterwards the industrial production of synthetic rubies was established; today this method is still followed throughout the world in some 20 factories with an estimated 1000 growing machines in nearly the same manner as invented by Verneuil. These figures prove that the Verneuil method is well in advance of all other growth methods.

The old method of sublimation was used by Durocher (1849; 1851) to prepare crystals of transition metal sulphides, passing hydrogen sulphide over the corresponding chlorides; even at that time he used the expression "transportation." Nowadays it is often Lorenz (1891)—who reproduced and discussed many of the older experiments—who was credited with having reinvented the sublimation technique for crystal growth technology. Concerning the growth methods from solution, G. Wulff (1895) was credited with the first construction of an apparatus with a rotating vessel, in this way breaking with the principle of avoiding any movement in crystallization experiments. But he had, in fact, already had a precursor in the less-known L. Wulff (1886). Johnsen (1915) invented a vertical setup with a rotating crystal. The two-tank technique was patented in 1910 by Krüger and Finke, but it had already been described accurately as early as 1852 by Payen. Following some precursors in experiments with high pressure hydrothermal solutions, Spezia (1905; 1906; 1909) succeeded first in the hydrothermal synthesis of larger sized quartz crystals.

Concerning the development of melt growth, Nacken (1915; 1916) grew single crystals from the melt on a cooled rod or seed crystal dipped into the melt. This method was modified by Kyropoulos (1926), who additionally slowly raised the growing crystal. Thus, his set-up became similar to Czochralski's (1918) method of pulling crystals from the melt. With the aim of growing metal crystals with a constant diameter, Gomperz (1921) put a platelet of mica onto the melt surface and pulled the crystal through a hole in the center of this platelet: in this way the technique of pulling profiled crystals was born; in the fifties this technique was reinvented by Stepanov (1959) and his co-workers. Concerning the freezing in crucibles, in 1914 Tammann had performed the growth of metal single crystals in small vertically arranged tubes by directional solidification; this method was made more sophisticated by Obreimov and Schubnikov (1924), who cooled the tip of the crucible blowing air onto it. With the same purpose Bridgman (1923; 1925) introduced the technique of lowering the crucible. In the 30s his method was applied by Stockbarger (1936) to grow large alkali halide crystals. The gradient freezing of large crystals in a resting crucible was performed by Stöber (1925). Somewhat later, Kapitza (1928) used the horizontal gradient technique in an open boat to grow bismuth crystals; in addition, he attempted to grow bismuth crystals by vertical zone melting too, the invention of which as a crystal growth technique must be credited to him. Finally we mention that in this period the preparation of aluminum crystals by the strain-anneal technique was performed by Sauveur (1912; cf. Carpenter 1922; 1926), the recrystallization of tungsten by the Pintsch technique (1916; cf. Böttger 1917) and the deposition of tungsten and other refractory metals from the vapor phase of halide compounds onto a hot wire by Koref (1922) and by van Arkel (1923; 1925).

Returning again to the theory of crystal growth, it was Volmer (1922) who introduced the adsorption of growth units onto the crystal surface, their diffusion along the surface and the generation of two-dimensional nuclei. Then, Volmer and Weber (1926) extracted from a thermodynamic treatment the basic expression of the rate of nucleation. This expression gives an exponential dependence of the nucleation rate on the work of nucleation ("*Keimbildungsarbeit*"). Due to its kinetic peculiarity, in this expression the preexponential factor still remains undetermined. The first kinetic approach to nucleation was given by Farkas (1927). At the same time the molecular kinetic theory of crystal growth was founded by Kossel (1927), introducing the half crystal position ("*Halbkristallage*") and by Stranski (1928), introducing the detachment energy and somewhat later (1931; 1932; 1934; 1935) in common with Kaischev the average detachment energy. After this Becker and Döring (1935) published their kinetic theory of nucleation.

Concerning the relationship between crystal structure and habit Donnay and Harker (1937) extended the principle of Bravais: they considered the influence of screw axis and glide mirror planes on the density of lattice points at the particular crystal faces and derived a morphological aspect for each space group. In the fifties Hartmann and Perdok (1955), regarding the actual structure of a crystal, introduced the concept of PBC-vectors that denote the chains of strongest bonds in a crystal—a concept that has proven fruitful with respect both to crystal habit and growth. Somewhat before this, Burton (1949), Cabrerar (1949) and Frank (1949) founded the well-known theory of spiral growth, the nowadays so-called BCF theory, solving a hitherto marked discrepancy between growth theories and measurements of actual growth rates. Also in the early fifties, Burton, Prim and Slichter (1953) derived their frequently cited expression for the effective distribution coefficient, and Rutter and Chalmers (1953) described the phenomenon of constitutional supercooling that is caused by a drop in melt temperature due to the enrichment of impurities adjacent to the surface of a growing crystal. Both papers became very important with regard to the practical aspects of crystal growth.

The onset of the modern development of crystal growth technology, dating from the Second World War, was boosted mainly by the demand for crystals for electronics, optics, and scientific instrumentation. Starting mostly from long-known growth methods, the growth technologies had to be raised to a very high and advanced level to fulfill the increasing demands in crystal size and quality. Frequently, for economic as well as for political reasons, similar developments were performed independently at several places, an example of which was the hydrothermal synthesis of quartz crystals in the forties. This also holds for the production of semiconductor crystals, starting from the fifties. The main progress in the latter field was marked by the adaptation of the Czochralski method to grow germanium crystals by Teal and Little (1950) and by Roth and Taylor (1952), by the zone melting invented by Pfann (1952; 1953) and subsequently by the floating zone technique for silicium, invented by Keck and Golay (1953; 1954) and by Emeis (1954). Finally, the old dream of crystal growers, the synthesis of man-made diamonds, perhaps the most spectacular event in the history of crystal growth, was published first by Bundy, Hall, Strong and Wentorf (1955). But it was realized in at least three places in the world independently and at the very same time, all using surprisingly similar apparata, that means high pressure equipment of the belt type. Nowadays about half of the diamonds used for industry are produced synthetically.

The increasing investigations and efforts in the field of crystal growth have shown up in the literature as well as at relevant conferences. Nowadays, most of the important papers are concentrated in the two leading journals "Journal of Crystal Growth" and "Crystal Research and Technology." The latter was founded in 1966 as "Kristall und Technik," the "Journal of Crystal Growth" was founded one year after in 1967. But now as before papers on crystal growth appear in many other periodicals, too. The first conference dedicated especially to crystal growth was held at Bristol by the Faraday Society in 1949; a further one was held at Cooperstown (N.Y.) in 1958. The Soviet Union Conferences started as early as in 1956. The International Conferences on Crystal Growth were founded in 1966 in Boston. Since 1976, in Zürich, there have been European Conferences, too. Besides other regional conferences not mentioned here, we have the well-known Hungarian Conferences on Crystal Growth to the third of which this paper was dedicated.

[Ed. We thank Acta Physica Hungarica and the author for allowing us to reprint this article which appeared in vol. 57 (3–4), pp. 161–178 (1985)].

Appendix A. The History of Crystal Growth—Chronology

1540	Birringuccio	Recrystallization of saltpeter
1556	Agricola	Production of various salts
1602	Caesalpinus	Typical forms of solution from crystal species
1611	Kepler	Structure of snow crystals
1658	Glauber	Crystallization of Glauber's salt
1665	Hooke	Structure of crystals
1666	Boyle	Influence of impurities and growth rate on crystal forms
1669	Steno	Law of constancy of crystal angles; crystal growth via addition of material from outside
1669	Bartholinus	Birefringence of calcite
1685	Leeuwenhoek	Description of crystals, also by microscopic observation
1688	Guglielmini	Correspondence of crystal forms and chemical species
1690	Huygens	Structural interpretation of birefringence
1692	Homberg	Crystal growth via addition of material from outside
1698	Hottinger	
1724	Fahrenheit	Supercooling of water
1767	Westfeld	
1773	Bergman	Building of crystals from small growth units (cleavage nuclei)
1782	Haüy	
1783	Romé de l'Isle	Description of crystals; change in nature of rocksalt by means of urea
1795	Lowitz	Supersaturation and crystallization of salt solutions; seeding
1813	Schweigger	Minimum size of crystal nuclei
1815	Fuchs	"Vicariates"
1815	Weiss	Crystal systems
1816	Weiss	Law of rational intercepts
1819	Mitscherlich	Isomorphism; epitaxy
1822	Mitscherlich	Polymorphism (dimorphism)
1824	Seeber	Lattice structure of crystals
1830	Hessel	Crystal classes
1837	Gaudin	Ruby from high-temperature solution
1839	Böttger; Elsner	
1839	Miller	Miller's indices
1849	Bravais	Lattice types; correspondence of lattice type and crystal form
1851	Durocher	Vapor growth of sulphide crystals; "transportation"
1852	Payen	Solution growth by the two-tank technique
1857	Löwel	Solubilities in the Na_2SO_4–H_2O system; metastable solutions
1865	Gernez	Reciprocal pairs of salts
1865	Marangoni	Liquid surface phenomena
1872	de Coppet	Spontaneous nucleation
1876	Sohncke	Groups of motion
1878	Gibbs	Heterogeneous phase equilibria
1880	Hannay	Man-made diamonds
1882	Gernez	Measurements of growth rates
1883	Barlow	Sphere packings
1885	Curie	Minimum surface energy of growth forms
1886	L. Wulff	Solution growth in a rotating vessel
1889	Stefan	Stefan's problems
1891	Schoenflies	
1891	Federov	Space groups
1893	Ostwald	Metastable region of supersaturation; critical size of nuclei
1894	Becke	Kinematic development of forms
1895	G. Wulff	Wulff's rule of construction of crystal forms; solution growth in a rotating vessel
1896	Liesegang	Liesegang's rings (crystal growth in a gel)
1897	Ostwald	Step rule
1897	Noyes & Whitney	Dissolution of crystals controlled by diffusion
1898	Tammann	Measurements of nucleation and growth rates
1900	Ostwald	Dependence of solubility on particle size
1902	Verneuil	Flame fusion technique (ruby)
1904	Nernst and Brunner	Diffusion layer on a crystal surface

1905	Spezia	Hydrothermal synthesis of quartz
1910	Johnsen	Kinematic development of forms
1910	Krüger and Finke	Solution growth by the two-tank technique
1911	Artemiew	Growth experiments with crystal spheres
1912	Sauveur	Strain-anneal technique
1912	Laue, Friedrich Knipping	X-ray diffraction by crystals
1913	Bragg	X-ray crystal structure determination
1914	Johnston	Diffusion technique to grow lowly soluble compounds
1914	Tammann	Directional solidification of metals
1915	Johnsen	Solution growth using a rotating seed
1915	Nacken	Melt growth using a cooled seed
1916	Schaller and Orbig	Recrystallization of a tungsten wire (Pintsch technique)
1918	Czochralski	Pulling of metal crystals from their melt
1921	Gomperz	Pulling of profiled metal crystals from the melt using a mica orifice
1922	Volmer	Adsorption and surface diffusion; two-dimensional nuclei
1922	Koref	
1923	Van Arkel	Vapor phase deposition (hot wire technique)
1923	Bridgman	Melt growth by lowering the crucible
1925	Stöber	Melt growth by the gradient technique
1926	Kyropoulos	Melt growth by using a cooled seed
1926	Volmer and Weber	Thermodynamic theory of nucleation
1927	Farkas	Kinetic approach to nucleation
1927	Kossel	Half crystal position
1928	Stranski	Detachment energy (both kinetic theory of crystal growth)
1928	Kapitz	Crystal growth by zone melting and by the horizontal gradient technique
1935	Becker and Döring	Kinetic theory of nucleation
1936	Stockbarger	Melt growth by lowering the crucible
1937	Donnay and Harker	Morphological aspect
1949	Burton, Cabrera and Frank	Spiral growth (BCF theory)
1950	Teal and Little	Czochralski growth of germanium
1952	Pfann	Zone melting
1953	Keck & Golay	Floating-zone technique
1953	Rutter and Chalmers	Constitutional supercooling
1953	Burton, Prim and Slichter	Effective distribution coefficient
1955	Hartmann & Perdok	"PBC-Vectors"
1955	Bundy, Hall, Strong and Wentorf	High pressure synthesis of diamond

References

Adams, F.D., 1938 and 1954. The Birth and Development of Geological Sciences. Dover, New York.

Bohm, J., 1981. Die historische Entwicklung der Kristallzüchtung. Crystal Research & Technol. 16, 275–292.

Buckley, H.E., 1951. Crystal Growth. John Wiley, New York; Chapman and Hall, London.

Burke, J.G., 1966. Origins of the Science of Crystals. University of California Press, Berkeley, Los Angeles.

Caldwell, H.B., 1935. Crystallization. Chem. and Met. Eng. 42, 213–214.

Darmstaedter, E., 1969. Die Alchemie des Geber. Springer, Berlin. (Repr. Sändig, Wiesbaden, 1922.)

Eitel, W., 1926. 1. Mineralsynthese, 2. Kristallzüchtung. In: Tiede, E., Richter, F. (Hrsg.), Handb. d. Arbeitsmethoden in der anorg. Chemie, Bd. 4. de Gruyter, Berlin, Leipzig, S. 391–447 u. 448–473.

Elwell, D., Scheel, H.J., 1975. Crystal Growth from High-Temperature Solutions. Academic Press, London, New York, San Francisco.

Groth, P., 1926. Entwicklungsgeschichte der mineralog. Wissenschaften. Springer, Berlin.

van Hook, A., 1961. Crystallization. Reinhold, New York; Chapman and Hall, London.

Kobell, F.V., 1864. Geschichte der Mineralogie von 1650–1860. J.G. Cottasche Buchhandlung, München.

Leicester, H.M., Klickstein, H.S., 1952. A Source Book in Chemistry. McGraw-Hill, New York, Toronto, London.

Lenz, H.O., 1861. Mineralogie der alten Griechen und Römer. E.F. Thienemann, Gotha. (Repr. Sändig, Wiesbaden, 1966.)

Marx, C.M., 1825. Geschichte der Crystalkunde. D.R. Marx'sche Buchhandlung, Carlsruhe und Baden.

Niggli, P., 1946. Die Krystallologia von Johann Heinrich Hottinger (1698). Sauerländer, Aarau.

Ray, P., 1956. History of Chemistry in Ancient and Medieval India. Indian Chem. Soc. Calcutta. (Cited after Elwell u. Scheel 1975.)

Schneer, C.J. (Ed.), 1977. Crystal Form and Structure. Dowden, Hutchinson and Ross, Stroudsburg (Penns. USA).

Schoen, H.M., Grove, C.S., Palermo, J.A., 1956. The early history of crystallization. J. Chem. Education 33, 373–375.

Shafranovskij, I.I., 1978 and 1980. Istorija Kristallografii. I u. II. Nauka, Leningrad.

Sohncke, L., 1879. Entwickelung einer Theorie der Krystallstruktur. B.G. Teubner, Leipzig.

Spangenberg, K., 1934. Wachstum und Auflösung der Kristalle. In: Handwörterbuch der Naturwissenschaften, vol. 10. Gustav Fischer, Jena, pp. 362–401.

Wilke, K.-Th., 1973. Kristallzüchtung. Dtsch. Verlag Wiss., Berlin.

For further references please see original article.

Dr. A.V.L. Verneuil and the synthesis of ruby and sapphire

K. Nassau

Bell Telephone Laboratories, Incorporated, Murray Hill, New Jersey

J. Nassau

Bernardsville, New Jersey

1. The early years—Frémy

1.1. Introduction

In many processes and products the actual originator is often difficult to locate. Frequently, various concepts of different origins are involved, with perhaps many improvements after these concepts have been assembled. Yet in the synthesis of red ruby and blue sapphire Verneuil was clearly "the father."

Others performed significant work on the synthesis of ruby both before and after him. Some had worked as long and some with more acclaim. There is M. Gaudin (Marc Antoine Augustin Gaudin, 1804–1880) whose work covered at least thirty-three years from 1837 to 1870; there is E. Frémy who directed first Feil and later Verneuil from before 1876 until 1892 and who wrote the book "Synthése due Rubis" in 1891; there are always the unknown producers of the "Geneva" synthetic rubies in the period 1886 to about 1905 (whose product sometimes has been incorrectly called "reconstructed" ruby); and many others.

Yet after all this is said, one inevitably returns to Professor Auguste Victor Louis Verneuil, Doctor of Science. He was involved in the final stages of Frémy's flux work. Although perhaps not the first to use a flame-fusion type of process for the growth of

ruby, he discovered the highly efficient flame-fusion technique known under his name which has been used for over 75 years essentially without change. He was also the first to uncover the secret of blue sapphire, recognizing that both iron and titanium were necessary to produce the correct shade of blue. He was instrumental in founding the synthetic corundum industry, which today has a production rate of about 1 billion carats (200 tons) of ruby and sapphire a year. A photograph of Verneuil taken in 1911 at the age of 55 is shown in Fig. 1.

Nevertheless, there has been close to a conspiracy of silence on the life and work of Professor Verneuil. Many biographical encyclopedias list Frémy, but none list Verneuil. Neither is Verneuil mentioned in Partington's four volume History of Chemistry or other such works. We have not been able to locate any detailed biographical account since an obituary collection in the Bulletin of the Association of the Students of Frémy an excerpt of which was also published elsewhere [3]. An article by Professor Lafuma [4] contains some additional details, but also inaccuracies in dates, etc. Very useful was a set of extended notes most helpfully provided by Professor Verneuil's nephew, Mr. A. Verneuil of Marseille, France [5]. Helpful information on the association of Verneuil with L. Heller and Son Co. was obtained from members of the Heller family. In

Previously published in AACG Newsletter Vol. 20(1,3) Spring and Winter 1990, Vol. 2(1) Spring 1991.

Fig. 1.

Fig. 2.

the absence of any additional significant accounts, we have had to go largely by these sources, by statements in Verneuil's own publications, as well as other references and accounts scattered throughout the published literature, with cross-checking for consistency.

In this account we intend to give an outline of Verneuil's life, concentrating on the details of his activities in connection with ruby and sapphire. Since there are many inaccurate and inconsistent statements on the early history of the synthesis of ruby [1], our account is based only on reliable primary data which we have been able to verify ourselves.

A detailed bibliographical listing of Verneuil's 19 published items on ruby and sapphire is given in Table I. These will be referred to as (A1) to (A12); we have examined closely every one of these items in its various versions. Since almost all the sources are in French, all quotations we give were translated by us.

1.2. Early years (1856 to 1875)

Auguste Victor Louis Verneuil was born in Dunkirk, France, on November 3, 1856, the third of three brothers. His grandfather, Jacques Auguste Verneuil, and his father, Auguste Marie Verneuil (1812–1904), were both mechanic-watch makers. One day his father saw a man standing behind a box on a tripod, his head covered with a black cloth, in the street outside his store. Being a friendly person, he went over to investigate. The man was Mr. Daguerre (Louis Jacques Mande, 1787–1851) operating his recently discovered photographic process. Daguerre was a painter and physicist who developed his photographic process in 1839, receiving for his achievement the Legion of Honor Award. A conversation ensued and Verneuil's father put up Daguerre for the night. As a result of this chance encounter, Verneuil's father changed his profession and opened a successful photography studio in Paris near the Pont Neuf. Helping his father work with the chemical processes used in Daguerrotype photography, young Auguste became interested in chemistry [5]. The photograph shown in Fig. 2 of a debonair young Verneuil, age 19, was undoubtedly taken by his father.

At the age of 17, Verneuil was accepted as laboratory assistant by Dr. Edmond Frémy (1814–1894), Professor of Chemistry and head of the chemistry laboratory at the Museum of Natural History in Paris. It was Frémy's practice to employ a number of promising young men as general laboratory assistants. There seems to have been considerable fellowships in this group, shown by the existence of the "Association of the Students of Frémy," with elected officers including a treasurer. It was this group that published a monthly bulletin, in which appeared an account of Verneuil's funeral, including several obituary

Table I
Bibliography of publications by A.V. L. Verneuil on Ruby and Sapphire

I. The Frémy–Verneuil work:	II. The flame fusion discovery:	III. The blue sapphire investigations:

I. The Frémy–Verneuil work:

A1. *The Action of Fluorides on Alumina*: Frémy and Verneuil, Action des fluorures sur l'alumina;
    a. Paris Acad. Sci., Comptes rendu 104, 738–740 (1887);
    b. Jl. Pharm. 15, 401–403 (1887) (identical with a).

A2. *The Artificial Production of Rhombohedral Ruby Crystals*: E. Frémy and A. Verneuil, Production artificielle des Cristraux de rubis rhomboédriques; Paris Acad. Sci., Comptes rendu 106, 565–567 (1888).

A3. *New Research on the Synthesis of Ruby*: E. Frémy and A. Verneuil, Nouvelles recherches sur la synthése du rubis; Paris Acad. Sci., Comptes rendu 111, 667–669 (1890).

II. The flame fusion discovery:

A4. Sealed Documents: *On a new process for the Fusion and Refinement of Chromium-containing Alumina and the Production of a Material having the Composition, Hardness, and the Density of Ruby (1891 and 1892, opened 1910)*: A. Verneuil, Plis Cachetés, Sur un nouveau proédé de fusion et d'affinage de l'alumine chromée et la production d'une matiere possedant la composition, la dureté et la densité du rubis, No. 4752, 23 décembre 1891 et No. 4849, 19 décembre 1892, ouverts 11 juillet 1910; Paris Acad. Sci., Comptes rendu 151, 131–132 (1910).

A5. *The Artificial Production of Ruby by Fusion*:
A. Verneuil, Production Artificielle du rubis par fusion;
    a. Paris Acad. Sci., Comptes rendu 135, 791–794 (1902);
    b. Rev. Ind. 33, 469–470 (1902) (with minor changes);
    c. Cosmos 936, 11–12 (1903) (with minor changes).

A6. *The Artificial Reproduction of Ruby by Fusion*:
A. Verneuil, Reproduction artificielle du rubis par fusion;
    a. La Nature 32, No. 1650, 177–178 (1904);
    b. Rev. Ind. 35, 448–449 (1904) (without the illustration and with minor changes);
    c. Scientific American Supplement 1535, 24594 (1905) translated into English under the title "The Artificial Production of Rubies," (but without the author's name!).

A7. Report on the Artificial Reproduction of Ruby by Fusion:
A. Verneuil, Mémoire sur la reproduction du rubis par fusion; An. De. Chim. et de Phys., Series 8, 3, 20–48 (1904).

III. The blue sapphire investigations:

A8. *Observations on a Note of Mr. L. Paris, on the Reproduction of the Blue Color of Oriental Sapphire*:
A Verneuil, Observation sur une Note de M. L. Paris, sur la reproduction de la coloration bleue du sapphir oriental; Paris Acad. Sci., Comptes rendu 147, 1059–1061 (1907).

A9. *On the Synthetic Reproduction of Sapphire by the Fusion Technique*:
A. Verneuil, Sur la reproduction synthétique du sapphir par la methode de fusion; Paris Acad. Sci., Comptes rendu 150, 185–187 (1909).

A10. *On the Nature of Oxides which Color Oriental Sapphire*:
A. Verneuil, Sur la nature des oxydes qui coloren le saphir oriental; Paris Acad. Sci., Comptes rendu 151, 1063–1066 (1910).

A11. *Process of Producing Synthetic Sapphires*:
A.V.L. Verneuil, U.S. Patent No. 988,230, March 28, 1911; Application May 10, 1910; Assigned to L. Heller and Son, New York, NY.

A12. *Synthetic Sapphire*:
A.V.L. Verneuil, U.S. Patent No. 1,004,505, September 26, 1911; Application June 28, 1911; Assigned to L. Heller and Son, New York, N.Y.

---

speeches [2]. Among Verneuil's friends, also working under Frémy, was Henri Moissan (1852–1902), well known for his work with diamonds and the electric arc furnace, done about 1896.

While working for Frémy at the Museum of Natural History, Verneuil also engaged in a series of studies which gained him the Bachelor's degree after two years, in 1875, the Master's degree five years later, in 1880, and finally the Doctor of Science degree in 1886. Verneuil's first publication was with L. Bourgeois in 1880 on the preparation of crystalline hydrous iron arsenate $Fe_2As_2O_8 \cdot 4H_2O$. By showing the identity of this with the natural mineral scorodite, the composition of this mineral was definitely established [6].

Professor Frémy was a chemist of widely varied interests. His work covered many different fields, including inorganic, organic, biochemical, industrial, and mineralogical problems. Among the latter was the synthesis of ruby, which had been intensively pursued by many chemists since 1837 [1]. Frémy, with his personal assistant C. Feil, succeeded in the first synthesis of clear red ruby which was published in 1877 [7]. The process used involved heating alumina, potassium dichromates lead oxide, and silica from the fireclay crucible used, but produced only small, thin, fragile ruby plates.

1.3. Verneuil and Frémy (1876–1892)

When Feil died, Verneuil became Frémy's personal assistant in 1876, with a private laboratory of his own. He also acted as demonstrator in Frémy's course on Mineral Chemistry. The Frémy–Verneuil work continued for some 16 years with three joint publications in 1887 (A1), 1888 (A2), and 1890 (A3).

Frémy summarized his extensive work with Feil and with Verneuil in his 1891 book [8]. The final process worked out by Frémy and Verneuil involved the recrystallization of alumina with a small amount of added potassium dichromate, by the use of potassium hydroxide and barium fluoride. The reaction was carried out in a ceramic crucible at the relatively high temperature of 1500 °C. The diffusion of humid air through the porous wall of the crucible was found to be an essential part of the process.

The product consisted of rhombohedral ruby crystals of great clarity. The color varied from colorless to red, violet, and blue, an occasional crystal being red on one side and blue on the other. The crystals were up to about 1/8 inch in diameter and up to one-third carat in weight. A photograph of some of these crystals and a view of a storage cabinet in Frémy's laboratory are given in Figs. 3 and 4.

At first they used crucibles up to 800 cc. in capacity and a small blast oven fueled with coke, both obtained from the Saint-Gobain glass factory. Temperature control was poor, however, and it was felt that larger quantities would be helpful. Subsequent work was done in large gas-fired ovens at the firm of Messrs. Apart, another glass manufacturer. In attempts to

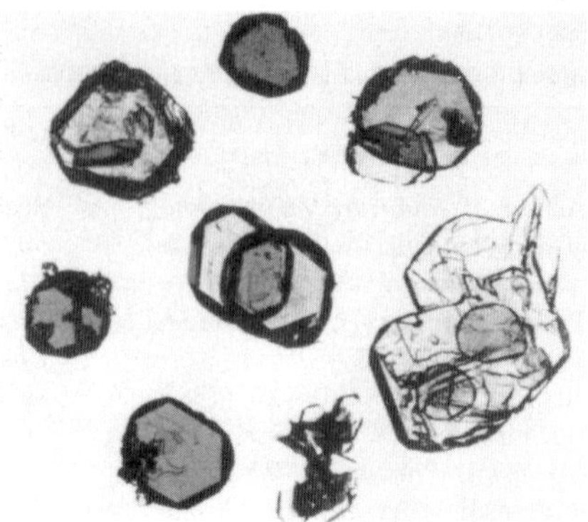

Fig. 3.

Fig. 4.

obtain larger crystals, ever larger and larger quantities were used, ultimately in crucibles having a capacity as, large as 50 liters (1 gallon)! But none of this helped increase the crystals' size appreciably. A large 12 liter run might yield up to 24,000 crystals weighing a total of 1,200 gms. Some of these ruby crystals lining crucibles can be seen on display at the Museum of Natural History, Jardin des Plants, Paris, V.

A consideration of the probable growth mechanism indicates the reason for the difficulties with crystal size. The mechanism can be assumed to involve the reaction of alumina with the barium fluoride to give gaseous aluminum fluoride. This then reacts with

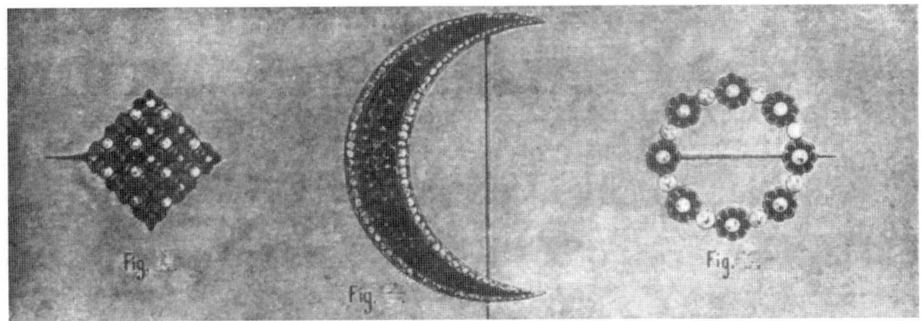

Fig. 5.

moist air to re-form alumina at an essentially constant temperature:

$$Al_2O_3 + 3BaF_2 \rightarrow 2AlF_3 + 3BaO$$

$$2AlF_3 + 3H_2O \rightarrow Al_2O_3 + 6HF$$

Since all of this occurs in a multiphase medium, the resultant vapor-phase nucleation and growth occurs in many small local cavities scattered throughout the porous mass filling the crucible. Accordingly, multiple nucleation is involved with only a small amount of growth on each nucleus. The growth technique would not be classified as flux-growth today, although it is often referred to in this way.

Figure 5 taken from Frémy's book [8] illustrates some jewelry made from these rubies by Jeweller Taub of Paris [8,9]. In most of this jewelry the ruby crystals were used without being faceted, and some crystals were also used as watch jewels (bearings). All of this was on an experimental basis and in very small quantities.

Frémy thought most highly of Verneuil. In his book [8] he speaks of Verneuil's ardor, talent, perseverance, and remarkable observational talents. He states that he had been "truly fortunate to find a co-worker such as Mr. Verneuil."

1.4. Other studies: (1873–1905)

Verneuil had been active on many other projects while working with Frémy on ruby. There was for example a series of more than a dozen reports on studies of the chemistry of selenium, the phosphorescence of zinc blende and other sulfide compounds, and so on. During this period he also served as a consultant to the glycerin manufacturers Clolus, Viandey, Linget,

and Co. of Billancourt, France. He developed for them a process for removing sulfur and arsenic from their glycerin so that it could be used for the production of dynamite.

From 1886 until his death, he also served as research director to Feil, Mantois, Parra Mantois, and Co., the major manufacturers of quality glass for optical instruments in France. In this capacity he developed new glass compositions of high refractive index, some of which were used by the well known firm Zeiss of Jena. For his work on the fabrication of large telescope objectives he was awarded a gold medal at the Paris Universal Exposition of 1900. From 1891 to 1900 he also served as inspector of historical monuments in Paris.

In 1892 Frémy retired (he died two years later) and his laboratory disbanded. Verneuil now moved to the chair of Applied Chemistry, which was part of the Organic Chemistry section of the Museum of Natural History in Paris, where he remained for 13 years (1892–1905). He took with him as his assistant Marc Pacquier, who had been one of Frémy's assistants. He also begin his teaching career, which was to continue until his death. He taught at various times during these early years at the Polytechnic Association (1879–1887), at the College Rollins (1880–1886), College Chaptal, and the College de France; these were senior high school to junior college level institutions.

If we take a single point in time, the year 1887 for example, we find Verneuil engaged simultaneously in an astounding number of projects:

  i with Frémy on ruby;
  ii work on the flame fusion growth of ruby;
  iii work on the phosphorescence of zinc blende;
  iv work on glycerin for Clolus;

v   work on glass for Feil; and
vi   teaching at the Polytechnic Association.

The accounts [2,3,5] also indicate that he did have other consulting activities not known by his colleagues owing to his discretion; nor have we come across any further records besides the studies on dental fillings with Doctor Cournand, a lifelong friend of Verneuil, which were never published [5].

Starting with the year 1895 he published 19 articles with G.N. Vyrubov (Wyrouboff) on the chemistry of the rare earth elements for which they were jointly awarded the prize "La Caze" by the Paris Academy in 1901; the academy had also awarded Verneuil the prize "Jecker" in 1889 for his doctoral thesis work on the chemistry of selenium. There was a paper with A.L. Arnaud on the extraction of rubber, and a paper on the action of sulfuric acid on charcoal which appeared in at least five different journals.

We can see in retrospect that this early work of Verneuil, important though it was in its own rights, was only the prelude for what was to follow. Without it he might not have been prepared for the complicated path leading to his crowning achievement: the development of the flame fusion technique for the growth of ruby and sapphire.

2. The flame fusion technique and its discovery

Even today one can still imagine the excitement surrounding the discovery of the synthesis of ruby. Towards the end of the Nineteenth Century it had been realized that there was nothing magical about diamond, ruby, and sapphire, and that the synthesis of these precious gemstones needed only careful control of a suitable crystallizing process. As of that time, however, none of the precious gemstones had been successfully produced in a form usable in jewelry.

We know now that the first usable synthetic gemstone was the "Geneva Ruby," sometimes mislabeled "reconstructed" ruby, first manufactured in 1886 by unknown producers and marketed in small quantities until about 1905 [1]. Since the claim that the "Geneva" product was made by the "reconstruction" of chips of natural ruby without destroying their identity was generally uncritically accepted at the time, these rubies were not recognized to be man-made. Other early synthetic rubies had been too small and imperfect to be useful. So that when Verneuil announced his flame fusion synthesis of ruby in 1902, an excited world hailed this as the remarkable achievement it was.

In his 1902 publication (A5a) Verneuil gave a brief account of the production of ruby by a fusion process utilizing an oxygen-hydrogen blow torch. This was reprinted twice (A5b and c) and followed by a similar account in 1904 (A6a and b) which also appeared translated into English in the Scientific American (A6c) (but without the author's name!).

The full experimental details of the process were finally published in a lengthy paper in 1904 (A7), and with this publication the flame fusion synthesis of ruby was public knowledge. Anyone with a reasonable amount of technical competence could do it himself. Admittedly, considerable skill was required and it would take an appreciable investment and at least several months of hard work to solve the many operational problems. The boules would surely crack if one attempted to grow them too large, but the way was clear for anyone willing to try. Several firms such as the Paris branch of the New York firm L. Heller and Son and the H. Djevahirdpian Co. immediately did try, succeeded, and proceeded with the large scale manufacture of synthetic ruby. Only three years later, in 1907, Verneuil reported (A8) an annual ruby production of 5 million carats.

Yet an account such as that of the previous paragraphs is deceptive in its simplicity. A careful search of the literature [1] shows that Verneuil had actually begun work on the fusion growth of ruby 16 years before his first public announcement in 1902, and had solved all the essential problems in six years with his second sealed note of 1892 (A4). We do not know the reason for this long delay in publication. It also appears to have been the only time in his career that Verneuil used sealed notes to establish priority of invention. The origin of this work presents a fascinating section of history.

2.1. "Geneva" synthetic ruby

In 1886 P.M.E. Jannettaz [10], a mineralogist and gem expert at the Paris Museum of Natural History, was shown by dealers some small rubies for which a natural origin was at first claimed. These were in fact the "Geneva" rubies which were later also erroneously called "reconstructed" or "reconstituted" rubies [1]. Jannettaz agreed with M. Friedel (a professor at the Sorbonne) and M. Vanderheym (President of the Syndicate of Diamonds and Precious Stones), who had also examined such rubies [1] that the spherical bubbles they contained indicated a synthetic origin, probably by fusion. George F. Kunz in his report to the New York Academy of Sciences, came to a similar conclusion [11].

Accordingly, Jannettaz discussed the matter with his associates at the Museum to see who might have the appropriate equipment to confirm this conclusion. Verneuil and Claire Auguste Terreil, a chemist, used an oxygen–hydrogen torch in the laboratory of Alexandre Leon Etard to fuse some powdered alumina containing a little chromium. They obtained only tiny specimens, the size of the head of a pin, but Jannettaz nevertheless was able to demonstrate that these were single crystals and gave the same strong fluorescence in Crooke's tube (cathodoluminescence) as the "Geneva" rubies. This was significantly different from the weak cathodoluminescence of natural rubies, thus confirming the fusion origin of the "Geneva" specimens.

Jannettaz "left to these gentlemen the task of reporting themselves how they had performed" these experiments [10], but they never did so. It appears that this chance request led Verneuil at the age of 30 to try a second approach to the synthesis of ruby, since at the time he was still working actively with Frémy on their joint experiments.

A recent study [1,12] has revealed the previously unknown process by which the "Geneva" rubies had been made. This was in fact a flame–fusion process, using the complex three-step technique illustrated in Figure 6. Other characteristics differing from the later Verneuil technique were the rotation of the growing boule and, at least in the last step, the use of two torches and the delivery of the feed powder down a platinum tube [1] in between the torches. It is possible that such a tube was also used in the earlier steps A and D of Figure 6.

The complexity of this process undoubtedly originated in the realization that cracking of the completed boule arises from intimate contact with the support if the process of step A of Figure 6 is continued to form a large crystal. As will be seen, Verneuil faced the same problem in 1891, but found a different and much more practical solution.

2.2. Verneuil's flame fusion experiments

Having thus been introduced to the concept of making ruby by the fusion of purified alumina containing

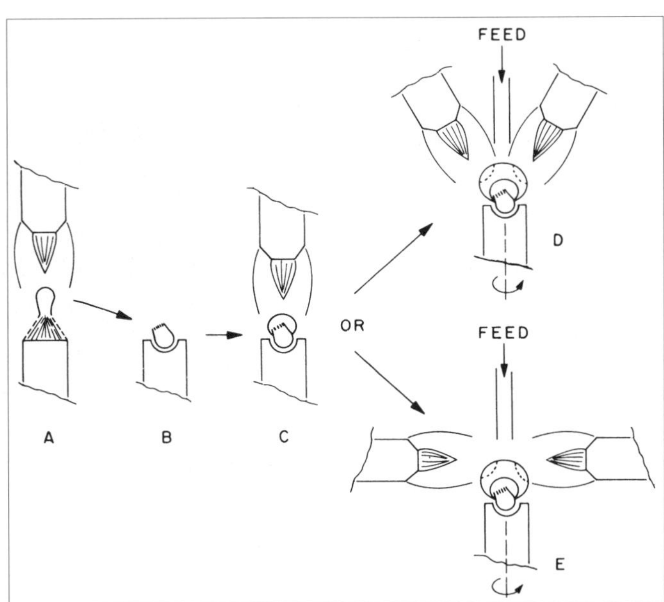

Fig. 6. Flame fusion technique used for the growth of "Geneva" synthetic rubies.

chromium oxide, Verneuil re-examined (A7) the publications of Gaudin (Marc Antaine Augustin, 1804–1880) who had attempted for more than 33 years to synthesize ruby. Gaudin had come to the erroneous conclusion that he had produced a ruby glass by the solidification of the molten material, and that the cracking which he saw represented the devitrification of the glass.

With the added clue from his "Geneva" ruby examination, Verneuil now recognized Gaudin's misinterpretation and set himself to investigating solidification of the melt as an alternative to the multiphase Frémy–Verneuil growth, which by 1886 had become stalled. Even very large scale experiments had not succeeded in increasing the size of the crystals appreciably. At the time the only controllable source of heat to reach 2050 °C, the melting point of Al_2O_3, was the hydrogen–oxygen or gas–oxygen blow torch ("chalumeau" in French). It was to this that Verneuil now turned. In this new approach to the synthesis of ruby, with which Verneuil was to occupy himself for the rest of his life, he was assisted by Marc Pacquier, and he was "happy to thank my student Mr. Marc Pacquier for the most active assistance which he has given during this long work" (A5).

By 1891 they had progressed sufficiently that Verneuil wrote down the major details of the flamefusion process as we know it, sealed the document and deposited it with the Paris Academy of Science. The only problem with the process was severe cracking of the ruby, which he could not control. Within a year he had solved this problem also by making the contact area between the support and the growing boule as small as possible, as described in his second sealed note of 1892. Both documents were opened at Verneuil's request in 1910 and were published in summary form (A4).

A Mr. Pacquier exhibited ruby crystals at the Paris World's Fair in 1900 and these were examined by Friedlander [13]. A close study [1] of this report indicates that the rubies were made by the newer Verneuil flame fusion process and not by the "Geneva" process. Friedlander states that the rubies were manufactured in Paris and he implies that such rubies reached the jewelry retailers as natural rubies. Friedlander also reported that "the unknown manufacturer was able to sell all of his production in France and Germany," and that "he deserves this reward since it took him

a very long time to accomplish this." We cannot be certain if this Mr. Pacquier was Verneuil's assistant Marc Pacquier, or whether others had learned of Verneuil's work and had independently begun commercial production.

2.3. *The 1904 flame fusion report*

Descriptions of the Verneuil technique usually outline the modern process which yields boules perhaps 3/4″ in diameter and several inches long, as for example the highly automated Russian unit of Popov [14]. The process was described in detail by Verneuil in 1904 (A7), the only time when he gave full construction and operation details. The apparatus used was quite small and is shown in the drawings of Figure 7. The actual apparatus shown in Figure 8 is still in existence and may be seen in Paris at the Conservatior des arts et metiers, together with the specimens shown in Figure 9 and others.

A satisfactory crystal growth run according to Verneuil would yield a 2 1/2 to 3 gm ruby of 5–6 mm diameter (12–15 carats, a little less than 1/4″ diameter). A set of Verneuil's boules at various stages of the process is shown in Figure 9. After a growth period of a little over two hours the gas and oxygen

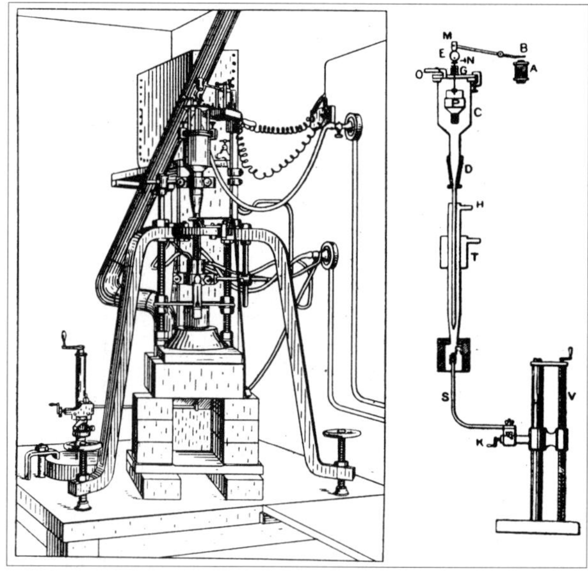

Fig. 7. Drawings of Verneuil's flame fusion apparatus (from Ref. A7).

were shut off abruptly, and after 10 minutes the boule, still attached to the alumina support rod, could be removed, to be separated only when cold. If the highly strained boule did not split into half by itself, a slight blow with a hammer would achieve this result and the quite strain-free halves were ready for delivery to the lapidary for faceting.

Among the interesting features not now used was a cover on the viewing window on the muffle R. The window was 15 mm high by 8 mm wide ($6/10''$ by $1/3''$) and was normally covered by a metal slide.

Fig. 8. Verneuil's flame fusion apparatus, now in the Museum of Arts and Sciences in Paris (from Ref. [4]).

On raising the slide a mica window in the slide would then permit viewing the growing boule. When closed the mica was protected from the flame so that it would not deteriorate. The part of the apparatus that appears to have given Verneuil most trouble was the tapping mechanism A, B, M, etc. There is a detailed description of the electro-mechanical device illustrated in Figures 7 and 8, but a little later Verneuil was to use a purely mechanical system much like the one in use today.

An oxygen-illuminating gas (coal gas) combination was used and a small water jacket T high up on the torch was adequate to cool the blow torch in this small (by present day standards) apparatus. The inside dimensions of the muffle (R of Figure 7B) were 105 mm high and 25 mm diameter ($4'' \times 1''$). The pedestal on which the boule grew was 3–4 mm in diameter (about $1/8''$) and consisted of an alumina rod held by a platinum tube attached to the bent iron rod S which was manipulated by the adjustments K and V, the latter with a 2 mm pitch screw thread. The gas flow was kept constant, but the oxygen flow was carefully regulated according to an elaborate schedule. A total of 100 to 110 liters of oxygen was used for each carat of ruby product.

Verneuil's feed powder was a mixture of ammonium alum and chromium alum heated to red-heat and ground and screened. It contained 2.5% chromium oxide. It was necessary to purify the alum by recrystallizing four to five times from water solution to avoid an orange cast in the ruby.

Verneuil saw three major factors to be observed: (1) to use a flame rich in hydrogen and carbon to prevent bubbling of the molten ruby and avoid the

Fig. 9. Set of specimens showing sequence of growth of a boule of ruby (from Ref. [4]).

introduction of gas bubbles; (2) to effect a gradual solidification of thin layers from the bottom upward to maintain transparency; and (3) to maintain a minimum contact area between the support and the boule so as to minimize cracking. By careful control of the gas flow and feed rate he was able to obtain crystals free of bubbles, but not free of the curved growth rings, which he reported as being the feature to distinguish his rubies from natural ones.

3. Blue sapphire and the last years

3.1. The middle years (to 1905)

Verneuil lived at the family residence at 25 Rue Humboldt in Paris while he was working on the flame fusion technique and until 1904. The three brothers, Auguste, Emile (two years older), and Ernest (four years older) had attended the same school, the Institution Leroy on the Boulevard Arago. The experience of having suffered the parsimonious conditions at "father and mother Leroy" helped to bind them together into a closely knit group. Among Auguste's classmates was the young Millerand (Etienne Alexandre, 1859–1943) who was to be the President of France from 1920 to 1924.

Music was one of the frequent diversions in the Verneuil household with August at the piano, Emile at the cello, and Ernest playing the flute. Emile, a jeweler, died in 1888 at the age of 30. Ernest studied electrical engineering and was able to help Auguste with some of his electrical equipment problems. He outlived Verneuil and died in 1924, leaving three children. Another member of the household was "Coco," a small monkey who entertained everyone with his antics. Since the father and the three brothers all had different professions and moved in different circles, the conversation at dinner was always lively and varied, with a true cosmopolitan flavor.

With the death of the father in 1904, the household broke up. Auguste inherited many of the pictures, statues, and other objects of art which his father had collected over the years and moved into his own apartment at 80 Boulevard St.-Germain. With him went the elderly Lucienne Tarone, who had been servant to the family for many years.

3.2. Professor Verneuil and blue sapphire (1905–1913)

In 1905, at the age of 49, Verneuil was appointed Professor at the National Conservatory of Arts and Sciences in Paris, the "Sorbonne of the Industrial Sciences" as it has been called. He succeeded Professor V.H. de Luynes (1829–1904) in the chair of "Lime, Cement, Ceramics, and Glassmaking." Much of his time was now taken up with teaching.

His course in Industrial Chemistry consisted of a three-year program of 120 lectures. Verneuil prepared his lectures carefully, with a separate folder containing his notes for each lecture. Since many industrial chemists attended these lectures to improve their knowledge and keep up to date with technical advances, the classes started at 9 p.m. There were usually 100 to 150 attendees. Verneuil became friends with many of his students, and after class a small group would usually join Verneuil at a local bistro to continue technical and undoubtedly also other discussions, well past midnight.

He nevertheless continued with his industrial consulting work, but most of his spare time was taken up with the problem of the nature of blue sapphire. At first it had been thought that a lower valence state of chromium was involved, since in his work with Frémy some violet and blue crystals had been obtained under what were thought to be partially reducing circumstances (A8). This was, however, due to the accidental presence of iron and other contaminants (A7). Later Mr. Paris of the Pasteur Institute concluded that cobalt oxide gave the blue of sapphire. However, calcium and magnesium oxides had also been added [15,16]. As Verneuil pointed out (A7), the addition of MgO changes corundum to spinel. Now cobalt does give a blue coloration to spinel, having a somewhat different shade from the blue of sapphire; cobalt by itself, however, does not impart any color to otherwise pure corundum.

In 1909 Verneuil added to his various activities by becoming chief chemist to the firm of L. Heller and Son, of New York and Paris. In a series of papers and patents (A8, A11 and A12) he worked out the full details of the manufacture of blue sapphire, using a combination of iron and titanium oxides and showed by the careful analysis of sapphires from Australia, Burma, and Montana (A10) that all of these contained iron oxide and, in addition, previously unsuspected titanium oxide. To duplicate the color of natural blue sapphire it was necessary to add about 1.5% iron oxide and 0.5% titanium oxide or their equivalents to the alumina feed powder.

Verneuil believed that ferrous iron was produced in the reducing part of the flame, and that this was subsequently oxidized to ferric iron by the titanium, which accordingly achieved a lower valence state which produced the color. Interestingly enough it is only very recently that the mechanism producing this color has been fully explained; Verneuil had actually come very close to the correct explanation. The process of "intervalence charge transfer" is at work which, in this case, involves Fe and Ti, each of which can be in two valence states. The transition involves the energy levels of the charge transfer process

$$Fe^{2+}-O-Ti^{4+} \Leftrightarrow Fe^{3+}-O-Ti^{3+}$$

These transitions occur with the absorption of light and are the ones that give the intense blue color in Al_2O_3–Fe–Ti [17].

This work was done for the New York City based firm of Lazarus Heller and Son (later to become the Heller–Hope Co.) which had a branch in Paris. They were dealers in doublets, imitation pearls, and various other imitation stones shortly after the turn of the century. They had felt the need for a synthetic blue sapphire to join synthetic ruby in rounding out their line and accordingly employed Verneuil as research director of the laboratory they established in Paris. A chemistry instructor from the City College of New York, I.H. Levin, was brought in to act as the intermediary between the Hellers and Verneuil. He supervised the Paris laboratory, views of which are shown in Fig. 10 and Fig. 11.

Abraham A. Heller, the son of Lazarus, was in charge of the Paris office of the Heller company from 1903 to 1910, and can be seen in Fig. 11; he lived on a farm, in Bernardsville, New Jersey, shortly after returning from France. Also in the photograph are H.W. Friedland, who was in charge of the Paris branch from 1912 on, and Mr. Spec-Torsky both relatives of the Heller family; the latter had a firm "Pierres Fine Reconstitutées" in Paris which cut and sold synthetics and imitation stones.

One of the problems in Verneuil's work for Heller was the large quantity of oxygen needed for the

Fig. 10. Prof. Verneuil (left) and Mr. Spec-Torsky in front of a flame fusion apparatus about 1910 (from Ref. [2]).

many burners seen in Fig. 11. I.H. Levin worked on this problem and developed a special electrolysis apparatus, the "Levin cell" for decomposing water into hydrogen gas and oxygen gas. This proved an important process for the International Oxygen Co. (now part of Union Carbide) for whom Levin next worked back in the U.S. He reported an annual production rate of 10 million carats ruby and 6 million carats sapphire in 1913 [18].

With the successful completion of the sapphire work and the patent applications of 1910 and 1911, the Heller laboratory in Paris was disbanded and the apparatus and stock were taken over by the Baikowsky Co., now at Annecy, Savoie, France. The Heller company reportedly had intended to set up manufacturing facilities in the U.S.A., but found that in 1911, as today, the labor costs were high and that it was more economical to import the faceted stones from Europe.

It is instructive to examine the difference between the 1910 U.S. Patent No. 988,230 (A11) and the 1911 U.S. Patent No. 1,004,505 (A12). Both were assigned to L. Heller and Son of New York, NY, and the draw-

Fig. 11. View of the laboratory in Paris about 1910. (After an old post card, courtesy of Mr. H.W. Friedland.)

ing sheet of the second is shown in Fig. 12. The first was based on a description which stated that the sapphire produced by following the patent revelation was "exactly similar in appearance to the blue sapphire found in nature." It apparently was soon realized that this was much too restrictive a claim and would not give adequate patent protection. The following year the second patent, identical in almost all other respects, speaks instead of a product "as closely resembling the natural sapphire as possible," and in each of its claims includes the phrase "having beneath its surface bubble-like spots bounded with rounded walls." This is a much more realistic description of Verneuil-grown sapphire than the previous one. Curved growth lines and curved cracks are also mentioned as distinguishing features compared to natural sapphire.

3.3. The last years

Throughout his life Verneuil had been interested in the arts as well as in the sciences. He had a profound interest in music, playing the piano, in painting, and in the plastic arts, particularly antique pottery. One of Verneuil's last three projects, published only in abstract form, dealt with his successful duplication [19] of the iron-containing black glaze, showing green reflections, which had been used on old Greco–Italian pottery, a subject which had long interested him. The

second dealt with a subject always of interest to a Frenchman, the blending of brandy [20].

The third described a novel gas-fired muffle furnace in which the flame was injected tangentially between the concentric furnace housing and the crucible [21].

With his many activities of teaching, research, and industrial consulting, exhaustion set in, complicated by diabetes. He went on several trips to rest at the resort of Nice in 1912 but this did not appear to help. He continued to teach as long as possible, but by January 1913, he was no longer able to leave his room. Nevertheless he still continued to serve as consultant from the couch in his living room. He died on April 13, 1913, at the age of 57, in the arms of his brother Ernest.

His funeral was held on April 30, 1913, in the local parish church, and obituary speeches were made by a number of his friends and associates: Professor Fleurent of the National Conservatory of Arts and Sciences; L. Lindet, President of the Society for the Encouragement of National Industry; E.D. De Laire, President of the Chemical Society of France; and Professor Maquenne of the Museum of Natural History of Paris. The obituary proceedings were published as a special issue of the Bulletin of the Students of Frémy [2].

From the age of 17 on Verneuil had worked his own way, acquiring degrees and other honors along the way. For his period this was not the usual path, nor

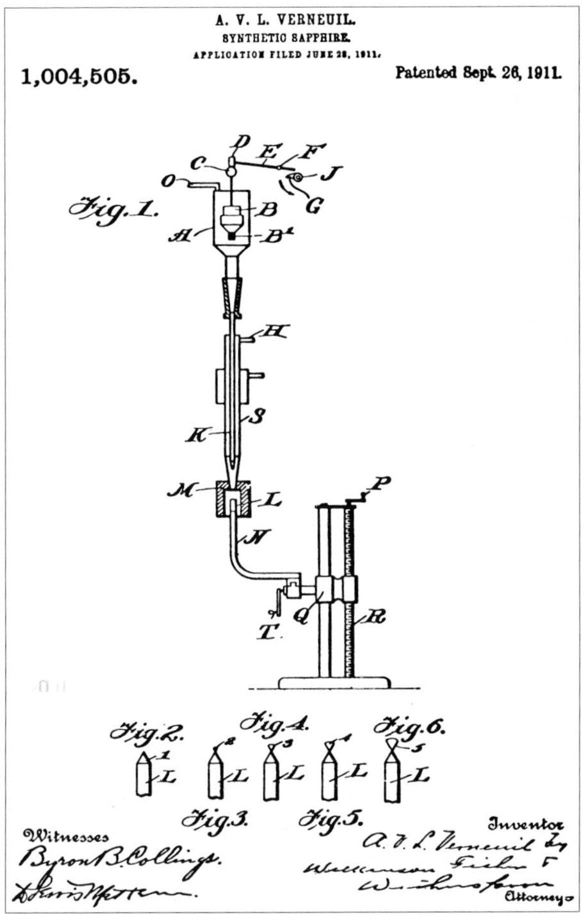

Fig. 12. Drawing page of the Verneuil Patent of 1911.

the relatively easy process it has become today. There is in fact some evidence in the lack of recognition of his achievements by his peers and successors, that the self-made practical man was not as highly regarded as the pure academic by the "establishment" science of that day.

During his 40 years of professional work, Verneuil's name appeared on some 70 publications and two U.S. patents. Particularly if one makes allowance for much unpublished work performed for industrial concerns, a body of achievements of most respectable proportion is evident. In addition we must not forget his teaching, at which he was excellent: he inspired both respect and affection in his students. Yet when all else has been said, Verneuil will always be remembered as "The Father of Synthetic Ruby."

Acknowledgements

The authors are particularly grateful to Mr. A. Verneuil, Professor Verneuil's nephew, of Marseilles, France, for much unpublished information; also for information, photographs, etc. to Mr. E.S. Heller, Mrs. G. Heller, Mr. J. Heller, Mr. H.W. Friedland and the many others acknowledged in Reference [1].

References

[1] References to the history and the current state of the synthesis of ruby will be found in the series by K. Nassau and R. Crowningshield, Lapidary Journal 23, 114–119 (April 1969); 313–314, 334–338 (May 1969); 440–446 (June 1969); 621 (July 1969).

[2] "Auguste Verneuil, 1856–1913"; Extrait du Bulletin de l'Association des Eleves de Frémy; about 1914; 39 pages and 3 plates. Contents: Auguste Verneuil p. 3, Discours de M. Fleurent p. 7, Discours de M. Léon Lindet p. 15, Discours de M. Maquenne p. 21, Discours de M. Barthélemy p. 25, Notice etc. par M.L. Maquenne p. 29, Liste Chronolique des Travaux d'Auguste Verneuil p. 37.

[3] L. Maguenne, Soc., Chim. Bull. 13, Obituary Section 1913, 7 pages text, 1 portrait, and 3 pages references.

[4] H. Lafuma, L'oeuvre d'Auguste Verneuil, presented at the conference of July 3, 1954 on the occasion of the 40th Anniversary of the founding of the H. Djévahirdjian Company of Monthey, Switzerland.

[5] A. Verneuil, unpublished information (1968).

[6] A. Verneuil, M.L. Bourgeois, Paris Acad. Sci., Comptes rendu 90 (1880) 223–225.

[7] E. Frémy, C. Feil, Paris Acad. Sci., Comptes rendu 85 (1877) 1029–1035.

[8] E. Frémy, Synthése du Rubis, Librarie des Corps Nationaux, Paris, 1891.

[9] H. Michel, Künstliche Edelsteine, W. Diebener, Leipzig, 1914, p. 13.

[10] E. Jannettaz, Bull. Soc. franc, mineralogie, Paris 9 (1886) 321–322.

[11] G.F. Kunz, Transactions of the New York Academy of Sciences, October 4, 1886, 3–10 (1886).

[12] K. Nassau, J. Cryst. Growth 5 (1969) 338–344.

[13] I. Friedlander, Verh. Ver. Beford, Gewfleiss, Sitz B, 71–80 (1901).

[14] S.K. Popov, in: Growth of Crystals, Vol. 2, Consultants Bureau, New York, 1962, p. 103 (translated from The Russian "Rost Kristalov").

[15] L. Paris, Paris Acad. Sci., Comptes rendu 147 (1908) 933.

[16] Anon., Sci. Am. (December, 1908) 452.

[17] M.G. Townsend, Solid State Comm. 6 (1968) 81.

[18] I.H. Levin, J. Ind. Eng. Chem. 5 (1913) 495–500.

[19] A. Verneuil, Chem. Ztg. 35 (1911) 259.

[20] A. Verneuil, Rev. Viticulture 37 (1911) 131–135.

[21] A. Verneuil, 1912: Chemical Abstracts 7, 876 (1913).

Remembrances of flame fusion

Leon Merker

The events that influenced the path that led me to crystal growth were more public, I would imagine, than those of most of us. In the Spring of 1938, in the second year of my chemistry studies at the University of Vienna, German fascism penetrated our country with the Anschluss. Within days, I and other Jewish students and teachers were forced from academic life. I had already made initial application for a visa from the United States' embassy, and I settled down to a summer of tennis while friends with Nazi connections kept me advised on our situation. By September, things had seriously deteriorated, and I left for Milan, Italy to see which way history would go.

The Italians, to their credit, had not included anti-Semitism in their brand of fascism, and we felt safe for many months. Meanwhile my American visa application continued to progress, and I was invited to the consulate in Naples for an interview and routine medical examination. My waiting would soon be over, I thought. All came to naught when the consulate physician pronounced my eyes incurably diseased with trachoma. Even with the statements of eminent Italian physicians to the contrary, the consul told me during a later appeal, "I believe that you are right, but no authority can contradict the consulate doctor." The entry of German troops into Italy changed the mood of that country, and I moved to a refugee camp in England.

After a few months there, I became aware of a scholarship that was being offered at the University of Michigan with funding by B'nai B'rith and the Jewish Refugee Committee and housing and meals supplied by Pi Lambda Phi. At first my hopes were small;

three hundred of us applied. But my optimism grew when I learned that the judge of the applications was the renowned chemist, Prof. Fajans. Only twelve of the applicants were chemists. My mentor in Vienna, Prof. Mark, who I had remembered as a passionate anti-Nazi, had fled to Canada. It was there that my letter asking for his support reached him. He wrote such a glowing recommendation that I got the award. I made arrangements to go to the embassy in London, now as a traveling scholar, and to my utter chagrin found the same doctor that had seen me in Italy behind the dispensary door. He didn't recognize me at first, and when I reminded him of our earlier meeting, he said, "Well you're better now." He obviously had acted earlier according to orders from his superiors because trachoma was at that time considered incurable without surgery.

I studied at Michigan for two years and completed a Masters in Chemistry. Near the end of my research, one of the boys at the fraternity said that I should meet his grand uncle, a Mr. Heller from France, and further that he was looking for someone to join one of his ventures. (The Heller family is named in Verneuil's patents as assignees. They financed his work.) Mr. Heller was in his 70's in 1940, a gracious businessman who wanted to re-establish his material business and help the war effort. Jewel bearings were an important part of timekeeping and navigational instruments. After we got to know each other, Mr. Heller explained the Verneuil process that had been running for forty years and stated that he had $100,000 to invest in a new flame fusion venture in North America. That was quite a sum in those days. His initial goal was

Previously published in AACG Newsletter Vol. 21(2) Autumn 1991.

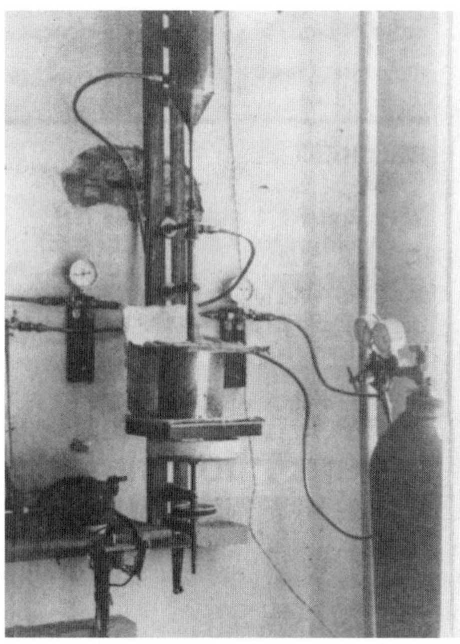

Fig. 1. The 'hand-made' burner used to produce our first usable ruby boule in 1941 at the General Synthetics Corp. in Newark, New Jersey.

to produce useful material within a year. If we were successful, we would easily expand the operation; if not, we would shake hands and call the effort a good try. I accepted and we set up shop in Newark, New Jersey where we began work. Ruby and sapphire had been grown in Europe.

We knew in very general terms what to do, and soon had the equipment assembled. In eight months we gave Mr. Heller an acceptable 1.2 cm diameter by 3 cm long boule of ruby and with the help of the War Materials Board and his investors, we set up the General Synthetics Corporation in 1941.

In that first year, we installed twenty production machines. The military claimed all of our whole boules (red to be seen in the miniature movements); the cracked ones we sold in the gem trade. This dual role of maker of materials for technology and pleasing gems continued throughout my work. While we had some descriptions of the process, we put much of our own thinking into the equipment design. One day, early on, Mr. Heller stood on the production floor and said that something was wrong. The small strikers used to vibrate the feed hoppers above the furnaces didn't sound at all like his remembrances

of Verneuil's machines. I assured him that everything was all right and explained our changes. Our work continued throughout the war years, and he always complained that the machines didn't sound right.

It was at this time that flame fusion ruby production facilities were also set up by the Bulova Watch Company on Long Island and by the Linde Company in Chicago. Linde's then director of technology, Dr. Leo Dana, died recently in Canton, Ohio. The Bulova facility had two hundred burners. Linde developed a process for making spectacular ruby rods at that time, surely a precursor to their laser work. We grew ruby for two years, 1941–1943 and then with other North American ruby operators well established, we went on to sapphire and blue spinels. That little company in Newark was truly a child of Verneuil. The same family owned it as owned the original facility, and it was built from Verneuil's drawings. The plant closed at war's end in the face of aggressive competition from Linde.

Crystal growth as a serious profession, and more so Materials Science, was in those days not yet established in the minds of knowledgeable technical people. I recall once talking with a Swiss engineer from their ruby factory who had come to North America to build the Bulova and Linde plants. He expressed surprise when I mentioned that I hoped to make a career of crystals. "No future in it", he offered. "They've all been grown." History is full of these insights. I didn't take his advice.

With the war over and my employer shutting down, I signed on with the Commerce Department as an adviser to McCloy's occupation government in Germany as a member of 'Branch FIAT', the Field Industrial Agency Technical. This was a rather sizable group of American scientists and technologists who were there to understand technology developed by the Axis powers during the war. My group interviewed Prof. Nacken in Berlin and Prof. Kyropolous in Frankfurt. Our status as adjuncts to the military government had its interesting moments. We were often "commissioned for a day" as officers with sufficient rank to equal that of people with whom we might deal. The other three great powers were there and it was common for us to interact with their military. I particularly remember the beautiful spinels grown by the people of Wiede Karbidwerke in Bavaria, and the offhand refusal of the Russians to permit us to visit the

I.G. Farben works at Bitterfield, a place in recent news for its polluted environment. We were actually trained to drink vodka, this to minimize the Russian's standard tactic of applying a "morning after" to our interviewing whenever they could. We worked on our FIAT projects for a year, and then I returned to the States. I toyed with synthetic emeralds at the suggestions of another member of the Heller family until one day in 1948 when a friend passed on a brief news item describing attempts to grow rutile at a company called National Lead. I wrote them immediately and was soon talking with Dr. Charles Moore who was in charge of their crystal research.

NL had (and has, I presume) a strong interest in pigments, a carryover from the days when white lead oxide, litharge, was common in paints. Titanium oxide has now replaced its use and in the late 1940's, NL and Dr. Moore were interested in learning as much as they could about this material. While its pigment use was as a fine powder, the company wanted to have a large pure single crystal to support, at the least, a determination of its optical properties. They had begun an attempt to produce a flame fusion boule, using hydrolized titanium dioxide material and a burner with a very high flame velocity. We changed both these approaches and produced a fine crystal in six months.

That first work highlighted two aspects of flame fusion that may not be generally appreciated. It is important that the powder that is dropped into the top of the flame be completely melted by the time it falls into the liquid cap of the growing boule. With the time of exposure of the particles to the radiant gases of the flame limited by the flame length above the growing boule and the flame velocity and heat conduction within the particle supported by only moderate thermal conductivity, most flame fusion processes demand particles with maximum diameters all less than a critical value. Another way to say this is that flame fusion feed materials must have very high surface areas and appear to be fluffy and light. An example of this is the Al_2O_3, feed material used for now nearly a century of sapphire production. Its surface area is so high that you can touch it as it is removed from 1000-degree calciners; the particles that are touched haven't enough heat to appreciably warm the skin. The discovering of ways to make high surface area powders has been the unpublicized half of the flame fusion story. Against the advice of the resident chemists, we invested weeks of effort to find a suitable material. We were first successful with titanium ammonium sulfate that we 'roasted' in air to produce TiO_2. We looked for a molecule where titanium shared space with large neighbors that could be removed by a gaseous reaction. We played a developer's hunch, and it worked.

Another part of that work that was typical of flame fusion problems was the design of the burner and its resulting envelope of streaming gasses. Temperature gradients, gas phase chemistry, and windage effects on the liquid cap needed to be tailored for each problem. In this case we knew that rutile grown in reducing conditions would crumble. We devised an easily maintainable burner with three concentric gas flows: oxygen in the middle and outside and hydrogen in between. This is commonly called a tricone burner, and it proved successful for us in late 1948. The high index of refraction and strong dispersion of this crystal made it a desirable gem stone. We produced it for jewelry in the four years that followed. Successful as we were, the jewelers were never truly happy with rutile's off-white appearance. Many attempts to eliminate this undesirable color tint via doping failed.

With some measure of financial success achieved with our rutile work, we were given the freedom to select other materials for our process. Both the needs of technology and our friends, the jewelers, drove our decisions. Barium titanate was known then as a strong ferro-electric and we attempted this material several times, never achieving success. The boules uniformly broke up on cooling. It wasn't until Joseph Remeika gave us his beautiful butterfly twinned crystals that barium titanate was solved.

As I mentioned earlier, our work and that of other crystal growers wasn't appreciated at that time. Virtually no crystals, other than the polycrystals used in infra-red optics, were used in technology. The revolution of the transistor lay ahead. An example of this might be interesting. In 1948, the Mineralogy Department at Michigan held a colloquim on oxides. Bill Bauer of Rutgers and I both submitted papers on crystal growth. I looked forward to returning to my North American alma mater and was surprised to learn that both our papers were accepted by the program committee only by the narrowest of margins. Crystal growth wasn't yet understood.

Fig. 2. A typical strontium titanate boule after growth and cooling ready to be harvested. National Lead Company (ca. 1955).

After barium titanate, we turned to strontium titanate with much more success. As before, material preparation occupied our attention in the early stages of the work. We developed a double salt precipitation method to fix the stoichiometry of our feed material. In 1952, we achieved material of exceptional beauty. It was quickly accepted by the jewelers, and its hardness, high index, and infra-red transmission led to its use in the heat seeking device on the sidewinder missile. It enjoys current interest as a substrate for high-T_c superconducting films for low frequency applications where its high dielectric constant isn't important. While our material was well accepted, our process could not deliver long boules. All attempts to produce long crystals failed with cracking and inclusions. After we observed rutile needles in a boule and realized that strontium oxide must be preferentially vaporized from the liquid cap and the melting of our rigorously stoichiometric feed particles, we soon established that an additional 3% of strontium oxide allowed us to grow beautiful boules as long as we wanted. Linde, to their credit,

brought the rutile needles under control and enjoyed a long success with their star sapphire gems.

Again, with strontium titanate as well as with rutile, the jeweler's initial acceptance waned in the absence of the elusive 'water white' crystal. All our trace chemical analysis failed to show anything that might be the cause of faint colorations. One day, almost by accident, we noted that the crystals changed color on UV exposure. We knew that photo-chromic effects were probably impurity related, and I prepared a recommendation to NL management for the pressurization of the entire building and what is now known as clean room facilities. They laughed. With the help of friends in the Chemistry Department at Wagner College, we prepared a batch of material under ultra-clean conditions and grew a boule that nearly satisfied the jewelers. Company management then renovated our building and our material improved measurably, but faint color tints still appeared occasionally in our boules. This kept us from reaching the mainstream of the market. Strontium titanate needed yet a little more work.

We speculated that tetravalent ions were at the root of our problems and that additions of penta- and hexavalents might help. Niobium and tantalum worked, but it was difficult to limit our additions to avoid a blue color that came with over-doping. We settled for a procedure in which we under-doped with tantalum or niobium and annealed our boules in steam to trim in the color. That proved to be the needed recipe and we operated twenty production burners continuously for many years. In fact, they're still operating in Florida under the care of Commercial Crystals, Inc. Strontium titanate was invented in 1953; its development and improvements, particularly with respect to color, size, internal quality, and strain continued to go on for many years. Because it was used in a great variety of applications (gems, optics, electronics, etc.), the market requirements greatly influenced our direction of work.

We worked for a short time with nickel titanate growing boules, but found no exceptional uses for them. Next we took on calcium titanate as a defense of our gem business. Our work on this did not appear until ten years later in the Journal of the American Ceramic Society. The acceptance of that paper was a far cry from the reluctance to recognize the scientific merits of our work fourteen years earlier.

The 'excuse' for attacking the growth of calcium titanate was to protect our profitable strontium titanate gem business, because it could be presumed that its optical properties might be similar to those of strontium titanate. My personal reason for initiating this project, difficult as it might be, was to show the then prevalent characterization of the flame fusion technique as "Not suited to grow a decent quality crystal" was largely based on ignorance.

Calcium titanate, the classical Perovskite, has a slightly distorted cubic structure. At that time, no other crystal growth technique had produced single crystals of a sufficient quality to permit even the identification of its crystallographic parameters with accuracy. 'Standard Verneuil conditions' produced only badly fractured and severely twinned crystals. After we developed a controllable boule growing furnace based on silicon carbide heaters which permitted reduction of the temperature gradient during growth and programmed boule cooling afterwards instead of the usual quench, we were able to obtain large truly single crystals.

The results of this two-year effort are of great meaning to me; they reinforce my contention (which I have publicized for many years) that the flame fusion technique, although practiced on a commercial scale for about seventy years, is essentially still in its infancy. The work with calcium titanate serves to show what dramatic results may be obtained by even minor improvements of Verneuil's method. It may be assumed that further technological improvements along these lines should produce flame-fusion grown crystals of a quality similar, or even better, to those grown by other techniques. We worked with lanthanum aluminate and several other titanates, yttrium and zinc among others in the following years, but none of our work brought more pleasure than the first sound large calcium titanate boule.

The world's interest in flame fusion titanates waned during the 1970's, and NL left the business with the sale of the equipment. Recent interest in strontium titanate as a substrate is the exception. Consumption of other Verneuil materials, ruby, spinel, etc., is as strong as ever.

Thanks to Dr. Reed Kinloch, who also began his crystal growth work with Verneuil's method, for suggesting that this memoir be undertaken and for his help in its preparation.

Evolution and application of the Kyropoulos crystal growth method

David F. Bliss

Rome Laboratory, Hanscom, MA 01731

1. Introduction

Although the Kyropoulos growth method has been used for more than 65 years to produce single crystals of many types, it has not aroused the scientific interest devoted to other techniques such as Bridgman or Czochralski growth. The story of Dr. Kyropoulos and the evolution of his growth technique have been overlooked in most studies of crystal growth. Despite such neglect, his method has survived and evolved over the years, wherever growth under low thermal stress is a requirement. Today, as these requirements become more stringent, the Kyropoulos method may now experience renewed activity and study.

Spyro Kyropoulos first proposed this technique in 1926 at the Physical Institute in Gottingen Germany, as a means of producing large single crystals of alkali halides free of cracks and damage due to confinement. Until then, the only known method for producing "large" single crystals was the Bridgman technique. Czochralski growth, invented in 1917, was used exclusively to grow thin metal whisker-like wires. Kyropoulos wanted to grow crack-free alkali halide crystals for precision optics. The Bridgman method was unsuitable because the container caused stress upon cooling, which resulted in flaws or cracks in the crystal.

Since Kyropoulos' original work, the method has evolved, expanding on the original concept to meet the demand for growth of brittle crystals which must be

Fig. 1. Spyro Kyropoulos in the library. Photograph courtesy of the Beckman Institute, California Institute of Technology Archives.

grown in a relatively stress-free environment. Some large twelve-inch diameter single crystals of alkali halides were grown in factories in Germany (1940s) and the U.S.A. (1950s) using the Kyropoulos technique. The technique is well-suited to growth of materials having a low thermal conductivity and a high thermal expansion coefficient. These materials are subject to fracture and slip unless they are grown in

Previously published in AACG Newsletter Vol. 23(2) Autumn 1993.

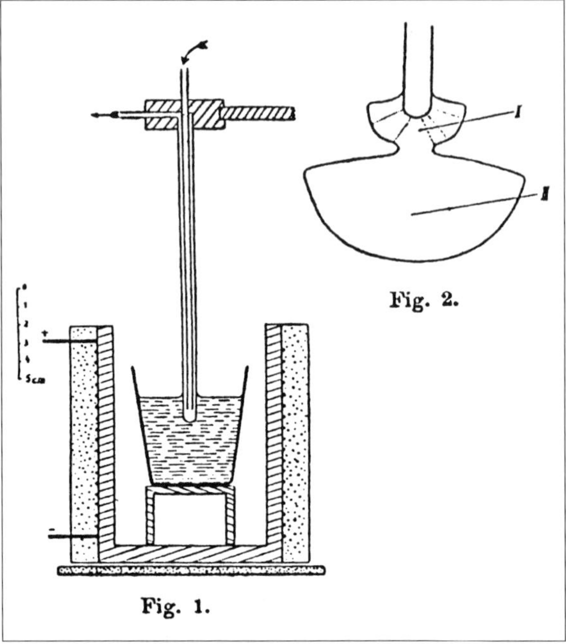

Fig. 2. Schematic of Kyropoulos growth method from his original paper (Ref. [1]).

a low-stress environment. During the last five years, perhaps because of increased demand for such materials, there has been a renewed research interest in Kyropoulos growth, with applications as diverse as oxide superconductors, compound semiconductors, and incongruent-melting peritectic substances.

This article reviews the evolution of the Kyropoulos method as it has been developed by crystal growth technologists to meet advancing materials needs. The requirement for new materials and materials with improved physical properties has caused many laboratories, including the author's, to draw on history to find a solution. For structural perfection and impurity control exceeding those which are available today from pulled crystals, the Kyropoulos method offers a useful alternative.

2. Historical development

While he was an Assistant Professor of physics at Gottingen University in the 1920's, Spyro Kyropoulos reported on a method of growing "large" crystals from the melt [1]. Departing from earlier work by

Tammann, Kyropoulos was able to demonstrate single crystal growth of many alkali halides from the molten salts. He dipped an air-cooled shaft into the melt to nucleate a clump of multi-crystalline material and then raised the shaft until only the bottom of the clump remained in the melt. Crystallization at the remaining point of contact favored the fastest growing orientation, so that a single crystal seed was obtained. Then by cooling the melt slowly he obtained a large single crystal. When the crystal achieved a dimension of several centimeters, he pulled it out of the melt. His professor, Tammann, was famous for a single crystal seeding technique which relied on controlled supercooling of the melt within a capillary tube. The main advantage of his technique over Tammann's was that crystals grown in this manner were free from the possibility of fracture due to confinement of the crystal. He also claimed good visibility and control of heat extraction through the air-cooled shaft. Since the apparatus did not provide for rotation of either the crucible or the shaft, the crystal shape was generally asymmetric. Kyropoulos' method required careful observation to determine whether the crystal was single or not; in effect, he created a new seed with every growth. Four years later, in 1930 when he was a fellow in fluid mechanics, Kyropoulos published the dielectric constants of the sodium, potassium and lithium halides grown in the original work as well as thallium and rubidium halide crystals [2]. The original apparatus was used with the addition of a micrometer to adjust the height of the seed shaft. The Gottingen school remained active in Kyropoulos growth for several years. In 1933, for example, Korth described the growth of 6 cm × 8 cm single crystals of KBr and KI [3]. Korth used a seed attached to a water-cooled shaft, and a water-cooled coil above the melt to increase axial heat flow. Kyropoulos was promoted to the position of assistant professor (Privatdozent) in Physical Technology in 1931, under the sponsorship of Prandtl with whom he studied fluid mechanics. He continued his work at Gottingen until 1936 when after a protracted trial he was dismissed from office by the Nazi government.

Further refinements were made to the Kyropoulos method during the 1930's and 1940's. Katherine Chamberlain [4] at Wayne University (now Wayne State) in Detroit, Michigan reported on growth of KBr crystals five inches in diameter weighing 6–7 lbs. in

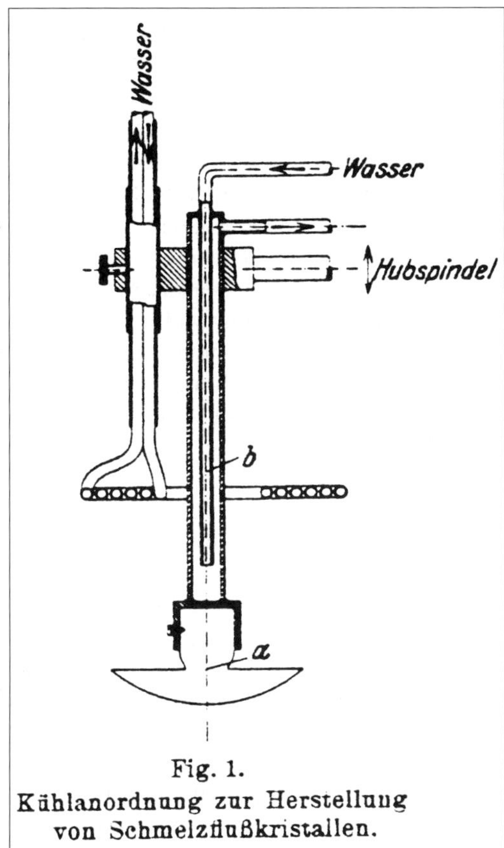

Fig. 1.
Kühlanordnung zur Herstellung
von Schmelzflußkristallen.

Fig. 3. Modification of Kyropoulos growth by Korth from the Institute at Gottingen, Germany (Ref. [3]).

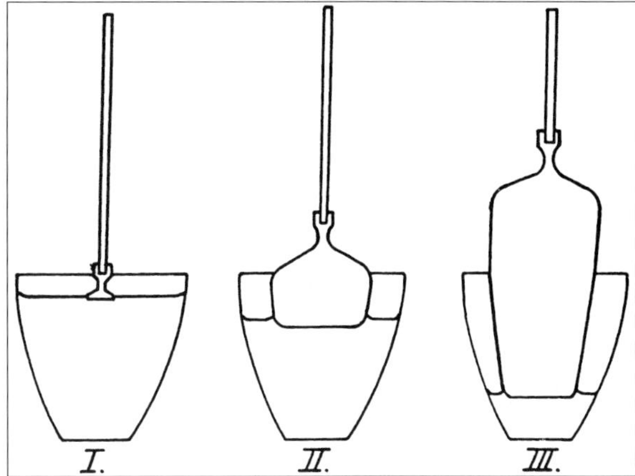

Fig. 4. "Three stages of Kyropoulos growth" from paper by Morgenstern (Ref. [5]).

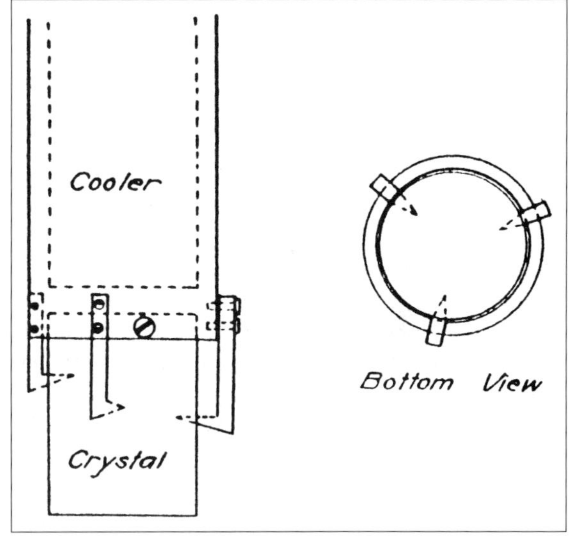

Fig. 5. Katherine Chamberlain's adaptation of the seed holder for Kyropoulos growth (Ref. [4]).

1938. She developed a seed clamp to hold a machined seed, but for crystals of this weight, seed breakage becomes a serious problem. In 1938 at the Physical Institute in the free city of Danzig (now Gdansk, Poland) Morgenstern [5] described Kyropoulos growth as follows: "The vertical growth from the melt as shown below has three definite states. The rod is dipped into the melt and withdrawn immediately. Part of the melt adheres to the rod and draws itself into a thin thread, which through a necking procedure similar to Bridgman growth, becomes a single crystal. The rod with the attached crystal slowly gets pulled upward. The single crystal begins to grow and gets a tetragonal cross-section. The third stage becomes the finished single crystal."

During the 1940's the Kyropoulos method was used for commercial production of prism crystals. Dr. Korber at I.G. Farben grew 12-in diameter, 6 in high NaCl single crystals. He used slow seed rotation to produce large cylindrical crystals. In a survey of crystal growth from 1931 to 1946, Wells [6] referred to three general methods for obtaining single crystals: Bridgman growth, Kyropoulos growth, and solid state recrystallization. Not mentioned was the Czochralski technique, which until the 1950's was used for growing thin metal wires. The Faraday Society dedicated their annual meeting in April, 1949 to a general discussion on crystal growth, with many international guests present. Menzies and Skinner [7] reported on the Kyropoulos method, pointing out that crystal size and shape are limited by the amount of heat extracted through the seed. They recommended a metal seed-chuck for maximum cooling. In Buckley's [8] book "Crystal Growth" the Kyropoulos method is described as a means of growing large crystals for infra-red lenses and polarizers, but with the limitation that it was not suitable for anisotropic crystals (unless growth is in the direction of maximum heat flow).

With the advent of silicon crystal pulling in the 1950's, Czochralski growth attained hegemony as the preferred method of producing large single crystals. Since silicon and germanium have thermal conductivities an order of magnitude larger than most of the alkali halides, Czochralski growth is well suited to these elemental semiconductors. Research in Kyropoulos growth declined temporarily during the next two decades. Bonner and Van Uitert [9] used the Kyropoulos method to grow $MgPb_3Nb_2O_9$ from a wide range of melt compositions. Weller and Grandits [10] grew the tungsten bronze Na_xWO_3 by an electrolytic adaptation of Kyropoulos growth, producing single crystal cubes of 1 cm. Nicklaus and Fischer [11] controlled the radial gradient with a ring furnace above the melt to grow BaFCl, a very anisotropic crystal. Large single crystals of $KNbO_3$ with low defect density were grown by Fukuda et al. [12]. A model for Kyropoulos growth based on a simple mass transfer approach was proposed by Singh et al. [13], which concluded that the shape of the crystal was dependent on heat flow through the seed. More advanced modeling was done by Schonherr [14] for KCl and Duseaux [15] for GaAs. Jacob [16] grew GaAs by the liquid encapsulated Kyropoulos (LEK) method to produce 6-inch diameter crystals with very low thermal stress and low dislocation density. However, the deep level defect EL2 concentration was very non-uniform. Ahern et al. [17] used magnetically stabilized LEK to grow 2-inch InP. Zhang and Shen [18] grew homogeneous crystals of $NaNb_3O_8$ with low stress by Kyropoulos growth.

Five patents have recently been issued for modifications of the original Kyropoulos method, four in Japan and one in the USSR. The materials claimed in the patents range from compound semiconductors like GaAs to oxide superconductors, to incongruent melting solids like potassium tantalate niobate grown from a peritectic solution. Dr. Kyropoulos never patented his original invention. He fled from Germany in 1936 to America where he taught at Cal Tech from 1937 to 1957, later becoming a consultant to the Air Force Missile Development lab at Holloman Air Force Base. At Caltech and at Holloman he became internationally recognized as an authority on lubrication and bearing problems, a study which he began at Gottingen under Prandtl. He jokingly recounted his life under the "three dictators," Tammann, Hitler, and R.A. Millikan, the first Chairman of Caltech. Known for his acid wit, his mastery of many languages, and his spirited commentary on current affairs, Kyropoulos was a contributor in the broadest sense to the development of Physical Science.

3. Early modeling of Kyropoulos growth

Early attempts to characterize the shape of crystals grown by the Kyropoulos technique were based on a simple heat or mass transfer approach. The shape of the growing crystal can be described as an ellipsoid of rotation, a shape for which there is a relatively simple analytical solution (numerical solutions were not considered at that time) to the heat flow equation under certain boundary conditions. In the case where the submerged end of the seed is held at constant temperature understationary conditions, the shape is explained by heat flow through the seed. Although the amount of heat lost by conduction through the seed is relatively small compared to the heat lost to the ambient above the melt, the direction of heat flow is determined by the seed temperature T_0. The actual crystal shape is determined by a competition between the two heat transfer mechanisms: conduction through the seed and convection/radiation at the melt surface. The simple conduction model says that when the seed temperature T_0 is held very low, the crystal approaches a spherical shape. As the seed temperature is raised, the interface becomes flatter. However, conductive heat flow reaches a saturation point very quickly when the crystal radius reaches 6–8 mm. The model fails to take into account the effect of convection, either in the melt or in the ambient gas. In reality, convective heat flow from the melt surface begins to dominate the growth process. Convective heat losses depend upon the pressure, the aspect ratio and size of the container in which the crystal is being grown.

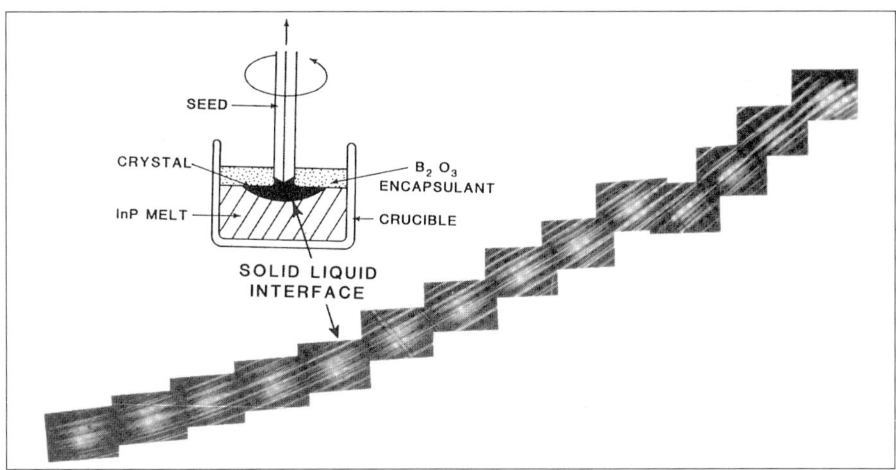

Fig. 6. Schematic showing MLEK growth of InP and NIR striagraph showing solid–liquid interface shape (Ref. [20]).

4. Kyropoulos growth of indium phosphide

In our laboratory we are exploiting the advantages of Kyropoulos growth to produce large twin-free indium phosphide (InP) crystals with low dislocation densities. In the past, Czochralski-pulled InP grown in the (100) direction has suffered from a high incidence of twinning; Hurle [19] has speculated that the angle of the Czochralski seed-cone is a major factor in twin formation. We have used a variation called magnetic liquid encapsulated Kyropoulos (MLEK) growth [20]. In MLEK growth there is no seed-cone; by growing a flat crown it is possible to grow twin-free (100) oriented crystals. In our adaptation we use a magnetic field to stabilize the crystal growth environment by reducing turbulence in the melt and by increasing the radial temperature gradient in the melt. Both effects contribute to the stability of crystal growth. Since the Kyropoulos method relies on heat extraction through the seed to initiate single crystal growth, a steep radial gradient assures heat flow in the direction of the seed. Unlike Czochralski growth, in which the solid-liquid interface remains at the melt surface while the seed is pulled upwards, the Kyropoulos crystal grows below the melt surface after the initial stages of seeding and growth.

Experiments were conducted to measure the melt temperature with and without an applied magnetic field. Using silica-encapsulated thermocouples, it was found that the axial gradient near the top of the melt is about 19 °C/cm, compared to the measured gradient in the B₂O₃ layer of 135 °C/cm. When the magnetic field is off, time-dependent fluctuations ranging up to 6 °C occurred every 1–2 seconds. These fluctuations stopped when a 1000 gauss field was applied. During the seeding and crowning process the crystal is subject to melt-back and regrowth conditions due to the thermal field. An InP crystal rotating through an asymmetric or time-varying thermal field is often subject to twinning. Infrared microscopy of axial slices from twinned crystals grown without an applied magnetic field has shown overlapping striations indicative of melt-back and regrowth. We have found it much easier to grow twin-free (100) InP with magnetic stabilization to suppress turbulent melt flow.

Controlling the thermal geometry of the crystal growth environment makes it possible to grow (100) InP by the MLEK process. There are several advantages for MLEK growth of twin-free InP crystals. (1) Random temperature fluctuations are reduced to less than 1 °C when the applied magnetic field exceeds 1 kg. (2) The magnetic field imposes a steeper radial gradient than exists in the non-magnetically stabilized melt. (3) Heat flow from the melt can be directed vertically by controlling the seed temperature. These three factors contribute to a solid-liquid interface marked by regular rotational striations and a smooth ellipsoidal shape.

5. Future research in Kyropoulos growth

As new applications for bulk materials in the 1990's demand crystals which are not readily produced by standard commercial techniques, research must resume on methods like Kyropoulos growth, which allow unconfined crystal growth in a low-stress environment. Such crystals are the high temperature superconducting allows, ternary semiconductors, and photorefractive materials like the sillenites. Some inherent advantages of MLEK growth are the ease of seeding and the low thermal stress during growth. Because the crown is flat, the operator can see what is going on and, if a twin nucleates, can restart the process. A low stress is made possible by the low axial thermal gradient in the melt, and the lack of confinement of the crystal within a container that has a different thermal expansion coefficient.

Despite its advantages, Kyropoulos growth has not had the years of intensive research and developments of the Czochralski and Bridgman methods. The following process improvements would propel Kyropoulos growth to the high degree of maturity attained by other growth methods:

- Improved understanding of the effects of crucible and crystal rotation and applied magnetic field strength on dopant uniformity and dislocation density and distribution.
- Improved understanding/control of convection and radiation as modes of heat transport in the high pressure growth furnace.
- Computer aided design of thermal environment to understand thermal stresses and interface shape so that they can be controlled.
- Computer control and process control monitoring of the key crystal growth parameters to maintain diameter and interface shape control.
- Computer modeling of fluid dynamics and heat transport to design large scale system for producing larger crystals.

Since its invention in 1926, Kyropoulos growth has succeeded in producing crystals not easily grown by other methods. Crystals which are either brittle or prone to plastic deformation are naturally suited to this technique. Research and development on Kyropoulos growth continued until the 1950's when Czochralski silicon growth dominated the research efforts of many labs, and it has only recently undergone a renaissance. Further development is needed to meet the need for new materials with less forgiving mechanical properties than silicon. For structural perfection and impurity control beyond that which is available today from Czochralski crystals, the Kyropoulos method offers a useful alternative.

At Rome Laboratory we have been developing the MLEK technique for growing InP (100) crystals with low dislocation density. The key difference between MLEK and commercial LEC growth of InP is the ability to control the crystal interface shape by controlling heat flows. Using MLEK, we have grown twin-free (100) crystals 75 mm in diameter. However, despite these accomplishments, the Kyropoulos growth method is still evolving and future work will focus on improved process control. Future research may promote the general understanding of Kyropoulos growth to a level of maturity on a par with Czochralski and Bridgman growth.

References

[1] S. Kyropoulos, Z. Anorg. Chem. 154 (1926) 308.
[2] S. Kyropoulos, Z. für Physik 63 (1930) 849.
[3] K. Korth, Z. für Physik 84 (1933) 677.
[4] K. Chamberlain, Rev. Sci. Instruments 9 (1938) 322.
[5] H. Morgenstern, Z. Kristallographie 100 (1938) 221.
[6] A.F. Wells, Chemical Society, Annual Reports, 1946, pp. 62–87.
[7] A.C. Menzies, J. Skinner, Faraday Society Annual Meeting, 1949.
[8] H.E. Buckley, Crystal Growth, Wiley and Sons, NY, 1951.
[9] W.A. Bonner, L.G. Van Uitert, Mater. Res. Bull. 2 (1967) 131.
[10] P.F. Weller, D.M. Grandits, J. Cryst. Growth 12 (1972) 63.
[11] E. Nicklaus, F. Fischer, J. Cryst. Growth 12 (1972) 337.
[12] T. Fukuda, Y. Uematsu, T. Ito, J. Cryst. Growth 24–25 (1974) 450.
[13] G. Singh, B. Ghosh, R.Y. Deshpande, Cryst. Res. and Tech. 16 (1981) 1239.
[14] E. Schonherr, J. Cryst. Growth 3–4 (1968) 265.
[15] M. Dusseaux, J. Cryst. Growth 61 (1983) 576.
[16] G. Jacob, J. Cryst. Growth 58 (1982) 455.
[17] Materials Letters 8 (1989) 486.
[18] W.J. Zhang, H.Y. Shen, J. Cryst. Growth 100 (1990) 655.
[19] D.T.J. Hurle, in: R.G. Chambers, et al. (Eds.), Sir Charles Frank 80th Birthday Tribute, Adam Hilger, London, 1991, pp. 188–205.
[20] D. Bliss, R. Hilton, S. Bachowski, J. Adamski, J. Electronic Materials 20 (1991) 967.

Fundamentals and applications: a 50-year retrospective of the Japanese crystal growth community

Ichiro Sunagawa *

Yamanashi Institute of Gemology and Jewelry Arts, Tokoji-machi 1955-1, Kofu 400-0808, Japan

Abstract

A review of the 50-year history of the Japanese crystal growth community can provide a good example of how science and technology, particularly in the field of crystal growth, can make significant progress when good cooperation and communication exists between fundamental scientists and device designers and engineers. This will be demonstrated both from a historical point of view and from the author's personal experiences and interest in crystal growth fundamentals, particularly the morphology of crystals.

© 2004 Elsevier B.V. All rights reserved.

Keywords: A1. Crystal morphology; A1. Hodankai; A1. Japanese crystal growth community; A3. Natural crystal growth; B1. Diamond

1. Introduction

The science of crystal growth has its roots in a treatise by N. Steno published in 1669. In this treatise, Steno argued that quartz crystals grew in hydrothermal solutions through an inorganic process, as a counter-argument against the general belief in those days that mineral crystals were formed through the action of bacteria in the earth. He also clearly stated that the hexagonal prismatic habit of quartz crystals and its other variations appear due to growth rate anisotropy, which is a very basic concept in present day crystal growth. The earliest success in growing single crystals for a specific application came in the middle of the 19th century when carat size emerald crystals were synthesized by the flux method. This was followed by the preparation of large single crystal boules of ruby

from the melt in 1902 by A. Verneuil. In spite of this progress, however, there was poor communication and cooperation between the crystal growth scientists and technologist of this day.

The earliest atomic level model of a crystal growth mechanism was put forth by Kossel and Stranski in the 1930s. It was followed over a decade later by the important spiral growth theory of Frank in 1949. Also around this later time period, a demand developed for using single crystals of piezoelectric and semiconductor materials for industrial purposes. This greatly helped promote the advancement of the technology for growing single crystals and later thin films, which in turn led to a better understanding of crystal growth mechanisms and defect formation. There followed an international consensus among scientists and engineers involved in crystal growth that we needed a forum to facilitate communication, and to facilitate cooperation between the fundamentals and applications workers in this field. For this purpose both the Inter-

---

* Corresponding author.

E-mail address: i.sunagawa@nifty.com (I. Sunagawa).

Originally published in
J. Crys. Growth 264 (4) (2004) 631–638.

national Organization for Crystal Growth was formed to arrange for triennial International Conferences on Crystal Growth (ICCG), and also the Journal of Crystal Growth was started with much a hope.

In the Japanese crystal growth community the situation was the same. A retrospective look at how the Japanese crystal growth community developed over the past 50 years, can serve as an example of the history of our science and technology, and may provide guidance for effective routes to our future development [1,2].

2. Before 1950

In Table 1, several Japanese individuals are listed who have made significant contributions to the crystal growth community in the early days. Among these, special mention is necessary concerning the works of Nakaya and Uyeda. Nakaya's haiku ("Snow crystals are the letters sent from the Sky") is particularly well known. Borrowing from Nakaya's haiku, Frank said later "Diamonds are letters sent from the depth of the Earth, more valuable to read than snow flakes, since we cannot reach there". Nakaya observed the morphological variations of natural snow crystals and those prepared in his laboratory in relation to temperature and supersaturation, and prepared the earliest "morphodrom" of snow crystals, the so-called Nakaya's diagram. His pupil observed how a circular platy ice crystal changed its form to petal-like mor-

phology and further to a dendritic form. This observation of morphological instability came much earlier than the Mullins–Sekerka's theory. Nakaya was an excellent essayist, published many essays on scientific topics, giving impact to and stimulating the scientific interests of the younger generation. Uyeda's contribution, though much later than Nakaya, was his ceaseless interest in using electron microscopy and diffraction methods to investigate ultra-fine particles of metals and compound crystals formed in evaporated smokes. Both were interested in the morphology of crystals.

My own curiosity about the variations in habit in natural pyrite (FeS_2) crystals started just after the end of the World War II. I wished to understand why the same crystal species could take on a wide variety of polyhedral forms. One of my interesting findings was that minute pyrite crystals impregnated within a handful-sized sample of clay had the simple cubic habit, whereas larger crystals showed a higher proportion of pyritehedral habit bounded by {2 1 0} faces (Fig. 1). The surface microtopography of {1 0 0} and {2 1 0} faces supported this observation, indicating that as crystals grew larger, striated {2 1 0} faces start

Table 1
Some Japanese names contributed to science and technology of crystal growth before 1950

Fundamentals	
T. Yamamoto	Habit of ionic crystals
M. Yamada	Theoretical work on equilibrium form
U. Nakaya	Snow crystals, earliest morphodrom. Morphological instability
S. Kaya, T. Fujiwara	Single crystal growth of metals for dislocation study
R. Uyeda	Ultra-fine particles
Application oriented	
T. Noda	Mica
G. Ohara, M. Kunitomi	Quartz, piezoelectric crystals
M. Hirose	Ruby
K. Sato	Dielectric materials
H. Inuzuka	Ge

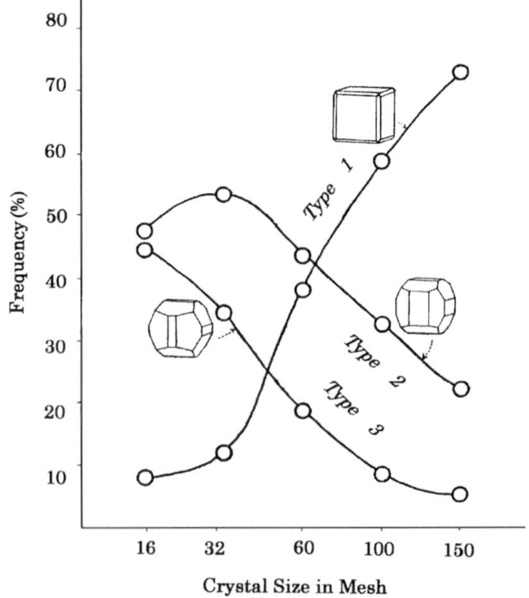

Fig. 1. Frequency of appearance (vertical axis, %) of different habits of pyrite impregnated in a handful clay sample, depending on crystal sizes (horizontal axis, in mesh).

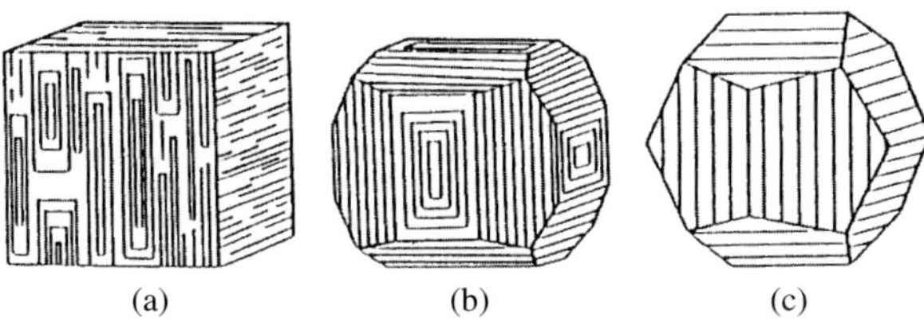

Fig. 2. Changes of surface microtopographs of {1 0 0} and {2 1 0} faces of pyrite as crystals grow larger.

to appear by the piling up of steps advancing on the {1 0 0} faces (Fig. 2). This observation was the start of my interest in crystal growth and the surface microtopography of crystal faces.

Around the same period, J. Nishizawa and I did a joint project on the types of thermo-electro-motive forces associated with natural pyrite crystals, and arrived at the conclusion that the stoichiometry, and thus crystal growth, determines the types found. We lived under the same roof in a high school dormitory in Sendai and were mutually stimulated. This cooperation formed the starting point for Nishizawa's concept of growing perfect crystals for industrial applications by controlling stoichiometry.

In the late 1950s through the early 1960s, I extended my research to the observation of the surface microtopography of natural hematite crystals. In this project I combined optical microscopy and multiple-beam interferometry. With these optical methods, it was possible to observe and measure step heights as small as 2.3 A on the growth spirals on the {0 0 0 1} faces of hematite crystals grown by a natural chemical transport process. Also it was possible to obtain a general view on what sort of surface features crystals exhibit when they grow under uncontrolled conditions. Although in recent years, AFM and STM have become the principal tools to investigate surface microtopography of crystal faces, I have an impression that most of the essential surface features relating to crystal growth and dissolution were already observed 40 years ago by optical and interferometric methods, except for two-dimensional nucleation on the terraces of spiral steps.

While progress was also being made in the growth of large single crystals for industrial purposes in this time period, including synthetic gemstones, piezoelectric and dielectric materials, mica and later on Ge (as mentioned in Table 1), there was no unified community for crystal growth. The fundamentalists and applications people were separated. Crystal growth was regarded either as a hobby science or a subsidiary service branch. Papers were presented sporadically at the mercy of major societies. There was no opportunity, therefore, for these two groups to communicate, and compare theory with experiments. The situation was nearly the same worldwide in those days.

3. Incubation period, 1950–1965

The period between the 1950s and 1966, when the First International Conference on Crystal Growth (ICCG-1) was organized, may be taken as an incubation period for crystal growth and before the realization that communication and cooperation between physical scientists and application engineers was essential. During this period, however, there were three important international meetings which focused on topics in crystal growth. These were the English Faraday Society Discussion No. 5 in 1949, the Cooperstown, New York Conference on the Growth and Perfection of Crystals in 1958, and the International Colloqium at Pont a Musson, near Nancy, France on "Adsorption et Croissance Cristalline" in 1965. In 1966, thanks to the effort of American colleagues, R.A. Laudise, K. Jackson and others, and particularly M. Schieber from Israel, the first international conference devoted to a range of topics in crystal growth was launched. This meeting was held in Boston, MA.

In Japan, a discussion group on the subject of artificial minerals was organized in 1956 within the Chemical Society of Japan. They also organized an annual symposium to communicate and discuss results on the synthesis of single crystal materials such as piezoelectrics, synthetic gemstones, mica, etc. This group later became an independent entity known as the Association of Synthetic Crystal Science and Technology (ASCST). In the same year, an ammonium dihydrogen phosphate (ADP, NH_4H_2PO4) Committee was organized by the Japan Society for the Promotion of Science. In 1970 this committee developed into the Crystal Technology Branch (CTB) of the Japan Society of Applied Physics (JSAP). While this branch did not hold independent annual meetings, papers were

Table 2
Crystal growth activities in Japan after 1950

Societies and Associations	
1956	Discussion group on artificial minerals, Chemical Society of Japan
	Later, Association of Synthetic Crystal Science and Technology (ASCST)
1956	ADP Committee, Japan Society for Promotion of Science
	Later, Crystal Technology Branch, Japan Society of Applied Physics (CTB-JSAP) in 1970
1969	NCCG-1, co-sponsored by 12 societies and associations
1973	1st "Hodankai"
1974	Japanese Association for Crystal Growth (JACG)
1985	National Committee for Crystal Growth, Science Council of Japan
Special research projects	
1972–1974	3-Year special research project on "Crystal Growth" (R.R. Hasiguti)
1991–1993	3-Year special research project "Crystal Growth Mechanism" (T. Nishinaga)
1996–2001	5-Year special research project on "Atomic Process of Crystal Growth" (T. Nishinaga)
Conferences	
1974	ICCG-4, Tokyo; ISSCG-2, Lake Kawaguchi
1989	ICCG-9, Sendai; ISSCG-7, Zao
2001	ICCG-13/ICVGE-11, Kyoto; ISSCG-11, Shiga

presented at sessions in the JSAP annual meetings. However, the group did organize independent seminars, short courses, schools, etc. on the synthesis of and industrial applications for single crystals and thin films. Table 2 lists some of the activities going on during this period.

The single crystal growth of semiconductor materials developed rapidly during this period, mainly owing to the demands of the electronic industries and the important role played by the CTB of the JSPS. Controlling perfection and homogeneity and growing large crystals were indispensable requirements for these industries. As a result of the intensive efforts launched by these groups the Si single crystal industry developed rapidly.

4. Crystal growth related activities in Japan after ICCG-1

The success of ICCG-1 encouraged the Japanese community to launch a unified crystal growth community. R.R. Hasiguti was the main organizer during this period. The First National Conference on Crystal Growth, NCCG-1, was organized in 1969 with the co-sponsorship of 12 societies and associations. After the formation of the Japanese Association for Crystal Growth (JACG) in 1974, the NCCG meetings were organized by the JACG and supported by 11 other societies.

In 1972–1974, we were successful in obtaining a large research grant (Special Research Grant) from Ministry of Education and Culture on "Crystal Growth". This helped greatly in promoting fundamental research activities and to build a bridge between theory and experiment and between fundamentals and applications. This led to unification of the crystal growth community in Japan. With a help of this grant, a group of theoreticians guided by A. Ookawa was organized. Ookawa believed that good communication with experimentalists was essential to the development of crystal growth theories, and spent the budget for travel expenses to visit experimental laboratories. Their first visit was to Sendai, where we were mainly interested in the morphology of mineral crystals. We organized a very informal discussion meeting, which was the first "Hodankai", about which an explanation will be given later.

Later, two more special research projects were started, one on "Crystal Growth Mechanisms" (1991–1993) and the other "Atomic Processes in Crystal Growth" (1996–2001). These two projects were organized by T. Nishinaga, who has been a regular participant in "Hodankai". These projects were in principle, fundamentally oriented, but were also coupled to applications. Both vapor phase and liquid phase epitaxy were the main focus of these projects. In addition to these projects, it should be noted that a few application oriented research projects were organized from time to time with the help of the Ministry of International Trades and Industries (MITI). A representative example of this was the establishment of a cooperative research effort (CRE) to develop methods for growing large, dislocation-free Si crystals. Before this, R and D was principally undertaken in individual company laboratories, and there was no intense cooperation among these researchers. In this CRE research program, staffs were sent from competitive companies, and worked together cooperatively to overcome the problems of interest. After the five-year project was completed the respective researchers went back to their own companies with new know-how. Cooperative research projects were also organized on compound semiconductors.

Another point to be mentioned was the formation in 1985 of the National Committee for Crystal Growth in the Science Council of Japan. The Science Council of Japan is the scientific policy making organization of Japan. The members of this Committee are elected from three associations representing the JACG, ASCST, CTB-JSAP. In Table 2, these activities are summarized.

5. Hodankai

The 1st "Hodankai" meeting was held in 1973 at a hot spring near the Zao volcano, when a newly organized group of theory people paid a visit to Sendai. Experimentalists interested in the morphology of crystals came from all over Japan to participate in this very informal discussion meeting. The 3-day meeting was held in an inexpensive hotel, and topics of mutual interest were discussed in a tatami-mat room. Extended time was required to build a common understanding of the technical terms involved, but this resulted in a constructive communication and mutual understanding. To keep the meeting informal and at the same time provide a stimulating atmosphere, all the sessions were presided over by one person. He was called the school master. This tradition continues even up to the present time. The success was proved by the fact that there were several papers presented in the next

year's NCCG conference to answer questions raised during the "Hodankai".

In 1974, when Japan hosted the ICCG-4 in Tokyo, there were comments from foreign participants the Japanese contributions were mainly industrially oriented, and not much concerned with the fundamental aspects of crystal growth. These comments motivated me and provided a driving force to continue "Hodankai", so that good communication between theory and experiment and between fundamental and applications may be realized. As a result "Hodankai's" became a regular meeting held immediately after each annual NCCG meeting. I served as school master of the "Hodankai" from 1973 to 1988, H. Komatsu from 1989 to 1999, and now M. Kasuga is serving. This year, we will have our 27th "Hodankai".

In 1989, when we hosted ICCG-9 in Sendai, A.A. Chernov commented at the banquet that he was most surprised to see that the Japanese contributions to the fundamentals of crystal growth had improved exponentially both in quantity and quality. His comments gave me a great satisfaction, and I felt I did at least make a useful contribution to our beloved science and technology of crystal growth.

6. Future research targets

During the past 50 years, our understanding of crystal growth fundamentals has advanced to such an extent that growth mechanisms, morphology, perfection and the homogeneity of bulk single crystals and thin films can be understood at the atomic level (at least in simple and single component system). This understanding has greatly helped advance the technology of growing crystals with a desired perfection and homogeneity. Although we may grow single crystals with the desired perfection and homogeneity through experience and personal know-how (even if we do not know the reason and the growth mechanism), it is certain that if we could understand the fundamentals, this will enrich ourselves and reduce the time to solve practical crystal growth problems. Both science and technology are mutually beneficial. The history of the crystal growth community worldwide has clearly demonstrated that close cooperation between fundamentalists and applications people has been essential in making significant advancement in this field. Both the ICCG and the NCCG-like national conferences in each country have provided a valuable forum for realizing such communication and cooperation. I certainly hope that this tradition will continue.

As future targets of research, I would like to mention the followings:

(1) Understanding of crystal growth mechanisms at the quantum mechanics level.
(2) Dislocation and homogeneity control in compound semiconductors and oxides, etc.
(3) Understanding the crystal growth and related problems involved in complex and compound systems, like earth and planetary materials and also the crystallization taking place in living organisms. We may be able to offer a new view-point to these fields, based on the understandings we achieved in the 20th century on simple and single component systems. Morphology will be the essential key code in solving these problems.

As an example of (3), I would like to explain in the next section what sorts of information we may be able to deduce in complicated and complex systems based on the knowledge we achieved over the past in 50 years. A similar methodology should be applicable to other systems such as biomineralization.

7. Diamonds, letters from the depth of the earth

As a final topic, I would like to briefly explain how the letters written in natural crystals can be decoded, and what sort of information we may obtain relating to the movement in the Earth. This can be done based on our present knowledge of morphology, perfection and the homogeneity of single crystals and polycrystalline aggregates, taking natural diamonds as a representative example.

It has been well established that:

(1) Natural diamonds grew in a solution phase in the Earth's mantle under high pressure, high temperature conditions, where diamond is the stable phase of carbon.
(2) Diamond crystals were brought up to the Earth's surface by the rapid ascent of kimberlite or lamproite magma and quenched metastably by volcanic eruption.
(3) There are three types of chemical environments necessary for diamond formation; ultramafic suite, eclogitic suite, and ultra-high pressure metamorphic rocks.
(4) During the ascent of kimberlite or lamproite magma, diamond crystals experienced partial dissolution and plastic deformation, and crystals were rounded and plastically deformed.

It has been also well established that:

(5) The morphological variation of single and polycrystalline diamonds has been understood in relation to the driving force conditions; under higher driving force conditions, polycrystalline aggregates, like ballas, borts, hailstone borts, etc., and cuboid appear, whereas under lower driving force conditions, single crystalline octahedral diamonds are formed.
(6) In a natural growth environment (i.e., silicate or carbonate solutions), only $\{1\,1\,1\}$ faces behave as smooth interfaces, whereas $\{1\,0\,0\}$ exclusively behave as rough interfaces.

In Fig. 3, X-ray topographs of an ordinary octahedral diamond crystal (a) and a round brilliant cut stone (b) are compared. Fig. 3a represents commonly encountered X-ray topographic images of single crystalline natural diamonds, which exhibit dislocation bundles with Burgers' vector of $\langle 1\,1\,0\rangle$, radiating from the center of a stone and running nearly perpendicularly to $\{1\,1\,1\}$ faces. Fig. 3b is an X-ray topograph of a pear-shaped brilliant cut stone. I was asked to investigate the origins of this stone, in particular, whether or not the stone was cut from the same rough stone as another round brilliant [3]. By correlating the X-ray topographs, the round and pear-shaped brilliant cut stones were proved to have been cut from a same rough octahedral crystal. At the centers of the two cut stones, X-ray topographs clearly indicated the presence of a square shaped core portion, and that the dislocations are seen to generate principally from the surface of the core portion. Fig. 4 shows a magnified X-ray topograph of the core portion. The dislocations have a Burgers' vector of $\langle 1\,0\,0\rangle$, which are different from the commonly observed $\langle 1\,1\,0\rangle$ Burger's vector

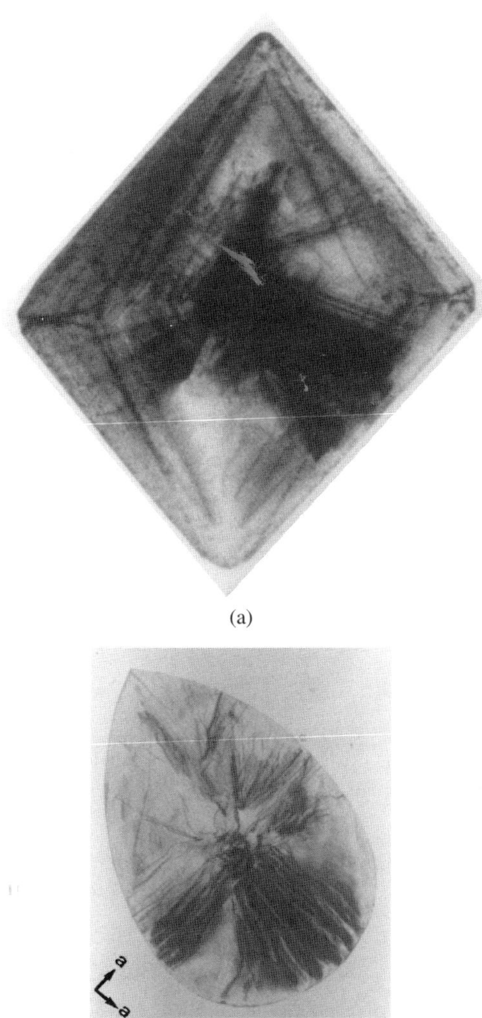

(a)

(b)

Fig. 3. Comparison of X-ray topographs of an ordinary octahedral diamond ((a); Burgers vector of dislocation bundles is $\langle 1\,1\,0 \rangle$) and a brilliant cut stone ((b); Burgers vector of dislocations is $\langle 1\,0\,0 \rangle$). X-ray topographs by T. Yasuda.

in ordinary octahedral single crystals (see Fig. 3a). These observations demonstrate that the core portion was formed elsewhere and had cuboid form which was brought into a new environmental phase and acted as a seed crystal on which new growth took place under much lower driving force conditions. It should be stressed that this was the first evidence proving the presence of seed crystals during natural diamond formation. Both X-ray topography and cathodoluminescence investigations of the two cut stones indicated

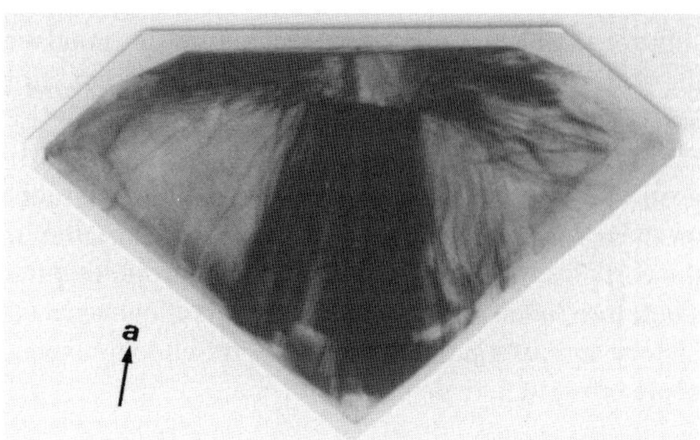

Fig. 4. Magnified X-ray topograph of the core portion, indicating that dislocations generate principally on the surface of the core portion. X-ray topograph by T. Yasuda.

that growth, in the new environment, was proceeded by the appearance and cooperation of $\{1\,1\,1\}$ microfacets. The growth environments were clearly different between the seed portion and the newly grown main portion.

In ultra-high pressure metamorphic rocks, micron size diamond crystals occur sporadically in garnet or zircon porphyroblastic crystals, and they principally take the cuboid form or polycrystalline aggregates. This indicates growth under higher driving force conditions. The content of diamonds in these rocks are much higher than in ultramafic rocks, going up to 2.0%. This is reasonable considering the fact that the ultra-high pressure metamorphic rocks originate from subducted oceanic sediments which contain organic carbon. Micro-diamonds were crystallized in liquid droplets formed by partial melting of C + silicate solid. If ultra-high pressure metamorphic rocks are further subducted deeper, they will be digested in a deep seated magma. The driving force conditions become much lower in such a magma. Micro-diamonds in ultra-high pressure metamorphic rocks will act as seeds for the further growth of diamond under much lower driving force conditions. The observation on the two brilliant cut stones suggests such a scenario.

It was surprising to see that the whole geological movement has been recorded in such small cut stones. It starts from the subduction of oceanic sediments and the formation of ultra-high pressure metamorphic rocks. This is followed by the partial melting and formation of liquid droplets of silicate + C solutions, and then the growth of micro-diamonds in the partially melted droplets. After further subduction and digestion of the ultra-high pressure metamorphic rocks containing micro-diamonds, there is further digestion of these rocks and further growth of diamond on these seeds. Finally there is rapid transportation to the earth's surface by kimberlite or lamproite magma. Therefore, even small stones have their own growth history and personality, and it is worthwhile to expend some effort to decode them.

References

[1] I. Sunagawa, Development of science, technology of crystal growth, in: Editorial Committee (Eds.), Crystallography in Japan—Its Historical Prospect, The Crystallographic Society of Japan, Tokyo, 1989, pp. 184–189 (in Japanese).

[2] I. Sunagawa, History and present state of crystal growth researches, in: Editorial Committee (Eds.), Handbook of Crystal Growth JACG, Kyoritsu, Tokyo, 1995, pp. 11–14 (in Japanese).

[3] I. Sunagawa, T. Yasuda, H. Fukushima, Gem and Gemology (Winter) 34 (1998) 270.

Reminiscences about the early background of the papers on "The distribution of solute in crystals grown from the melt"

J.A. Burton, W.P. Slichter

AT&T Bell Laboratories

The writers were more than a little surprised when AACG asked us to tell us about the history and times of the Burton–Prim–Slichter (BPS) paper in the Journal of Chemical Physics, 1953. That was 30 years ago, which suddenly seems like a very long time. The request from AACG set us to remembering some exciting days that were part of the history of semiconductor technology leading to today's sophistication of microcircuitry and surely extending to further advances beyond the imagination.

In the decade before World War II, an increasing appreciation developed of the existence of a distinct class of materials, now broadly called *semiconductors,* that possess electrical properties intermediate between those of the familiar metals and the dielectric insulators. These substances included elemental silicon and germanium, and also a wide number of oxides such as cupric oxide and ores such as galena. However, the successful use of these and other materials depended mostly on trial and error: you got a good detector or you didn't.

The metallurgical studies by J.H. Scaff and H.C. Theuerer at Bell Laboratories on polycrystalline ingots of silicon showed beyond doubt that impurities in these solids are basic to the electrical properties. Indeed, the terms n-type and p-type come from this work on silicon ingots. It became increasingly apparent that great variability in electrical resistivity accompanied the process of the growth of the polycrystalline ingots.

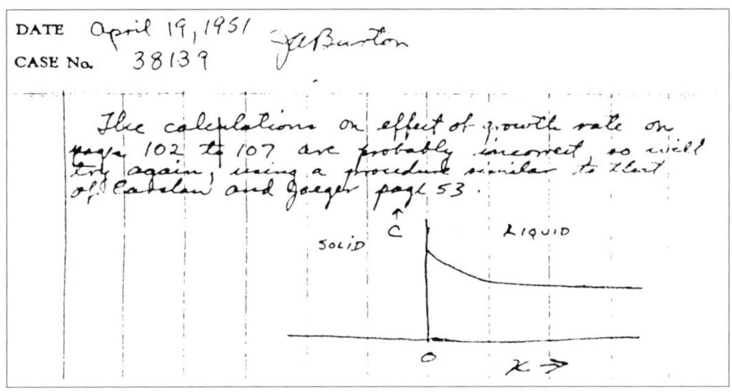

Previously published in AACG Newsletter Vol. 13(3) November 1983.

next, consider the boundary conditions:

Suppose for $x > l$ the liquid is thoroughly stirred so that

$$C = C_\infty \quad \text{at } x = l \qquad C_\infty = \text{constant}$$

(note: l can have any positive value; for $l = \infty$ we have the case of no stirring in the liquid.)

at $x = 0$ $C = C_0$ $\dfrac{\partial C}{\partial x} = \left(\dfrac{\partial C}{\partial x}\right)_0$

$$-D\left(\frac{\partial C}{\partial x}\right)_0 = (1-K)C_0 U$$

or finally:

$$\frac{C}{C_\infty} = \frac{\dfrac{K}{1-K} + e^{-Ux/D}}{\dfrac{K}{1-K} + e^{-Ul/D}}$$

This gives the steady state concentration as a function of x, the distance from the interface.

Immediately after the discovery of the transistor in 1947, great emphasis was given at Bell Laboratories to the new materials problems relevant to transistor technology. The need for reliable single crystals of germanium was widely recognized, and a method for growing such crystals from the melt was developed at Bell Laboratories by G.K. Teal and J.B. Little (1950), based upon the classic work of J. Czochralski (1918). The combined needs of single crystallinity and controlled purity were sought in this approach. It was also recognized that the accurate control of electronic impurities in semiconductors represented a new frontier in the identification and control of impurities at very low levels. One of us (J.A.B.), in collaboration with J.D. Struthers, was already using radioactive tracer techniques in research on vacuum tube materials. These methods seemed especially relevant to the chemistry of impurities in semiconductors at the parts per million (or billion) level. So in 1949 Burton and Struthers started a program using radioactive tracers to study the known donors and acceptors in germanium, and also other elements such as copper, nickel, and iron in the search for causes for low carrier lifetime and p-type thermal conversion. They measured solute concentrations, solid/liquid distribution coefficients, diffusion coefficients, and looked at solute non-uniformities with radioautographs of crystals and ingots. The radioautographs of pulled Ge crystals containing radioactive donors showed large fluctuations in concentration that corresponded to variations in resistivity and carrier lifetime (G.L. Pearson and J.R. Haynes). At about this time, Burton developed an equation describing the effect of growth rate on solute concentration, using a procedure similar to certain problems in heat flow. This equation was the same as the steady-state solution published in our later papers.

In this same time frame, A.H. White, a chemist who was later to head up most of the materials science and engineering at Bell Labs, was given responsibility for the different groups working on semiconductor chemistry and metallurgy. The two of us combined our scientific efforts under White's leadership. Slichter's emphasis was on the reproducible preparation of high quality crystals of germanium, suitable for transistor fabrication. He, with the collaboration of E.D. Kolb, undertook a large program to assess broadly the effects of such parameters in Czochralski growth as the crystal growth rate and the speed of rotation of the crystal in contact with the melt. This latter parameter was important, since the effect of the rotating crystal was to pump the melt, more or less effectively according to speed, to the growing solid-liquid interface, thereby affecting the availability of solute impurity during growth. We were joined by mathematician R.C. Prim (The "P" of BPS) in an extensive analysis of this problem including possible transient effects. Prim discovered and helped to apply the 1934 paper by W.G. Cochran (Proc. Cambrige Phil. Soc.) on the fluid flow produced by a rotating disk in a semi-infinite liquid. The resulting statement, sometimes generously referred to as the BPS (Burton–Prim–Slichter) equation, describes the rate of incorporation of impurity solute into a growing crystal which is rotating in contact with the surface of the melt. Two fluxes of solute atoms in the melt are visualized, one away from the growing crystal surface, described by the diffusion coefficient of solute in the melt; and one toward the crystal surface owing to the pumping action of the rotation. A zone of liquid next to the surface was postulated, within which the transport of solute is dominated by diffusion.

These times were exciting and often confusing, with people working day-by-day on new ideas, with conflicting results and puzzling phenomena popping up continually, and with constant pressures from the device development people who literally stood around waiting for slices of "good" crystals. We at Bell Laboratories were aware of competing workers at other institutions such as R.N. Hall at General Electric Laboratories and K.A. Lark-Horowitz at Purdue University. The environment and the times added greatly to the excitement. People would come back from technical meetings with intriguing rumors or hard news of progress. At home, the top managers such as W. Shockley, J.B. Fisk (then Director of Physics Research), and R. Bown (Vice President, Research), kept week-by-week contact with the progress. A good time was had by all!

In April 1952, our work on "The Distribution of Solute Elements: Steady State Growth and Transient Conditions" was presented at the "Symposium on Transistor Technology" where members of the technical staff of Bell Labs and Western Electric described their work in this field for the first time to a large audience of representatives of defense agencies and industrial companies. Our work was published as Chapters 5 and 6 in the two-volume proceedings entitled "Transistor Technology," Western Electric Co., July 1952. These volumes were government classified "Restricted—Security Information," and declassified in 1954. Eventually our July 1952 papers were republished in: "Transistor Technology" (3 volumes), D. Van Nostrand, 1958. In the meantime, our papers in *Journal of Chemical Physics,* Vol. 21, pp. 1987–1996, November 1953 by Burton, Prim, Slichter, Kolb, and Struthers were published.

Even after thirty long years, it still seems that the BPS equation gives a pretty good description of the distribution of solute in crystals grown from the melt. It is a pleasure to remember these happy and exciting events!

Notes on interface growth kinetics 50 years after Burton, Cabrera and Frank

A.A. Chernov *

*BAE Systems North America at US NASA Marshall Space Flight Center, NASA, Mail Code SD46,
Marshall Space Flight Center, AL 35812, USA*

Abstract

This overview is devoted to some unresolved basic problems in crystal growth kinetics. The density wave approach to propagation of a spatially diffuse interface between a growing crystal and its simple (e.g., metallic) melt is discussed in Section 2. This approach allows for the calculation of kinetic coefficients and is an alternative to the localized interface concept in which each atom belongs to either a solid or a liquid. Sections 3 and 4 deal mainly with layer growth from solution. Mutual retardation of the growth steps via their bulk and surface diffusion fields is the major subject. The influence of solution flow on step bunching (Section 4) suggests the essential influence of bulk diffusion on the surface morphology. The flow within the solution boundary layer enhances step–step interaction, influences the step bunching process and the resulting step pattern morphology on the growing surface. Recent experiments on the rates at which strongly polygonized steps on protein and small molecule crystals propagate during growth from solution are analyzed in Section 5. We have shown that the step segments may be "singular" and that "one-dimensional nucleation" may be the rate limiting stage for the segments that are shorter or comparable in length to the thermodynamic equilibrium interkink distance. In this case, the reciprocal dependence of the segment propagation rate on the segment length that follow from the Gibbs–Thomson law, should be replaced by an abrupt switch from zero to a finite constant velocity. Until recently, the Kossel crystal remained the only model used in crystal growth theory. In such Kossel crystals, all kinks at the steps are identical and the kink rate is a linear function of the supersaturation. In the non-Kossel crystals, there may be several kink configurations characterized by different geometries and energies. These configurations should appear in a specific sequence when each new lattice unit cell is filled. As a result of such a cooperative interaction within the unit cell, a non-linear dependence of the kink rate on the vapor pressure or solution concentration in excess over the equilibrium value should be expected.
© 2004 Elsevier B.V. All rights reserved.

PACS: 81.10.-h; 81.10.Aj; 87.15.Nn

Keywords: A1. Fluid flows; A1. Growth models; A1. Impurities; A1. Morphological stability; A1. Surface processes

1. Introduction

* Tel.: +1-256-544-9196; fax: +1-256-544-8762.
E-mail address: alex.chernov@msfc.nasa.gov
(A.A. Chernov).

The 1951 paper of Burton, Cabrera and Frank (BCF) [1] brought to crystal growth the spirit of contemporary physical theory and opened the possibility of a much stronger quantitative link between theory and experiment. The "giants on whose shoulders this paper stands" are numerous. Some of their breakthroughs are: Gibb's concepts of interfaces, equilibrium shape and nucleation [2], Volmer's finding of surface diffusion [3], Kossel's and Stranski's "half-crystal" position (kink) concept and related molecular kinetic growth theory [4–7], the method of average detachment work for equilibrium shapes by Stranski and Kaishev [8,9] and Frenkel's pioneering ideas on fluctuation-induced meandering of steps and interfaces [10]. On this foundation, Frank's enlightening insight into the role of dislocations in crystal growth, together with Burton and Cabrera's approach to surface fluctuations brought about a new level of understanding of crystal growth. Since BCF, interface structure and kinetics analysis has moved towards a much more intense, deeper quantitative physical insight [11–16].

The great demand for high-quality semiconductor, laser, non-linear optical, magnetic crystals, etc., for electronic, coherent optical and high-temperature structural (like sapphire) applications, has provided practical incentives for the technology-based industries to study crystal growth kinetics and numerous related problems concerning defects in crystals (i.e., internal stress, point defects and impurities, striations and inclusions [16–18]). The issues involved in single crystal perfection, together with long-standing metallurgical interests in dendrites and pattern formation, and geological interests in mineral diagnostics, led to the concepts of constitutional supercooling and morphological stability for rough [19–21] and faceted interfaces [22,23]. Later, roughening and stability issues attracted the serious attention of mathematical physicists.

The past half-century (since 1951) has shaped crystal growth into a discipline significant in its own right, while, at the same time, being closely linked with physics, chemistry and crystallography. Crystal growth remains an important part of materials science and applications and supplies many puzzles to surface science. Some branches have matured to the industrial level like, e.g., semiconductor bulk growth. Advances in computer modeling allows one to address numerous practical issues, beginning with melt growth and deposition from the gas phase. During the last several decades, crystal growth, along with many other areas of science and technology, have become highly specialized. This specialization demands not only the general framework theory that has been developed so far, but also a quantitative theory of multiparametric phenomena. Such a theory is still missing. Also, there are fundamental issues not clearly understood and general problems not solved. For instance, not many papers on melt growth mechanisms can be found in the literature nowadays. Theories concerning the trapping of vacancies and impurities that can affect defects in bulk crystals are far from complete. Also, there are no equations that predict the kinetic coefficients connecting interface supercooling with growth rate. True, these coefficients are usually high for metals so that the growth is limited by heat transport. However, this is not the case for more complex systems, such as dielectrics. Recent efforts on phase field modeling [24–27] have the potential to link growth rate with real crystal–melt rough interface structures. Therefore, in Section 2 earlier work on how to calculate kinetic coefficients for simple melts will be discussed.

Another issue is layer growth. Typically, single crystalline films and other semiconductor electronic structures grown from the gas phase or from solution grow layer-by-layer, by step propagation. The reason is that the solubility or vapor pressure is low and, thus, interfacial energies are high. High surface energies and low temperatures lead to a low density of kinks on steps, which is the case, e.g., in low-temperature epitaxy. Also, layer-by-layer growth is typical in conventional chemical vapor and molecular beam epitaxy [28]. Growth studies of non-linear optical materials from room-temperature aqueous solutions, culminated in the development of fast growth technologies which can yield up to half-a-meter sized KDP crystals of great perfection for frequency doubling of lasers for nuclear fusion [29–31]. The significance of these materials provides a practical incentive for paying closer attention to not only the layer growth mode kinetics, but also to the morphological stability of singular faces, i.e., the interaction between steps and the transport in surrounding solutions. Issues not fully understood in this area include: the importance of surface diffusion in solution growth, the rate and mechanisms of kink formation on steps and surfaces as compared to the rate of their annihilation during growth, the applicability of the Gibbs–Thomson law at moderate supersaturations and the applicability of all major classical concepts to crystals in which the unit cell includes several molecules in non-equivalent positions. An understanding of some of these issues was facilitated by the crystal growth of large (2–200 nm) biological macromolecules like proteins, nucleic acids and their complexes, which, via AFM, allow molecular-level resolution of surface features. These solution growth problems will be reviewed in Sections 3–5.

2. Diffuse interface approach: kinetic coefficient for melt growth

On the molecular level, BCF considered the boundary between the crystal and mother phase as a localized surface: each molecule or atom is assumed to belong to either the crystalline or to the surrounding gas, liquid or other solid phase. This is the molecular-kinetic theory approach of Kossel and Stranski rather than Gibb's vision of the interface as a layer of finite thickness. In this localized interface model, roughening means intense meandering of the interface localized on the molecular level. The longitudinal scale of meandering is a correlation length, which is close to the molecular dimensions above the roughening transition. Another view on rough interfaces is that positional and/or rotational order parameter(s), η, gradually decrease(s) from a fully ordered crystal bulk, $\eta = 1$, to a fluid, $\eta = 0$.

Assuming that cooperative interactions along the localized interface is ruined by thermal fluctuations and does not affect propagation rate, R, of this crystal–melt interface, one usually uses the classical relationship

$$R = (D/d)\{1 - \exp[-(\mu_l - \mu_s)/kT]\}. \tag{1}$$

It is linear in $\Delta\mu = \mu_l - \mu_s = \Delta s(T_m - T)$, at $\Delta\mu/kT \ll 1$. Here, μ_l and μ_s are the chemical potentials in the liquid (l) and solid (s) phases, T is the actual interface temperature, T_m the melting point, Δs the entropy of fusion and a the atomic or molecular size. Within the localized interface model, diffusivity D through the interface is assumed to be a Boltzman function of temperature with the activation energy related to the melt viscosity. In other words, D is assumed to be inversely proportional to the melt viscosity with an adjustment factor to fit the linear

dependence of the growth rate on the undercooling $T_m - T$. This approach provided a qualitatively correct description of the growth kinetics [32], although the absolute value for the kinetic coefficient and even the temperature dependence of the growth rate on undercooling, e.g., for cyclohexanole and succinonitrile, did not fit well [33]. Of no less importance, molecular dynamic (MD) simulations over a wide range of supercooling suggested that the kinetic coefficient does not depend exponentially on temperature, but rather as $\sqrt{T}$ [34]. The MD simulations demonstrated that the transition between a fully ordered crystal and a fully disordered liquid occurs within several molecular layers within which the thermal motion of the particles destroys the order [35,36]. Fig. 1 illustrates such a continuous spatial order–disorder transition. Fig. 1a shows, as an example, atoms arranged in a simple cu-

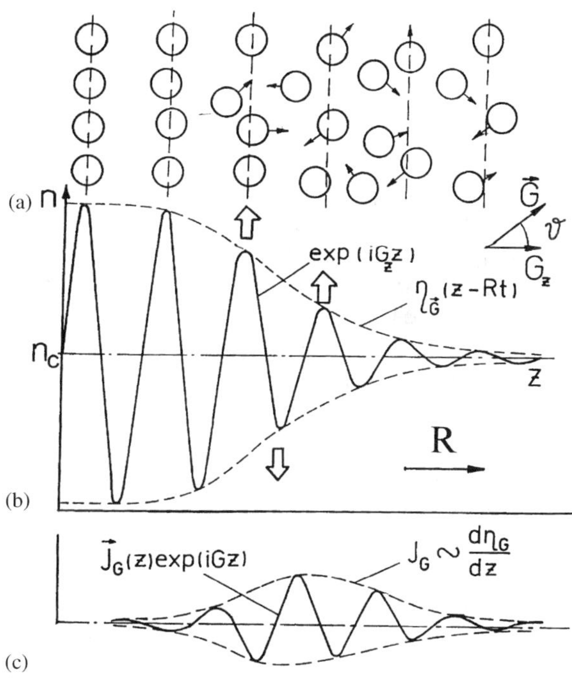

Fig. 1. Diffuse interface between crystal (left) and melt (right) propagating during growth. (a) Atoms are ordered in the crystal and disordered in the melt. Ordering process occurs simultaneously in several future lattice planes (...) in between; thin arrows symbolize atomic motion. (b) Number density $n(z)$ assuming average crystal density, n_c, equals that of the melt. As the interface proceeds to the right, the density amplitude, $\eta(z)$, taken as the order parameter, increases as shown by the bold arrows. The ordering flux, $j \sim d\eta/dz$, is maximal within the interface moving to the right at the rate R.

bic lattice on the left so that the mass density is a periodic function of the coordinate z normal to the $(0\,0\,1)$ crystallographic plane and to the crystal–melt interface. The simpler the Miller index of a crystallographic plane, the longer is the period and the larger the density amplitude. At $T = 0$ K, all such density waves are regular repetitions of narrow peaks of height $n = n_c + \eta$, where n_c follows from the average crystal density and the interplanar distance, while the density wave amplitude, η, is chosen as an order parameter. As we approach the crystal–melt interface from the bulk crystal, the density wave smears out, disappearing in the melt when $\eta = 0$ and n is a constant, assuming the latter to be not much different from the crystal density, n_c. In other words, the density distribution $n(\vec{z})$ (cm^{-3}) near the crystal–melt interface may be presented as

$$n(\vec{r}) = \sum_{\vec{G}} \eta(z - Rt) \exp(\mathrm{i}\vec{G}\vec{r}), \tag{2}$$

where the wave amplitude, $\eta(z)$ (cm^{-3}) decreases from 1 to 0 when the coordinate z normal to the interface increases from $-\infty$ in the crystal to $+\infty$ in the liquid. Practically, this occurs within a transition layer several interplanar spacings wide. In Eq. (2), R is the interface growth rate, t is time, $\vec{G}$ are reciprocal lattice vectors over which the sum is taken. These vectors for crystal and short-range order in the melt are assumed to be the same. At equilibrium, $R = 0$, the extension of crystalline order from the crystal to the liquid, i.e., the order inheritance, occurs via short-range atomic interactions. A crystal surface induces melt ordering in its vicinity, "exciting" the density waves most relevant to the surface and the liquid. Since the liquid short-range order potentially includes all density waves of the crystal, waves characterized preferentially by different $\vec{G}$ vectors are induced by each crystal surface. The thermal motion of atoms dynamically builds up and eventually destroys the planes at the interfaces. This motion is symbolized by the arrows in Fig. 1a. At a supercooled crystal–melt interface, there are more species that build up the crystal planes than there are leaving these planes. As a result, the whole crystal–melt transition layer propagates (to the right in Fig. 1). The ordering atomic flux building up the crystalline planes increases the density wave amplitude, η, as is symbolized by bold arrows in Fig. 1b. This flux, $(\mathrm{cm}^{-2}\,\mathrm{s}^{-1})$, is the net flux

of atoms from the less dense interplanar spacings to the more dense future crystalline planes. All $\vec{G}$ vectors contribute to this flux and all types of atomic planes are constructed simultaneously by the flows $\vec{J}_{\vec{G}}$ related to different $\vec{G}$'s. Since the average density is preserved, it can be shown [37] that for each $\vec{G}$ vector,

$$\vec{J}_{\vec{G}} = \mathrm{i}\left(R\vec{G}/G^2\right)\mathrm{d}\eta/\mathrm{d}z. \tag{3}$$

Indeed, as is clear from Figs. 1b and c, $R\,\mathrm{d}\eta/\mathrm{d}z$ is the rate at which the density wave amplitudes rise: it is zero within the crystal and liquid, and reaches a maximum in the middle of the smeared interface, where the wave envelope $\mathrm{d}\eta/\mathrm{d}z$ varies most quickly with z. The factor $\mathrm{i} = \sqrt{-1}$ comes from the phase shift between the $n(\vec{r})$ and $j(\vec{r})$—the flux is maximal at z between, not on, the planes under development, Figs. 1b and c.

As mentioned earlier, building a crystal lattice occurs not only via forming smectic planes parallel to the interface, but also via ordering within each plane. This is associated with the formation of a set of density waves normal to the $\vec{G}$ vectors in addition to that normal to the interface. Therefore, the total flux is the sum of all these fluxes. In the model considered so far [37], only the $\{1\,1\,1\}$ planes are presenting the largest interlayer spacing in the FCC lattice. Correspondingly, the fluxes generated by eight different $\langle 1\,1\,1 \rangle$ vectors have been considered. This set includes two opposite directions of each $\vec{G}$, because each atomic plane is built up from both sides.

The ordering is associated with the release and dissipation of crystallization heat, which is compensated for by a gain in chemical potential, $\Delta\mu$. The energy dissipation rate per particle is the particle average velocity, $\vec{J}_{\vec{G}}/n$, times the force $\vec{\nabla}\mu$ acting on this particle. The ordering flux, $\vec{J}_{\vec{G}}$, is driven by the local gradient of chemical potential:

$$\vec{J}_{\vec{G}} = -D_{\vec{G}}\nabla n = -(D_{\vec{G}}n_c/kT)\nabla\mu. \tag{4}$$

Therefore, the dissipation rate per particle is $\vec{J}_{\vec{G}}\vec{\nabla}\mu/n = kT j_{\vec{G}}^2/2D_{\vec{G}}$. Here, $D_{\vec{G}}$ is the diffusivity normal to the $\vec{G}$ wave. Dissipation occurs within the transition layer, induced by ordering along all the $\vec{G}$ vectors. Therefore, the energy dissipation balance per unit area is

$$(kT/2)\sum_{\vec{G}} \int_{-s}^{\infty} |\vec{J}|^2/D_{\vec{G}}n_c\,\mathrm{d}z = n_c R\Delta\mu. \tag{5}$$

The diffusivity, $D_{\vec{G}}$, characterizes the atomic arrangement rate in the $\vec{G}$ wave and can be expressed via a frequency of thermal motion of particles, $\omega_{\mathrm{H}}(G)$, in the $\vec{G}$ density wave [38]:

$$D_{\vec{G}} = \omega_{\mathrm{H}} S(G)/G^2 = (1/G)\big[S(G)(kT/m)\big]^{1/2}. \quad (6)$$

Here, $S(G)$ is the maximum of the static structure factor at the wavenumber G, while ω_{H} is the half width at half height of the dynamic structure factor, $S(k,\omega)$ at the same wavenumber $k = G$. This information has been provided by experiments in coherent thermal neutron scattering. The frequency $\omega_{\mathrm{H}} = \Delta E/\hbar$ is measured from the neutron energy transfer, ΔE, as a function of the inelastic neutron scattering angle, i.e., the wavenumber, k. It is measured from the half width at half height of the neutron scattering spectra. In Fig. 2 [38], the frequency ω_{H} is normalized by the time $t_a = (m/kT)^{1/2}a/2$, at which the particle of mass m moves over its typical diameter, a, at its thermal gas velocity, $(kT/m)^{1/2}$. It is known that the typical frequency, ω_{H}, reaches a minima, $\omega_{\mathrm{H}}(G)$, at the wave vectors $k = \vec{G}$, when the conventional static structure factor of the liquid, $S(k)$, reaches a maxima [38–40]. These minima come from smaller energy exchanges between the liquid particles and the neutrons reflecting off small close-packed "planes" within the short-range order liquid. Consequently, the absolute value of the minimal frequency, $\omega_{\mathrm{H}}(G_i)$, decreases as the liquid density increases:

$$\omega(G) = A(n)(2/a)(kT/m)^{1/2}, \quad (7)$$

where the function $A(n) \cong 4.09\text{–}5.45\,(na^3/\sqrt{2})$ approximates the experimental data for $0.3 < na^3/\sqrt{2} < 0.65$. The parameter $a^3\sqrt{2}$, calculated in Ref. [38] as the specific volume per particle in an equivalent close-packed hard sphere liquid, provides the same first peak of the structure factor, $S(k)$, as the real liquid density n.

The left-hand side of Eq. (5) is $\sim \Delta\mu^2$. Therefore, the kinetic coefficient with respect to $\Delta\mu/kT$ can be expressed via the structure of the crystal–melt boundary layer [37]:

$$\beta = 2S(G_1)\omega_{\mathrm{H}}\xi_{\mathrm{b}}\Big/ \sum_{\vec{G}_1}, \quad (8)$$

where the correlation length, $\xi_{\mathrm{b}} = 1.8a$ corresponds to the reciprocal width of the first maximum of the structure factor at $G_1 = \langle 1\,1\,1\rangle$, $S(G_1) = 2.85$ for liquid of hard spheres at the melting point [40]. If the density waves generated by all eight $\{1\,1\,1\}$ reciprocal vectors G_1 are selected, then the sum for the $\{1\,1\,1\}$ interface in Eq. (8) is

$$\sum_{\vec{G}} \equiv \sum_{|G|=G_1} 1/\cos\theta_{\vec{G}\vec{n}} = 8\sqrt{3}. \quad (9)$$

The most important feature of the kinetic coefficient given in Eq. (8) is its proportionality to the gas velocity $(kT/m)^{1/2}$. This comes from the dynamics of particle motion in the liquid via the frequency ω_{H} given in Eq. (7). For Pb, $\beta^T = 28$ cm/s K, while the experimental figure is amazingly close at 33 cm/s K [41].

Recent MD simulations confirm that the density wave approach described above, provides the correct hierarchy for the kinetic coefficients for the $(1\,1\,1)$, $(1\,0\,0)$ and $(1\,1\,0)$ faces of Au, Cu, Ni. However, absolute values for these coefficients, following from the density wave theory, are $\sim 50\%$ lower than those from simulation. This discrepancy may be understood and

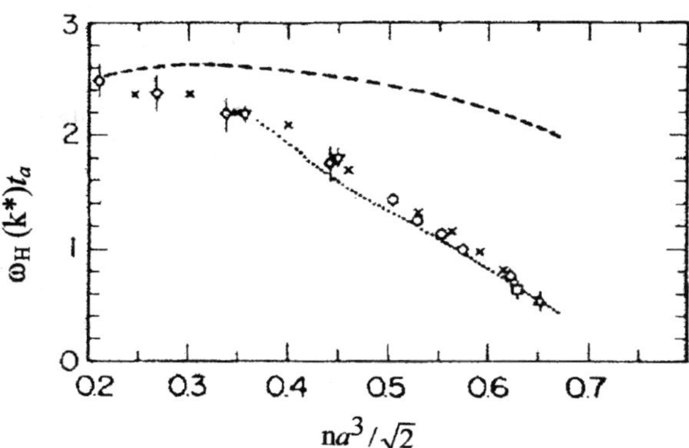

Fig. 2. Typical frequency ω_{H} of atomic motion in simple liquids as a function of the liquid density, n. The frequency ω_{H} is determined from experimental energy spectrum of thermal neutrons scattered by a simple liquid and resulting in the dynamic structure factor of this liquid. This dynamic factor shows sharp minimum at the neutron scattering directions where static structure factor is maximal. The ω_{H} is half width at half height of the minimum. Liquid density n is normalized by volume $a^3/\sqrt{2}$ per hard sphere of equivalent model liquid showing peak of the static structure factor at the same wave vector as the real fluid. The frequency ω_{H} is normalized by time t_a equal to the particle diameter over thermal velocity. Different experimental points correspond to Ar at various temperatures, Kr, Rb. Dotted line is theoretical. From Ref. [39].

Table 1
Kinetic coefficients of face, β^T, and step, β_{st}^T, in growth from pure melt

Material	β^T, β_{st}^T (cm/s K)
Succinonitrile [30], β^T	> 17
Cyclohexanol [30], β^T	1.1×10^{-2}
Tin [30], β^T	$> 3 \times 10^{-3}$
Lead [30], β^T	33
Silicon, steps on (1 1 1) [83], β_{st}^T	50

eliminated if one takes into account the vibration frequency, ω_H, in Eqs. (6) and (8), making the dissipation rate in Eq. (5) and the kinetic coefficient higher within the "semi-solid" diffuse interface as compared to the bulk liquid [26].

Some experimentally measured kinetic coefficients for melts are summarized in Table 1. The big difference between the β values for cyclohexanole on one hand, and succinonitrile and lead on the other, is not yet understood. The relatively low coefficient for tin might be associated with the tendency of tin to form chain clusters in the melt. This issue requires serious liquid structure studies similar to that for other complex liquids. It remains a challenge to check the density wave approach for other simple liquids and alloys and to modify this model for non-simple liquids having rough interfaces. The possible multicomponent nature of such liquids and the less evident direction of all $\vec{G}$ vectors, make obtaining and interpreting dynamic structure factors a big challenge.

Another important issue involves the motion of rough steps on smooth interfaces. In this case, the interstep terraces are smooth and, thus, these portions of the interface are localized. Therefore, the steps should only be smeared over several atomic/molecular rows, similar to Fig. 1. AFM observation suggests that this is not the case, at least for biomacromolecular crystals in their growth solutions. However, smectic layers of melt or solution adjacent to the smooth terraces between the steps are partly ordered [42,43] and, therefore, may also be enriched in crystallizing species (in solutions). This makes the liquid better prepared for crystallization during step motion. On the other hand, species mobility within the first smectic layers above the flat terraces may be lower and thus, make this crystallization slower.

3. Smooth interface: layer growth

The energies of unsaturated bonds of surface atoms are about an order of magnitude higher at the crystal–vacuum or gas interface as compared to the crystal–melt interface. Solutions present an intermediate case, though, even in this case, as compared with the crystal–melt interface, the ratio of excess energy per unsaturated bond to kT is usually close to unity or higher. Therefore, smooth, faceted interfaces are typical of the growth of thin films from molecular beams, chemical vapor deposition and the growth of bulk crystals from solutions. In this section, I focus on solution growth kinetics.

Contrary to the diffuse interface approach taken in Section 2, the localized interface model is considered in this section: each atom or molecule belongs to either the crystal or the mother phase in which the crystal may grow or dissolve. We assume first, following the BCF estimates, that the step is intensively meandering so that distance between kinks, λ_0, covers only several lattice spacings, a. Each isolated kink propagates along a step via statistically independent attachments and detachments of molecules or atoms. This simply means that there is no cooperative interaction between these species and that the detachment of each molecule from a kink requires the same work, no matter how many molecules are in the lattice unit cell: all these molecules are crystallographically equivalent and each may occupy a kink position. This is the Kossel model (see Section 6 for the non-Kossel model). In the Kossel model, the kink rate is just the difference between ingoing and outgoing components of the exchange flux between the crystal and its surroundings. Under these conditions, the step rate [44]

$$
\begin{aligned}
v &= (a^2/\lambda_0)\nu\big\{\exp[(\mu_1 - E)/kT] \\
&\quad - \exp[(\mu_s - E)/kT]\big\} \\
&= (a^2/\lambda_0)P\nu\exp(-E/kT)\omega C_e\sigma \\
&= \beta_{st}\omega(C - C_e), \\
\sigma &\equiv (C - C_e)/C_e.
\end{aligned}
\tag{10}
$$

In Eq. (10), ν is the typical frequency at which a molecule attempts to overcome potential barrier E to join the lattice at a kink, typically starting from a hydrated position nearby in solution, and ω is the specific molecular volume in the crystal. In this position, the molecule should already be in the correct orientation, for which the probability is P. The probability P may be close to unity for small molecules. However, it may be expected to be as low as 10^{-3} for the incorporation of big biological macromolecules (2–200 nm in diameter). Indeed, "docking" of such a molecule to the lattice at a kink requires that this molecule, despite Brownian liberation and rotation, approach the kink site in the orientation that is strictly in register with its future neighbors in the lattice. The typically complex shape of the large molecule prevents its close contact with the lattice molecules already in the kink—except along several random areas where, in addition, the interaction may be weak. Therefore, the torque applied to the big docking molecule by molecular forces from the kink is similar to that applied to small molecules. Consequently, rotational adjustment of the docking molecule, easy for small species, should be impossible for the much bigger asymmetric species. Depending on the nature of the contact to be established between the docking molecule and its neighbors in kink, the frequency ν in Eq. (10) may have a different origin: $\sim kT/\hbar$ if electronic bonding dominates or $\sim a(m/kT)^{1/2}$ if Brownian motion is the limiting stage. Elegant AFM experiments [45] demonstrated that the step kinetic coefficients are the same for apoferritin (molecular weight $M = 450$ kDa) and holoferritin ($M = 670$ kDa) (these molecules are both hollow spherical shells of the same diameter and structure, though in holoferritin the core is filled by ferruous oxide). This independence of kinetic coefficient on mass suggests that the Brownian motion in solution supplies enough species for future docking, a process which includes overcoming entropic and potential (probably dehydration) barriers.

In Eq. (10), the equilibrium, (e), and actual concentrations, C_e and C, should be taken immediately at the step and the activity coefficient is assumed to be unity. Eq. (10) also serves as the definition of step kinetic coefficient, β_{st}, or of a reciprocal resistance to crystallization at the step. With such kinetic coefficients, the mass balance at the growing step is [12]

$$D\partial C/\partial r = \beta_{st}(C - C_e), \qquad (11)$$

where r is the radial distance from the step. It is important to remember that the kinetic coefficients should relate the crystallization flux to the step ($D\partial C/\partial r$ or

$\omega\nu$) and to the supersaturation $C - C_e$ right at the step, rather than the supersaturation averaged out over the growing surface. Therefore, the introduction of kinetic coefficients as a factor in the already calculated step rate [1] will lead to the incorrect ultimate dependence of the step and face growth rate on supersaturation, particularly if the resistance for incorporation of a new species to the step is low or comparable to the resistance in other growth stages, i.e., surface diffusion or adsorption [12].

If the step rate, ν, and elementary step density on a vicinal crystal face, p/h (h is the step height and p is the local vicinal surface slope), are known, then the local propagation rate, R, of a face may be found from the evident kinematic relationship:

$$R = p\nu. \qquad (12)$$

If the steps are generated by a screw dislocation creating a vicinal hillock on the face, then p is just an average slope for this hillock $p = \bar{p} \cong 10^{-2}$–$10^{-3}$ (cf. Fig. 4). If the steps are generated by two-dimensional nucleation, then the local step density (local slope p) enters Eq. (12) [16]. This slope has a macroscopic sense if the nucleation occurs preferentially in regions of high supersaturation, e.g., near the crystal edge [16,46]. From Eq. (12), the simplest kinetic coefficient, β, for the face growth $\beta = \bar{p}\beta_{st}$.

Experimental data on $\beta_{st}\bar{p}$ and the growth modes are summarized in Table 2. The much lower values for β_{st} in proteins may be ascribed to the entropy constraints associated with the large molecular size and the asymmetric shape of the molecule described above. The area per intermolecular contact in proteins may reach 10^3 Å^2 or more and, thus, should also be associated with the high-potential barrier, E, in Eq. (10). Biomacromolecules are typically stable in a narrow range of temperature ($\sim$ 0–40 °C) and there are no direct reliable measurements of the activation energy.

In many cases, growth kinetics is treated in the simplest way, as a result of the direct incorporation of crystallizing species from solution at the kink sites at steps [12]. Alternatively, adsorption on interstep terraces and surface diffusion is considered, following BCF, even in solution growth [47]. The direct incorporation model works well in many cases of solution growth [31]. It is also known that the micromor-

Table 2
Kinetic coefficients β_{st} and vicinal slopes p

Substance, Face	β_{st} (cm/s)	p	$p\beta_{st}$ (cm/s)
ADP, KDP, DKDP $(1\,0\,0)$ 10^{-4}–10^{-3}	$(5\text{–}12) \times 10^{-2}$	3×10^{-4}–8×10^{-3}	
NH_4H2PO_4, KH_2PO_4			
ADP $(1\,0\,1)$	0.4×10^{-1}	10^{-4}–5×10^{-3}	4×10^{-5}–5×10^{-3}
$BaNO_3$ $(1\,1\,1)$	1.3×10^{-2}	$(3\text{–}15) \times 10^{-4}$	4×10^{-6}–2×10^{-5}
$KAI\,(SO_4)_2 \cdot 12H_2O$ $(1\,1\,1)$ alums	8×10^{-2}	$(0.4\text{–}3.5) \times 10^{-3}$	3×10^{-5}–3×10^{-4}
$Y_3Fe_5O_{12}$ $(1\,1\,0)$, $(2\,1\,1)$		$(0.3\text{–}3) \times 10^{-2}$	$(0.4\text{–}1) \times 10^{-3}$
$(YSm)_3(FeGa)_5O_{12}$ $(1\,1\,1)$ 10^{-2}	1.4×10^{-2}		
$(EuYb)_3Fe_5O_{12}$ $(1\,1\,1)$	$(0.1\text{–}3) \times 10^{-3}$		
(a) Lysozyme $(1\,0\,1)$ 14,300	$2\text{–}20 \times 10^{-4}$	$(1.1\text{–}1.5) \times 10^{-2}$	$3\text{–}30 \times 10^{-6}$
(a) Lysozyme $(1\,1\,0)$ 14,300	$2\text{–}3 \times 10^{-4}$	$(1.0\text{–}1.5) \times 10^{-2}$	$3\text{–}30 \times 10^{-6}$
(b) Canavalin 147,000	9×10^{-4}	9×10^{-3}	9×10^{-6}
(b) Thaumatin	22,000	2×10^{-4}	2D nucleation
(c) Catalase	25,000	3.2×10^{-5}	2D nucleation
(d) STMV	1,600,000	$(4\text{–}8) \times 10^{-4}$	2D nucleation
(e) Ferritin	480,000	6×10^{-4}	2D nucleation

References:
Inorganic:
A.A. Chernov, J. Crystal Growth 118 (1992) 333.
P.G. Vekilov, Yu.G. Kuznetsov, A.A. Chernov, J. Crystal Growth 121 (1992) 643.
Proteins:
(a) P.G. Vekilov, M. Ataka, T. Katsura, J. Crystal Growth 130 (1993) 317.
(b) Y.G. Kuznetsov, A.J. Malkin, A. Greenwood, A. McPherson, J. Struct. Biol. 114 (1995) 184.
(c) T.A. Land, J.J. DeYoreo, J.D. Lee, Surf. Sci. 84 (1997) 136.
(d) A.J. Malkin, et al., Phys. Rev. Lett. 75 (1995) 2781.
(e) S.-T. Yau, B.R. Thomas, P.G. Vekilov, Phys. Rev. Lett. 85 (2000) 356.

phology of a growing face depends on solution flow [48,49]. On the other hand, the average step rate, v, on the $(1\,0\,1)$ face of ADP ($NH_4\,KH_2\,PO_4$), found to depend on the average vicinal slope, $\bar{p}$ (cf. Fig. 4), may be stronger than expected from the model of direct incorporation by steps on a regular step train. Within the framework of the local kinetic coefficients, the velocity dependence on the local slope, p, was approximated by [46,50]

$$v(p) = v_0/(1 + Kp). \tag{13}$$

Here, v_0 is velocity of an isolated step. With the constant $K \sim 10^2$–10^3 and for typical p values to be $p = 10^{-2}$–10^{-3}, the product $Kp \sim 1$. For an infinite equidistant train of elementary steps of height, h, average vicinal slope, $\bar{p}$ and kinetic coefficient, β_{st}, in an infinite stagnant solution occupying semispace above the interface, Kp in Eq. (13) should be replaced by $(\beta_{st}h/D) \ln |\bar{p}|^{-1}$ and is practically independent on the vicinal slope, $\bar{p}$ [23]. Also, with $\beta_{st} = 0.1$ cm/s, $h = 10^{-7}$ cm, $\bar{p} \cong 10^{-2}$–10^{-3} and bulk diffusivity $D = 10^{-5}$ cm^2/s, one has $(\beta_{st}h/D) \ln |\bar{p}|^{-1} \cong 5 \times$

10^{-3}. The $v(p)$ dependence might be different if the train of elementary steps is replaced by train of step bunches (Section 4). Nevertheless, the possibility of combined surface and bulk diffusion [47] remains an option.

The dependence of $v(p)$ in Eq. (13) is in agreement with the dependence of the propagation rate, v, of a step bunch on the local slope, p, within the bunch. This can be measured with high-precision interferometry on step bunches propagating over the steep vicinal hillock slope on the $(1\,0\,1)$ face of a KDP (KH_2PO_4) crystal in a flowing solution [51,52]. The behavior of step bunches has shed light on the long-standing problem of bulk vs. surface diffusion. Step bunching also affects crystal quality, which makes it of practical significance and is discussed in the next section.

4. Step bunching

By a step bunch we mean an agglomeration of elementary steps approximately parallel to one another. A surface profile showing elementary steps and two step bunches, B_1 and B_2, are given in Fig. 4. Step bunches trap point defects in amounts different from that incorporated by single steps. This difference results in striations within the grown crystals [53] as seen in Fig. 4. Unlike striations induced by variations of external temperature, compositions and liquid flows, the bunch-induced striations are intrinsic phenomenon and have not been eliminated, so far, by stabilization of external conditions and convective flows. In particular, this problem is essential to the fast solution growth of large KDP family crystals (up to 50 cm) for nuclear fusion [29,30]. In these crystals, step bunches lead to microinclusions and other inhomogenieties compromising crystal performance. Bunch-induced striations were also found in protein crystals [54].

Step bunches are also well known in vapor growth and epitaxy and have been thoroughly studied [55–58]. Step–step interaction in vapor growth is very strong, since surface diffusion is its most important component. Indeed, between parallel steps, the surface diffusion field is one-dimensional so that the concentration of crystallizing species in the adsorbed layer exponentially changes with the distance from the step. The typical interaction distance is the path, λ_s, covered by an adsorbed molecule, atom or ion during its lifetime on the terrace. Typically, λ_s exceeds the lattice spacing by 2–3 orders of magnitude, depending on the adsorption energy and temperature. Steps separated by a distance $h/p \gg \lambda_s$ do not sense one another, while interstep distances of $h/p < \lambda_s$ result in a lower material supply to these steps and a lower step rate—as compared with that of widely separated steps. This strong interaction is mitigated if the incorporation rate at each step is low as compared to the surface diffusion rate, i.e., $\beta_{st} \ll D_s/\lambda_s$, where D_s is surface diffusivity [12]. This may be the case when chemical reactions occur at the steps. When surface diffusion is essential, the species supplied to a step from the lower and upper terraces are attached to the step at different rates (higher from the lower terrace) [59]. This asymmetry is essential for bunching in vapor growth.

Mutual retardation of step motion causes "traffic jams"—step bunches or "shock waves" of steps. The step bunching induced by surface diffusion is relatively well understood from studies of the vapor growth or MBE of simple substances, like silicon, or

metals where there is no resistance for incorporation at steps [55–58]. Facetting, following step bunching assisted by surface diffusion, has also been extensively modeled [55–57,60].

Besides bunches, facetted macrosteps with the macrostep riser built by a crystallographic face are sometimes also present on growing surfaces (Fig. 4). Step bunches and real macrosteps on crystal faces grown and growing from solutions are very common and may often be seen by the naked eye [61–63].

The step density within the bunch, p/h, Fig. 4, is still much lower than the atomic density on the terrace, $1/h$, i.e., the local bunch slope, $p \ll 1$ [51,64]. On the other hand, p may essentially exceed the average vicinal slope, $\bar{p}$. The riser of a real macrostep is often a narrow crystallographic face other than that on which the macrostep propagates. In other words, $p \cong 1$ at the riser and the riser on its own may grow via elementary steps running along this microfacet [12] (Fig. 4). Bulk diffusion around the macrosteps, their stability and the interaction between real macrosteps, were analytically addressed in Ref. [65]. Below, we consider step bunches only.

In solution growth, step bunching is much less understood than in vapor growth and related epitaxies since bulk diffusion and incorporation at the steps are more important. The roles of surface diffusion and the existence of Schwoebel asymmetry [59] are still not clear. A detailed understanding of bunching may allow insights into a number of molecular processes.

Earlier studies assumed that step bunches (and real macrosteps) might be considered as an entity of a height essentially exceeding the lattice parameter (i.e., elementary step height) and, thus, propagating at a velocity inversely proportional to its height (because of an overlapping of the bulk diffusion fields of the elementary steps and their mutual retardation [12]). Then, since bunch heights are not equal to one another, the bunches must coalesce. If splitting of the bunches is ignored, then the average height increases as $\sim x^{1/2}$, where x is the path covered by the sequence of bunches [12]. Step–step interactions in biomacromolecular crystals via their bulk and surface diffusion fields have been modeled [66], revealing enhanced instability only with respect to finite amplitude perturbations. The stability of an elementary step with respect to its meandering within the growth plane in a flowing solution has also been analyzed analytically [67].

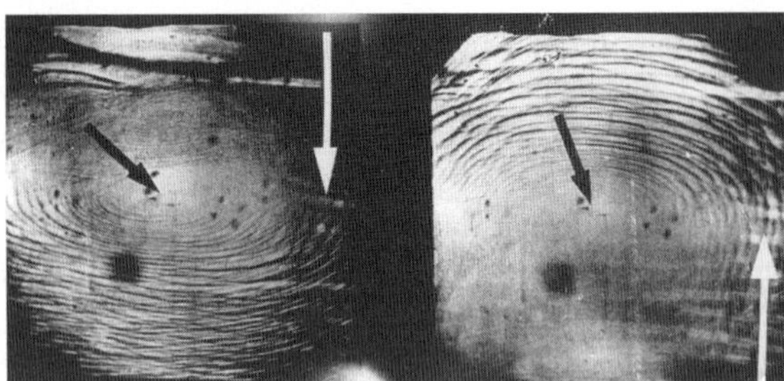

Fig. 3. Flow-dependent step bunching on the $(0\,0\,1)$ face of $NH_4H_2PO_4$ (ADP) crystal. One dislocation hillock, with apex shown by black arrow, occupies the whole face ~ 5 mm wide. White arrow shows solution flow velocity. Ripples are S step, bunches are reversible. The bunches are developed on the downstream slope of the hillock. When direction of solution flow changes to the opposite, bunches disappear on the now upstream slope and appear on the now downstream slope after ~ 1 min.

Bunching was also observed to be enhanced by impurities. One of the suggested mechanisms is that the impurities which hinder step propagation are adsorbed on interstep terraces: the wider the terrace, the longer it is exposed to impurity adsorption, the larger the density of these impurities, the stronger they diminish the bunch propagation rate, the higher are the bunches [68,69]. In this model, the average bunch/macro-step height should also increase infinitely with $\sim \ln x$, in contrast to the $x^{1/2}$ dependence mentioned earlier.

The importance of bulk diffusion for step bunching follows directly from the dependence of bunching behavior on solution flow. This dependence was found originally on the $(1\,0\,0)$ face of an ADP $(NH_4\,H_2\,PO_4)$ crystal growing from a solution flowing parallel to the face as shown in Fig. 3 by bold white arrows [48]. Fig. 3 shows one vicinal conical hillock occupying the whole $(1\,0\,0)$ face of ~ 8 mm long. The hillock was generated by a dislocation step source in the center shown by the black arrow. The hillock slope, where steps are moving in the same direction as the solution flows, is rippled, i.e., step bunches are formed. The opposite slope of the same hillock where the steps and solution flow in opposite directions does not show visible bunching. When the direction of solution flow is switched $180°$, the ripples disappear on the now upstream slope and develop on the now down-stream slope. This flow-induced morphology change is completely reversible [48]. This morphological instability phenomenon was understood in terms of solution flow within the diffusion boundary layer [49,70–73]. To

better explain this mechanism, a close-up of the corrugated crystal–solution interface is illustrated schematically in Fig. 4 as a cross-section normal to the surface and the steps on this surface. A weak corrugation may be induced by variations in step source activity, in solution supersaturation, by the presence of local step stopping impurities on the surface, or just by the random meandering of elementary steps. An inhomogeneous step distribution on the corrugated surface induces an inhomogeneous distribution of concentration over this surface: the concentration being lower when the density of elementary steps absorbing the crystallizing species is higher, i.e., over the bunches B in Fig. 4. Similarly, over the flatter, low step density segments, the concentration is higher, as illustrated in Fig. 4 by the darker and lighter regions at the terraces. If solution flows to the right, as shown in Fig. 4 with the arrow u, the enriched solution from the regions of lower step density comes first, and is discharged to the protrusions of the surface profile. This local increase in supersaturation increases the propagation rate of the steps forming these protrusions and, therefore the protrusion heights. Similarly, the solution depleted over the bunch, B, comes first to the depressions and aggravates these depressions. Thus, if the solution flows down the step flow, it indeed should cause instability. Analogously, if the step and solution flow directions are anti-parallel, the interface is stabilized.

The non-linear, S-shaped, dependence of face growth rate on supersaturation, typical if impurities are present in the system, enhances instability. Namely, step bunches on the $(1\,0\,0)$ ADP face appear preferentially within the supersaturation range near $\sigma \sim 1.5 \times 10^{-2}$, where the normal growth rate increases steeply. This bunching occurs independently of the solution flow direction and velocity [74]. Otherwise, the flow still influences the onset of bunching at other supersaturations [72]. Another type of step bunching on the $(1\,0\,0)$ face, probably also related to impurities, happens preferentially within two azimuthal sectors of a large growth hillock, the sectors being centered at the hillock apex. Bunches filling these two sectors are oriented along the direction of maximal growth rate, i.e., propagate in the direction of the minimal rate [31]. It was recently found that on vicinal hillocks on the $(1\,0\,1)$ KDP face, usually not very sensitive to impurities, the flow-induced bunching is reversed: bunches appear when the di-

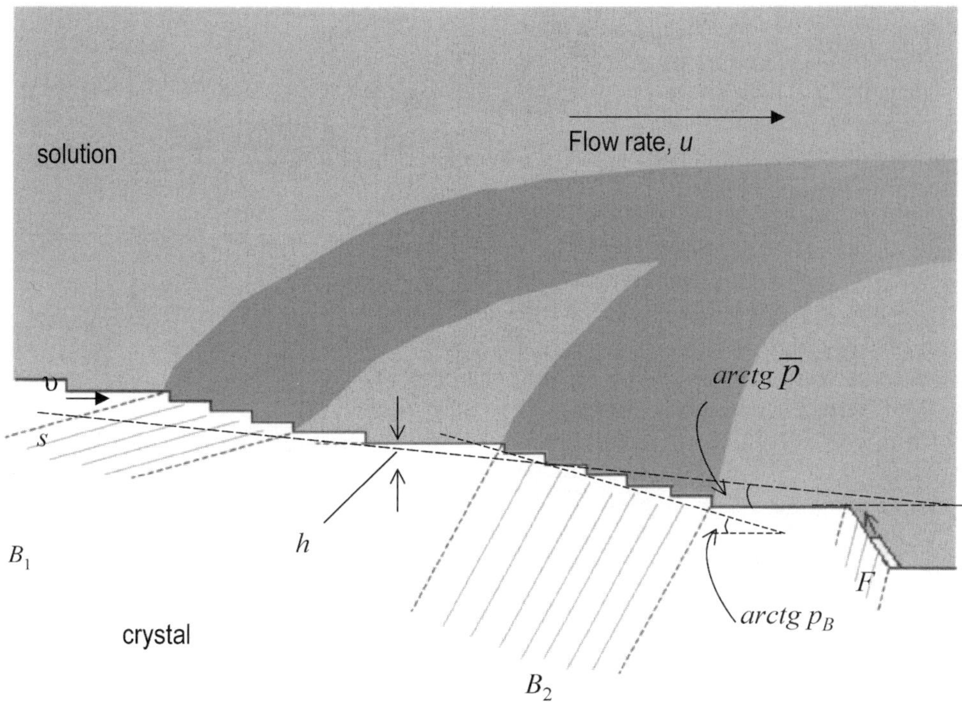

Fig. 4. Stepped surface profile with elementary steps, S, step bunches, B_1 and B_2 and faceted macrostep, F. The riser of the latter is formed by a crystallographic facet, unlike risers of the bunches B_1 and B_2, which are only slightly declined from the singular terrace orientation, $p = 0$, and from the average vicinal slope, p. Both the average vicinal slope, $\bar{p}$, and the local bunch riser slope, $\bar{p}$, are small: $\bar{p} = 10^{-2}$–10^{-3}. The shadows from bunches B_1 and B_2 symbolize the drag of the depleted solution down the solution flow and their mutual overlapping, resulting in enhanced interaction between the bunches.

rections of the step propagation and solution flow are anti-parallel. This happens, however, only at low supersaturations and also might be associated with impurities.

The impurity effect on bunching at least partly follows from the superlinear $R(\sigma)$ dependence. This originates from a steep superlinear increase in the step rate when either the solution is in the supersaturation range where the step breaks through the "fence" of step retarding impurities, or not enough time is available for the impurity species to be adsorbed on terraces and steps [72]. Indeed, a periodic perturbation of a flat vicinal, i.e., equidistant step train, leads to (periodic) local slope variation and, thus, to the corresponding perturbation in the concentration field above this growing surface. Within the supersaturation range where the $R(\sigma)$ dependence is strong, the amplitude of the perturbed concentration wave is larger than when the $R(\sigma)$ dependence is weaker. In this case, the impurity-induced bunching found in

experiments is not as sensitive to the solution flow as it is in the theory and it is observed at any flow direction. Away from the σ range where $R(\sigma)$ is non-linear, conventional behavior is observed [74].

The recently developed phase-shift interferometric technique [52] now allows us to reveal bunch formation to the accuracy of ~ 5 nm in bunch height. A growth hillock interferogram taken on the $(1\,0\,1)$ face of KDP is viewed in Fig. 5a [51]. It consists of three slopes of which the upper is the most shallow, while the lower is the steepest out of the three. Enlarged images from different portions of the vicinal slopes are presented in Figs. 5b–d. The lineage structures in c and d are step bunches. Recording the surface height z normal to the face allows us to quantitatively characterize the surface profile, $z(x)$ along the coordinate x normal to the mutually parallel bunches within the growth plane. The profile and its Fourier spectrum are presented in Fig. 6. First measurements show that the typical wavelength of ripples (recipro-

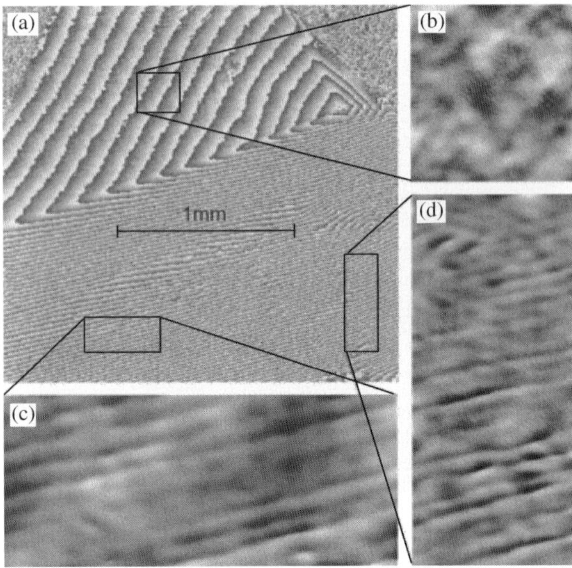

Fig. 5. High-precision phase shift interferometry of a dislocation growth hillock on the growing $(1\,0\,1)$ KH_2PO_4 (KDP) face (a). Solution flows down dislocation. The hillock has three different slopes: average step density is proportional to the interference fringe density. Solution flows down to the fringe flow on the steepest hillock. Zooming in for different areas are shown in (b), (c) and (d). Step bunches are seen on (c) and (d) as darker bands, but not on (b).

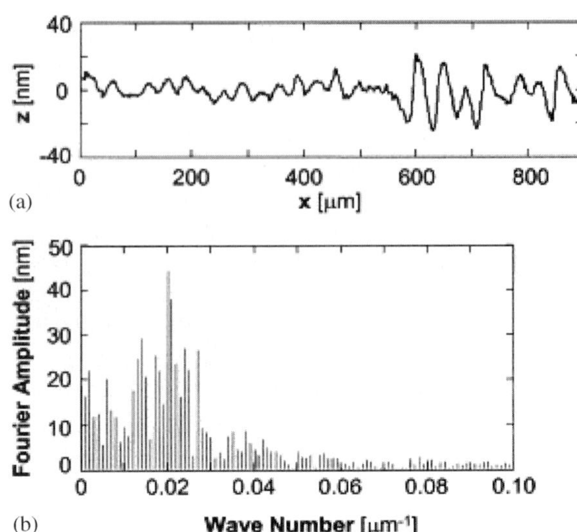

Fig. 6. Profile of the bunched slope (a) and Fourier spectrum of this profile (b).

cal wavenumber) decreases as the flow rate increases and Eq. (13) holds with p being local slope within the bunches.

The $v(p)$ dependence given in Eq. (13) has not yet been theoretically understood. It may come from both bulk and surface diffusion. In the case of bulk diffusion, an analogy between elementary [23] and macrostep trains may be expected. This analogy assumes that the bunch width, Λ, is (1) shorter than the interbunch distance, (2) that the local bunch slope, p, is proportional to the average slope, $\bar{p}$, and (3) that the stagnant boundary layer is thicker than the interbunch distance. Then, the elementary step height, h, should be replaced by the bunch height, Λp, so that $Kp \cong (\beta_{st}\Lambda p/D)\ln|p|^{-1}$. Alternatively, $\beta_{st}p$ may be considered as the kinetic coefficient of the bunch riser. Within these assumptions, $K \cong 10^2$–10^3 at $\Lambda = 20$–200 µm, $\beta_{st} = 0.1$ cm/s, $\bar{p} \cong 7 \times 10^{-3}$, $D = 10^{-5}$ cm^2/s. This estimate, while encouraging for the bulk diffusion model to explain the whole problem, still awaits a consistent analysis. The diffusion field around the step bunches, taking into account laminar or laminar/turbulent, rather than stagnant boundary

layers, needs to be found. The decrease in bunch width and height when the solution flow rate increases has no consistent explanation. However, we may speculate that the flow rate increase cuts off the bulk diffusion fields of the bunch and, thus, the interaction between ends of such a bunch becomes weak. Weak interaction facilitates bunch splitting. Also, increasing the flow rate enhances the mutual influence of steps within a bunch and may increase the difference in feeding between the upstream and downstream portion of a bunch, increasing its local slope. On the other hand, it is known that the higher step density, the less stable should be vicinal face [49,71–73]. The same tendency may be expected for a single bunch. Therefore, increasing free flow velocity, u, and thus the shear rate, $\partial u/\partial z$, may indeed reduce the stability of the bunch with respect to splitting into two or more bunches. This option is still to be analyzed. Also, the analyses in Refs. [71,73] were performed at low shear rates, $\partial u/\partial z < 1$ s^{-1}. If the solution flow is fast, up to 1 m/s, the shear rate may be estimated from turbulent boundary layer theory [75,76] to be $\sim 10^3$ s^{-1} so that the consideration should be generalized.

5. Steps with rare kinks and spiral growth

In the classical BCF, the thermal meandering of a step was considered to be so intense that the step roughness and step propagation rate only weakly depended on step azimuthal orientation within a growing face. In this case, screw dislocations generate rounded spiral steps. The radial distance between two neighboring turns equals $19\rho_c$, where $\rho_c = \omega\alpha/\Delta\mu$, is the radius of the two-dimensional nucleus. The coefficient 19 [77] replaces the original $4\pi \cong 12.6$ value from BCF. A similar consideration was applied to polygonized spirals, and the coefficient $\cong 19$ was confirmed for the interstep distance. These results follow from the Gibbs–Thomson dependence of the step rate on its curvature for rounded spirals and the step segment rate dependence on the segment length, L. However, recent AFM experiments have shown that the Gibbs–Thomson rule to evaluate the growth rate, v, of a straight step segment of length L as

$$v = v_\infty(1 - L_c/L), \tag{14}$$

does not provide the correct results [78–81]. In Eq. (14), v_∞ is the rate of an infinite step ($L \to \infty$), L_c is the side length of a polygonized critical two-dimensional nuclei. If this nuclei is a rectangle with the sides L_1 and L_2 bearing free step riser energies α_1 and α_2, respectively, then

$$L_{c1} = 2\omega\alpha_2/\Delta\mu, \qquad L_{c2} = 2\omega\alpha_1/\Delta\mu, \tag{15}$$

where $\Delta\mu = kT \ln C/C_e$ is the difference in the chemical potentials of the crystallizing molecules in the actual (concentration C) and equilibrium (concentration C_e) solutions and ω is the molecular volume. For definitiveness, we will take $\alpha_1 < \alpha_2$, i.e., $L_2 < L_1$.

Fig. 7 presents an AFM image of straight steps parallel to the c-axis on the (0 1 0) face of orthorhombic lysozyme crystal in aqueous solution. One end of the segment, B, is at the stacking fault normal to the segment and is unable to move except along this stacking fault. The opposite end is the corner where the step bends to acquire another orientation with a high kink density. This end can move along the segment, thus increasing the segment length. It turns out that the segments do not propagate normal to themselves until they reach a length $L_{c2} = 1140$ nm. With $\Delta\mu = kT \ln C/C_e = 4.10^{14}$ erg (at $C/C_e = 2.7$, estimated for the experiment) $\omega = 3 \times 10^{-20}$ cm^3 per one lysozyme molecule in the lattice, the second part of Eqs. (15) results in $\alpha_1 = 75$ erg/cm^2. An independent figure for the step riser energy, $\alpha = 1$ erg/cm^2, averaged over azimuthal orientations, comes from the face

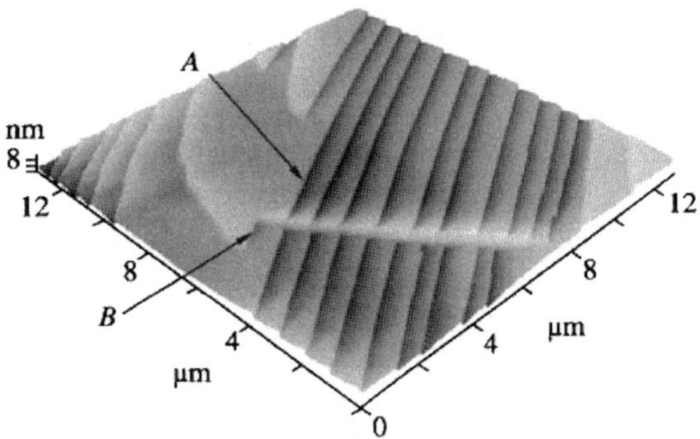

Fig. 7. Very straight steps propagating to the right and down from the invisible dislocation source located above the upper right corner of the image and not visible on the image. When a step meets stacking fault in the middle of the image, its longer portion (A) continues to propagate while the short segment (B) does not move until its length reaches ~ 1140 nm. See text for discussion. This segment is assumed to contain no kinks until it is sufficiently long.

growth rate dependence on supersaturation based on a disc-shaped 2D nucleus [82]. Assuming that the nucleation work is the same for both a rectangular and disk nuclei, the effective riser free energy, following from the growth rate, is $\alpha = [(4/\pi)\alpha_1\alpha_2]^{1/2}$. Taking for an estimate the ratio $\alpha_1/\alpha_2 = v_2/v_1 = 7$, the corresponding step rates ratio, we get for the larger step riser energy $\alpha_1 = 2.4$ erg/cm^2, i.e., about 30 times lower than the ~ 75 erg/cm^2 found above. No less important is that as soon as the segment L_2 begins to propagate normal to itself, its velocity, v_2, is independent of L_2. It, therefore, immediately acquires (to the accuracy of experiment) the velocity of the infinite step ($v_2 = 2.4$ nm/s). In an earlier work [83], AFM analysis of the same $\langle 0\,0\,1\rangle$ step on the (0 1 0) lysozyme face, though at a bit lower supersaturation, showed that kinks are extremely rare and that the average interkink distance was 575 nm. Thus, the step segment $L_{c2} = 1140$ nm should not include, on average, more than about two kinks. Therefore, there is no sense in an instantaneous velocity of such a short segment and the average velocity over numerous similar segments should be considered. To put it differently—step fluctuations might not be fast enough to ensure the applicability of equilibrium statistics and thus, of the classical BCF concepts at any moment of time [84].

The Gibbs–Thomson Equation (14) also does not describe the development of polygonized spirals on calcite [78] lysozyme [80] and potassium hydrogen phthalate crystals (KAP) [81]. Thus, steps with low kink density should be more carefully considered. This issue was first addressed theoretically by Voroknov [85,86] and continued recently on the basis of new AFM findings [78–81,83,84].

The present state of understanding may be summarized as follows. There are two interrelated modes in the propagation of step segments having low kink densities.

1. *Thermodynamic mode*: In this case, kink density on a singular step segment is low, but fluctuations are still fast enough to maintain Herring's equation

$$\Delta\mu = \omega K\big[\alpha(\varphi) + \partial^2\alpha/\partial\varphi^2\big] \equiv \omega K\tilde{\alpha}(\varphi), \tag{16}$$

on this step, not only at equilibrium, when the step is, on average, standing still, but also when the step is moving at a finite speed and, therefore, has a limited time to deposit each new row of molecules. In Eq. (16), φ is the azimuthal orientation of the step. It is well known [12] that at equilibrium, $\Delta\mu =$ constant over the whole step, so that the step curvature, K, is inversely proportional to the step stiffness $\tilde{\alpha} = \alpha + \partial^2\alpha/\partial\varphi^2$. For each singular orientation, in absence of kinks, $\alpha(\varphi)$ has a cusp minimum and, thus, $\partial^2\alpha/\partial\varphi^2 = \infty$. For a one-dimensional object, like a step, this singularity can only happen theoretically at $T = 0$. However, if the (kink energy)$/kT$ ratio is large, then kinks are extremely rare. For example, for steps on a (1 1 1) face of Si in a vacuum, during low-temperature epitaxy, say, at $500\,^\circ$C, the ratio is $\cong 18$. Then, the interkink distance is exponentially large, $\cong 10$ cm. Under these conditions, only the long time behavior of a finite step or properties of a sufficiently long step obey the conventional thermodynamics of large systems—since the thermodynamics find its way through the fluctuations, i.e., by the creation and diffusion of kinks. If time or step length are limited, we come to the kinetic mode.

2. *Kinetic mode*: In this mode, equilibrium with respect to kinks is not established, i.e., fluctuations have no time to be developed during the creation and elongation of the step segment. In this case, the direct kinetic approach should be applied.

Now, we may elaborate on some specific features of these two modes.

1. *Thermodynamic mode*: Voronkov's equation for the step rate propagation, v, along the normal to a singular orientation is

$$v = \beta_{\text{st}}(p)\big(1 + p^2\big)^{1/2}\Delta\mu_\infty/kT$$
$$+ \beta(p)(\omega\tilde{\alpha}/kT)\,\mathrm{d}p/\mathrm{d}x, \tag{17}$$

where the step kinetic coefficient $\beta_{\text{st}}(p)$ depends on the local step orientation $p = \text{tg}\,\varphi$, $\Delta\mu_\infty/kT$ is the supersaturation for an infinite step and $\tilde{\alpha}$ is the step stiffness defined by Eq. (16). Note that the step curvature $\mathrm{d}p/\mathrm{d}x < 0$. At and close to equilibrium, the kinetic coefficient, β_{st}, is directly proportional, while the step stiffness $\tilde{\alpha}$ is inversely proportional to the kink density [87]. Thus, the coefficient at $\mathrm{d}p/\mathrm{d}x$ in Eq. (17) is independent of p and Eq. (17) may be integrated at the steady-state step rate $v = $ const. The result describes the dependence of the step rate, v, on the segment length, L, and its value at a critical nuclei, L_c as [81]

$$(L/L_c) - 1 = (\pi kTa/2\omega\alpha)\big[1 - (v/v_\infty)^2\big]^{1/2}$$
$$\times (1 - 2/\pi \arcsin v/v_\infty), \tag{18}$$

where a is the kink depth. Eq. (18) replaces Eq. (14) for the polygonized steps if the kink fluctuations are fast enough to implement the equilibrium kink structure and, thus, Herring's Eq. (16). Eq. (18) gives a much steeper dependence of the segment rate v as a function of its length, L. The physical reason for this may be understood on the basis of "squeezing" the critical sized polygonized nucleus (which has very stiff straight edges ($\tilde{\alpha} \gg \alpha$) and rounded and thus "softer" corners) through a gate between, e.g., two dislocations of opposite signs or between a dislocation outcrop and next step segment of different orientation [84]. Eq. (18) works well for polygonized spirals on the (0 1 0) face of potassium hydrogen phtalate (KAP) [81], since experiments show an almost abrupt onset of v_∞ from $v = 0$ when the segment length L exceeds L_c:

$$v = \begin{cases} 0, & L < L_c \\ v_\infty, & L > L_c \end{cases}. \tag{19}$$

This function, independent of supersaturation σ, replaces the Gibbs–Thomson law, Eq. (16). Such an independence from supersaturation was found in AFM studies of calcite growth [78,81] while conventional Eq. (16) provides different $v(L)$ dependencies at different σ, since $L_c = 2\omega\alpha/kT \ln(1 + \sigma)$.

With Eq. (19), the period of the polygonized spiral "rotation," τ, and the distance, λ_i, between neighboring segments of the orientation are

$$\tau = \sum (L_{ci}/v_i) \sin\theta_{i,i+1}, \qquad \lambda_i = v_i\tau. \tag{20}$$

For a squared spiral, $\theta_{i,i+1} = \pi/2$, so that L_c, v and λ are independent of i:

$$\tau = 4L_c/v, \qquad \lambda = 4L_c. \tag{21}$$

On the other hand, the Gibbs–Thomson Eq. (16) results in $\lambda = 9.5L_c$, i.e., predicts a twice faster face rate on dislocations. The experiments [81] follow Eqs. (19) and (20). A similar conclusion follows from the kinetics of the polygonized spiral development on the $(1\,0\,1)$ monoclinic lysozyme face. Like on KAP, steps on monoclinic lysozyme include more kinks than the $\langle 0\,0\,1\rangle$ $(0\,1\,0)$ steps on the orthorhombic polymorph of lysozyme. Both on KAP and monoclinic lysozyme [80], the corners between segments of different orientations are also rounded to a scale easy detectable by AFM [80].

On monoclinic lysozyme, the time τ, Eq. (20), of the full cycle of the spiral development fluctuates strongly: $\tau = 94$ s for the first turn and 135 s for the next. The difference is mainly due to fluctuations in the length of one of the first segments ending on a screw dislocation. Unlike segments of the other three orientations, this segment is completely straight. The length fluctuations may be, of course, interpreted as a result of one impurity stone on the segment. However, the fluctuations may also happen because of a very fundamental reason—a long expectation time for kinks to appear on the segment. The latter option supports our guess that the creation of kinks is not fast enough. These fluctuations are connected to the kink dynamics.

2. *Kinetic mode*: There is no interaction energy between kinks on a step or ends of a step segment at least several lattice spacings apart. Therefore, the only way Gibbs–Thomson law can be implemented is by the exchange of kinks between the segment ends [84]. This entropic interaction is indeed possible because kinks diffuse along the step segment and may either reach the opposite ends or meet a kink of the opposite sign somewhere on the segment. In both cases, kinks annihilate and, thus, reduce the total surface energy. In other words, one segment end "feels" the presence of another only via a kink exchange. The kink behavior on the step may be summarized as follows. There are only two sources of kinks on a perfectly singular step segment: (a) one-dimensional nucleation of a new row along the segment, (b) splitting of kinks from the segment corners where the segment under consideration changes its orientation and is transferred into another singular segment, e.g., from $\langle 1\,0\rangle$ to $\langle 0\,1\rangle$. Mechanism (b) partly reflects the physics of an entropic interaction, though it may lead to segment dissolution only, even in a supersaturated solution. It can be shown that in this simplest case, if no 1D nucleation occurs, the dissolution rate equals

$$
\begin{aligned}
&-(w_+ - w_-)/\big[(w_+/w_-)^N - 1\big] \\
&\quad = -w_-\big[1 - \exp(\Delta\mu/kT)\big]/\big[\exp(N\Delta\mu/kT) - 1\big].
\end{aligned}
\tag{22}
$$

Here, w_+ and w_- are, respectively, attachment and detachment frequencies of growth species at the kink site and $2N$ is the segment length measured by the size of species. The negative sign in Eq. (22) automatically comes from the calculation of the net flux of kinks crossing any point on the segment for any w_+ and w_- and means dissolution. Dissolution occurs even in a supersaturated solution since kink annihilation leads to the step energy gain for any segment having a final length.

In a supersaturated solution, where $w_+ > w_-$, kinks that are split from the segment corners are pushed back to these corners so that the rounded corners have a smaller radius of curvature than at equilibrium. Also, the probability that a kink splits from a corner and diffuses along the segment reaching a distance n spacings long from the segment center is

$$\big[(w_+/w_-)^n - 1\big]/\big[(w_+/w_-)^N - 1\big] \cong (w_+/w_-)^{n-N}. \tag{23}$$

This probability is exponentially small at $n \ll N$ and $w_+ > w_-$. Therefore, the Gibbs–Thomson law is applicable if the segment half length, N, is less than or comparable with the ratio of kink diffusivity $(w_+ + w_-)/2$ to the kink rate, $w_+ - w_-$. This condition holds if the supersaturation

$$\sigma \equiv (w_+ - w_-)/w_- < (w_+ + w_-)/2w_- N \cong 1/N. \tag{24}$$

In the lysozyme case discussed at the beginning of this section [79], the immobile segment, L_c, was ~ 300 lattice cells (or 600 molecular sizes) long. In this case, Eq. (24) requires $\sigma < \sim 3 \times 10^{-3}$, while the actual $\sigma \sim 1$. Therefore, only 1D nucleation should control the kink density away from the regions which are $\sim 1/\sigma$ spacing long from each side of the segment. At $\sigma \sim 1$, this is essentially the whole segment. The estimate given above suggests that this is the expectation time for the creation of a 1D nuclei on a kink-free step segment, that controls the stagnation time of this segment as described in the beginning of this section. If this is the case, Eq. (18) is not applicable anymore, unless averaging over numerous cycles of spiral development provides the same result. Such an averaging has not been done so far.

We should not exclude the fact that the segment rate in the kinetic mode might be reduced to the same as that in the thermodynamic mode if an average is taken over numerous similar segments.

The term "1D nucleation" should not be confusing, even in the case of one crystallizing species per lattice unit cell because the irreversible creation of the nucleus—a stable pair of species—takes some time. This problem is more serious in complex lattices as discussed below.

6. Non-Kossel crystals

Practically, everything that has been done theoretically on interface kinetics so far has been done using the Kossel model. This model ignores the fact that numerous crystals have more than one molecule or atom in symmetrically non-equivalent positions within its unit cells. These crystals, unlike elements with simple cubic, FCC, BCC or HCP lattices, are topologically different from the Kossel crystals. In the non-Kossel crystal, identical molecules occupy non-equivalent positions within the lattice unit cell. Therefore, different kink configurations are inevitable and the energies required to detach a molecule from these different positions differ from one another. As a result, only the unit cell as a whole (though selected in any possible way) may present "repeatable units" in a kink. However, solutions or melts are typically composed of single atoms, molecules or ions rather than of unit cells. Therefore, building a new complex unit cell in the kink by the subsequent addition of new molecules (or groups of them) is similar to the growth of a crystal built of different chemical species. The driving force for crystallization is a deviation from the equilibrium

solubility product. The latter, in this case, is C_{e}^{m}, where m is the number of non-equivalent molecular positions per unit cell. In other words, the driving force for the whole unit cell should be used and equals $m \Delta\mu = kT \ln(C/C_{\mathrm{e}})^{m}$. This power dependence comes from cooperative effects. Indeed, the molecular positions in the complex unit cell should be filled in a specific sequence controlled by different energies and accessibilities of these positions, i.e., not independently from one another. On the other hand, since only one type of molecule is present, e.g., in molecular (or biomacro-molecular) solutions, the driving force is conventionally taken as C/C_{e} instead of the actual $(C/C_{\mathrm{e}})^{m}$. Therefore, following Eq. (10), instead of the conventional kink rate $v_{\mathrm{k}} \sim (C - C_{\mathrm{e}})$, one should expect $v_{\mathrm{k}} \sim [\exp(m \Delta\mu/kT) - 1] \sim C^{m} - C_{\mathrm{e}}^{m}$. This conclusion was also obtained from explicit kinetic considerations [84]. Still, at low supersaturations, $v_{\mathrm{k}} \sim (C - C_{\mathrm{e}})$, the kinetic coefficient should be different from the conventional expression, Eq. (10). Even more important is the non-linearity in the $v_{\mathrm{k}}(C/C_{\mathrm{e}})$ dependence where $C/C_{\mathrm{e}} \gg 1$, which is typical, e.g., in the case of protein crystal growth. Evidently, the non-linear dependence of kink rate on supersaturation should result in a non-linear dependence of the step and face rates, in addition to the conventional sources of non-linearity from dislocations and the 2D nucleation modes involved in step generation.

The non-linearity in the step rate dependence on the supersaturation C/C_{e} (or $\Delta\mu/kT$) for non-Kossel crystals may also come from 1D nucleation at the step riser. If the lattice unit cell includes m molecules in non-equivalent positions, then the 1D nucleus consists of $2m$ molecules, instead of 2 molecules for the conventional Kossel crystal. Consecutively, building of the nucleus, one by one, from $2m$ molecules in m different positions is first associated with an increase and then decrease of the Gibbs potential, for the same reason that leads to conventional potential barriers for 3D nucleation. However, in the case of non-Kossel crystals, the 1D critical nucleus cannot include more than $2m$ particles, no matter how low the supersaturation is. The tetragonal lysozyme structure is an example of a non-Kossel crystal [84], though no systematic data on non-Kossel crystal nucleation and growth are available.

7. Conclusions

The BCF approach brought contemporary physics to crystal growth theory and stimulated further development of crystal growth science. The half a century that has passed was a golden era in crystal growth that brought about a core understanding and rational approach to crystal growth. This, in turn, has resulted in a multibillion dollar industry based on electronic, optical and magnetic single crystals of the future. This fundamental development has provided a diagnostics framework to guide experiments. However, the chemical diversity of materials, especially in chemically complex systems, still limits our ability to make quantitative predictions. Also, challenges of fundamental significance still exist in basic crystal growth science. Some of them are mentioned in this paper and are summarized below.

First, we must be able to account for the real liquid structure at the crystal–liquid interface, rather than the lattice model for a mother liquor. This is a subject for MD simulations and may not be completely hopeless analytically.

Second, the pattern formations resulting from the interconnection between the interface and transport processes, and which results in dissipative structures, remains a challenge. This is especially true for systems with singular interfaces experiencing layer growth. In this case, species incorporation into the lattice occurs only at 10^{-2}–10^{-3} portion of the surface sites and is much slower than that with the normal growth of rough interfaces. As such, these processes are more important in comparison with transport phenomena than in pattern formation in simple melts and alloys. The interactions of solution flow with stepped singular interfaces raise new questions on dissipative step bunching structures and their development. The existence of surface diffusion at the crystal–liquid interface still needs to be addressed.

In addition to the above, the high kink density on steps, as noted by Frenkel and developed by BCF, turns out not always to be the case. Steps with low kink density raise the problems of whether kink nucleation and fluctuations are fast enough to compensate for kink annihilation during growth step propagation. The problem of how fluctuating kinks implement communication between two corners of a straight, nearly kink-free step segment should be addressed in more detail. This communication turns out to be effective only at very low supersaturation, and affects the form in which Gibbs–Thomson law should be applied to polygonized steps. The same problem for facetted crystals is even more interesting.

Finally, crystals in which the unit cells have several non-equivalent molecular positions may show strong cooperative interaction within the cell, inducing a real potential barrier for 1D nucleation and new non-linear dependencies of kink, step and face rates on supersaturation, i.e., new crystal growth physics and chemistry.

Acknowledgements

NASA support under grants NAG8-1748, NAG8-1454, NAG8-1829 and USRA/NASA MSFC CORBAMS contract are highly appreciated. I am thankful to Prof. L.N. Rashkovich, Prof. P.O. Vekilov and Dr. N.A. Booth for collaboration in obtaining some of the results described in Sections 4 and 5. I am indebted to Prof. R.S. Feigelson for his careful editorial work with the manuscript. Sincere thanks go to Ms. L.K. Lewis for her technical help in preparation of this manuscript.

References

[1] W.K. Burton, N. Cabrera, F.C. Frank, Philos. Trans. Roy. Soc. London Ser. A 243 (1951) 299.

[2] J.W. Gibbs, in: The Collected Works of J.W. Gibbs, vol. 1, Yale University Press, New Haven, 1961.

[3] M. Volmer, Kinetik der Phasenbildung, Steinkopff, Dresden, 1939.

[4] W. Kossel, Die molekularen Vorgange beim Kristallwachstum, Leipziger Vortrage, 1928.

[5] I.N. Stranski, Z. Phys. Chem. 136 (1928) 259.

[6] I.N. Stranski, Z. Phys. Chem. B 11 (1931) 342.

[7] I.N. Stranski, Z. Phys. Chem. B 17 (1932) 127.

[8] I.N. Stranski, R. Kaischew, Phys. Chem. B 26 (1934) 100.

[9] I.N. Stranski, R. Kaischew, Phys. Chem. B 26 (1934) 114.

[10] J. Frenkel, Phys. J. USSR 1 (1932) 498.

[11] K.A. Jackson, Interface structure, in: R.H. Doremus, B.W. Roberts, D. Turnbull (Eds.), Growth and Perfection of Crystals, Chapman and Hall, London, 1958, pp. 319–323.

[12] A.A. Chernov, Sov. Phys. Uspekhi 4 (1961) 116.

[13] S.T. Chui, J.D. Weeks, Phys. Rev. B 14 (1978) 4978.

[14] J.P. Van der Eerden, Crystal growth mechanisms, in: D. Hurle (Ed.), Handbook of Crystal Growth, vol. 1a, North Holland, Amsterdam, 1993, pp. 307–475.

[15] C. Godreche, in: C. Godreche (Ed.), Solids Far From Equilibrium, Cambridge University Press, New York, 1992.

[16] A.A. Chernov, Modern Crystallography III, Crystal Growth, Springer, Berlin, 1984.

[17] W.A. Tiller, The Science of Crystallization: Microscopic Interfacial Phenomena, Cambridge University Press, New York, 1991.

[18] W.A. Tiller, The Science of Crystallization: Macroscopic Phenomena and Defect Generation, Cambridge University Press, New York, 1992.

[19] W.W. Mullins, R.F. Sekerka, J. Appl. Phys. 34 (1963) 323.

[20] W.A. Tiller, K.A. Jackson, J.W. Rutter, B. Chalmers, Acta Metall. 1 (1953) 428.

[21] V.V. Voronkov, Sov. Phys. Solid State 6 (1965) 2378.

[22] A.A. Chernov, Sov. Phys. Crystallogr. 16 (1972) 734.

[23] A.A. Chernov, J. Crystal Growth 24/25 (1974) 11.

[24] R.F. Sekerka, Fundamentals of phase field theory, in: K. Sato, Y. Furukawa, K. Nakajima (Eds.), Advances in Crystal Growth Research, Elsevier, Amsterdam, 2001, pp. 21–41.

[25] A. Karma, Phys. Rev. Lett 87 (2001) 115701.

[26] J.J. Hoyt, M. Asta, A. Karma, Phys. Rev. Lett 86 (2001) 5530.

[27] R.L. Davidchack, B.B. Laird, Phys. Rev. Lett 85 (2000) 4751.

[28] G.B. Stringfellow, J. Crystal Growth, this issue; doi:10.1016/jcrysgro.2003.12.037.

[29] J.J. De Yoreo, Z.V. Reck, N. Zaitseva, B.W. Woods, J. Crystal Growth 166 (1995) 291.

[30] V.I. Bespalov, V.I. Bredikhin, V.I. Katsman, V.P. Ershov, L.A. Lavrov, J. Crystal Growth 82 (1987) 776.

[31] L.N. Rashkovich, KDP-Family Single Crystals, Adam Hilger, Bristol, 1991.

[32] K.A. Jackson, D.R. Uhlman, J.D. Hunt, J. Crystal Growth 1 (1967) 1.

[33] D.E. Ovsienko, G.A. Alfintsev, Crystal growth from the melt, in: Crystals. Growth, Properties and Applications, Springer, Berlin, 1982, pp. 119–169.

[34] J.Q. Broughton, G.H. Gilmer, K.A. Jackson, Phys. Rev. Lett. 49 (1982) 1496.

[35] J.P. Van der Eerden, Atomic models for crystal growth, in: J.P. Van der Eerden, O.S.L. Bruinsma (Eds.), Science and Technology of Crystal Growth, Kluwer, Dordrecht, 1995, pp. 15–26.

[36] A. Bonissent, Structure of solid-liquid interface, in: A.A. Chernov, H. Muller-Krumbhaar (Eds.), Crystals. Growth, Properties and Applications, Springer, Berlin, 1983, pp. 1–21.

[37] L.V. Mikheev, A.A. Chernov, J. Crystal Growth 112 (1991) 591.

[38] E.G.D. Cohen, P. Westerhuis, E.M.D. Shepper, Phys. Rev. Lett 59 (1987) 2872.

[39] P.G. de Gennes, Physica 25 (1959) 825.

[40] L.V. Mikheev, A.A. Chernov, Sov. Phys. JETP 65 (1987) 971.

[41] G.H. Roadway, J.D. Hunt, J. Crystal Growth 112 (1991) 554.

[42] T. Nakada, S. Miyashita, G. Sazaki, H. Komatsu, A.A. Chernov, Jpn. J. Appl. Phys. 35 (1996) L52.

[43] W.J. Huisman, J.F. Peters, M.J. Zwanenburg, S.A. de Vries, T.F. Derry, D. Abernathy, J.F. van der Veen, Nature 290 (1997) 379.

[44] A.A. Chernov, H. Komatsu, Topics in crystal growth kinetics, in: J.P. van der Eerden, O.S.L. Bruinsma (Eds.), Science and Technology of Crystal Growth, Kluwer Academic Publishers, Dordrecht, 1995, pp. 67–80.

[45] D.N. Petsev, K. Chen, O. Gliko, P.G. Vekilov, Proc. Natl. Acad. Sci. USA 100 (2003) 792.

[46] P.G. Vekilov, A.A. Chernov, Solid State Phys. 57 (2002) 1.

[47] G.H. Gilmer, R. Ghez, N. Cabrera, J. Crystal Growth 8 (1971) 79.

[48] A.A. Chernov, Y.G. Kuznetsov, I.L. Smol'sky, V.N. Rozhanskii, Sov. Phys. Crystallogr. 31 (1986) 705.

[49] A.A. Chernov, J. Crystal Growth 118 (1992) 333.

[50] P.G. Vekilov, Y.G. Kuznetsov, A.A. Chernov, J. Crystal Growth 121 (1992) 643.

[51] N.A. Booth, A.A. Chernov, P.G. Vekilov, J. Crystal Growth 237–239 (2002) 1818.

[52] N.A. Booth, B. Stanojev, A.A. Chernov, P.G. Vekilov, Rev. Sci. Instrum. 73 (2002) 3540.

[53] E. Bauser, Atomic mechanisms in semiconductor liquid phase epitaxy, in: D.T.J. Hurle (Ed.), Handbook of Crystal Growth, North-Holland, Amsterdam, 1994, pp. 879–911.

[54] P.G. Vekilov, F. Rosenberger, Phys. Rev. E 57 (1998) 6979.

[55] M. Sato, M. Uwaha, Y. Saito, Phys. Rev. B 62 (2000) 8452.

[56] M. Uwaha, Y. Saito, M. Sato, J. Crystal Growth 146 (1995) 164.

[57] F. Liu, H. Metiu, Phys. Rev. E 49 (1994) 2601.

[58] K. Fujita, M. Ichikawa, S. Stoyanov, Phys. Rev. B 60 (1999) 16006.

[59] R.L. Schwoebel, E.J. Shipsey, J. Appl. Phys. 37 (1966) 3682.

[60] D.G. Vlachos, L.D. Schmidt, R. Aris, Phys. Rev. B 47 (1993) 4896.

[61] I. Sunagawa, Morphology of minerals, in: I. Sunagawa (Ed.), Morphology of Crystals, Part B, Terra Science, Riedel, Dordrecht, 1987, pp. 509–587.

[62] I. Sunagawa, Morphology of crystal faces, in: I. Sunagawa (Ed.), Morphology of Crystals, Part A, Terra Science, Riedel, Dordrecht, 1987, pp. 321–365.

[63] D. Elwell, H.J. Scheel, Crystal Growth from High Temperature Solutions, Academic Press, London, 1975.

[64] A.A. Chernov, T. Nishinaga, Growth shapes and their stability at anisotropic interface kinetics: theoretical aspects for solution growth, in: I. Sunagawa (Ed.), Morphology of Crystals, Terra Science Publishers, Tokyo, 1987, pp. 207–267.

[65] S.Y. Potapenko, J. Crystal Growth 147 (1995) 223.

[66] F. Rosenberger, H. Lin, P.O. Vekilov, Phys. Rev. E 59 (1999) 3155.

[67] S.Y. Potapenko, J. Crystal Growth 158 (1996) 346.

[68] J.P. Van der Eerden, H. Mueller-Krumbhaar, Phys. Rev. Lett. 57 (1986) 2431.

[69] J.P. Van der Eerden, H. Mueller-Krumbhaar, Electrochem. Acta 31 (1986) 1007.

[70] B.T. Murray, S.R. Coriell, A.A. Chernov, G.B. McFadden, J. Crystal Growth 218 (2000) 434.

[71] S.R. Coriell, B.T. Murrary, A.A. Chernov, G.B. McFadden, J. Crystal Growth 169 (1996) 773.

[72] S.R. Coriell, A.A. Chernov, B.T. Murray, G.B. McFadden, J. Crystal Growth 183 (1998) 669.

[73] A.A. Chernov, S.R. Coriell, B.T. Murray, J. Crystal Growth 132 (1993) 405.

[74] L.N. Rashkovich, B.Y. Shekunov, J. Crystal Growth 100 (1990) 133.

[75] H. Schlichting, Boundary Layer Theory, sixth ed., McGraw-Hill, New York, 1968.

[76] H. Schlichting, K. Gersten, Boundary Layer Theory, eighth ed., Springer, Berlin, 2000.

[77] N. Cabrera, M. Levine, Philos. Mag. 1 (1956) 450.

[78] H.H. Teng, P.M. Dove, C.A. Orme, J.J. De Yoreo, Science 282 (1998) 724.

[79] L.N. Rashkovich, N.V. Gvozdev, M.I. Sil'nikova, I.V. Yaminski, A.A. Chernov, Crystal Rep. 46 (2001) 860.

[80] L.N. Rashkovich, N.V. Gvozdev, M.I. Sil'nikova, A.A. Chernov, Crystal Rep. 47 (2002) 859.

[81] L.N. Rashkovich, E.V. Petrova, O.A. Shuston, T.G. Chernevich, Phys. Solid State 45 (2002) 400.

[82] A.A. Chernov, Phys. Rep. 288 (1997) 61.

[83] A.A. Chernov, L.N. Rashkovich, I.V. Yamlinski, N.V. Gvozdev, J. Phys.: Condens. Matter 11 (1999) 9969.

[84] A.A. Chernov, J. Materials Science, Mater. Electron. 12 (2001) 437.

[85] V.V. Voronkov, Sov. Phys. Crystallogr. 15 (1970) 8.

[86] V.V. Voronkov, Sov. Phys. Crystallogr. 18 (1973).

[87] V.V. Voronkov, in: B. Springer (Ed.), Crystals. Growth, Properties and Applications, Springer, Berlin, 1983, pp. 93–96.

How the constitutional supercooling formula was developed

William A. Tiller

Stanford University, Stanford, CA

As new Engineering Physics graduates of the University of Toronto in 1952, Ken Jackson and I went to work on Masters degrees in Physical Metallurgy with Bruce Chalmers in the general area of solidification. Ken had been set to work on an investigation of the origins of those dislocations leading to the crystal substructure called "striations." I had been set to work to understand the origins of the solidification phenomenon called "banding," which involved abrupt solute fluctuations in the crystal along the locus of the interface.

I sat in the same laboratory with John Rutter, who had just finished his Ph.D. research on the cellular substructure in crystals and was continuing, with Bruce Chalmers, as a Postdoc. The laboratory, located in the Mining building at the University of Toronto, was old but large and had an end wall covered with blackboard. This made our lab the natural place in which to congregate for coffee and bull sessions. It became a regular practice that Bruce and his solidification group (about eight of us) would meet in that lab for coffee each morning around 10 a.m. There, we would have great and "strong" discussions on what might be going on to explain the various phenomena under scrutiny. The fine hand and personality of Bruce Chalmers stimulated and guided us all.

After a few months of settling in, a sub group, composed of Ken, John, Bruce and myself, became strongly interested in solute partitioning during freezing; I was the most interested since it was the topic of my thesis. Our initial discussions, with much "hand-waving," centered about the qualitative understanding of the general class of phenomena associated with solute redistribution. After several days of vigorous discussion, we thought we could give an acceptable qualitative explanation for banding and the cellular interface phenomenon. Earlier, Bruce and John had proposed the concept of constitutional supercooling and it fitted well with our qualitative explanation.

Central to both of the explanations was the actual form for the solute distribution in a liquid ahead of an interface advancing at constant velocity. Our qualitative discussions had led us to realize the importance of the steady-state distribution. Being somewhat mathematically inclined, I tried to set up the problem in a coordinate system moving with the interface and was quickly successful in solving for the steady-state solution. The excitement of our blackboard sessions went up a notch since the results predicted, surprisingly to many at the time, the existence of a very thin layer of strongly solute-rich liquid at the interface. Very quickly, thereafter, we gained approximate solutions for the initial, intermediate and terminal transients. Following this, it was very easy for us to write down the quantitative criterion for the onset of constitutional supercooling and thus the onset of cellular interface development for metallic systems. At this point our sense of joy and satisfaction was unbounded.

Bruce suggested that I might as well write it up for a paper while I was still filled with such enthusiasm. Being very excited about it all and fairly articulate with words, I wrote the paper (my first) that weekend

Previously published in AACG Newsletter Vol. 13(1) March 1983.

and, with a minimum of haggling and polishing, he accepted it for publication in the second issue of *Acta Metallurgica*.

It was a wonderful time! On reflection, it seems that our accomplishment was primarily a matter of being at the right place at the right time with the right tools. Because of the strong semiconductor interest, the world seemed poised and ready for this new contribution. The new quantitative understanding expanded the dimension of our qualitative appreciation allowing us to begin to explain a whole host of solidification and casting phenomena in a fairly precise way. The initial successes involving controlled materials preparation for the semiconductor industry spilled over into the empirical field of casting and foundry technology. The beginnings of a science-based technology of the freezing process began to form which is still continuing to grow today.

Constitutional supercooling surface roughening

Kenneth A. Jackson *

University of Arizona, 600 WindSpirit Circle, Prescott, AZ 86303, USA

Abstract

This paper consists of my recollections about the studies of crystal growth which I have been involved with over the past decades. I am deeply indebted to many outstanding co-workers who have collaborated with me. The descriptions are drawn from my memory of events, and subject to all of the failing which that might entail. I only regret that I will not be able to mention all of my collaborators over the years, due to a limitation of space and a faulty memory. This paper includes sections on interface instabilities, on the surface roughening transition, on the Jackson alpha factor, on transparent analogues of metals, on analytical models for crystallization, and finally, a few remarks about on-going and future of studies of crystal growth processes.
© 2004 Published by Elsevier B.V.

Keywords: A1. Crystal morphology; A1. Dendrites; A1. Eutectics; A1. Interfaces; A1. Roughening; A1. Solidification

1. The past

The shape and form of crystals has been a topic of interest for centuries. The modern understanding of crystal growth processes began over a century ago, with the work of Knudsen [1] and Wilson [2], and continued into the last century with such great workers as Frenkel [3], Stranski [4], Volmer [5] and Tamman [6]. The current understanding of crystal growth rests on the shoulders of these giants.

My introduction to crystal growth came from my mentor, Bruce Chalmers, and my introduction to some of the giants in the field was at the Cooperstown Conference in 1958 [7]. Fig. 1 is a reproduction of the center part of the front row of the group photo which was taken at that conference.

Fig. 1. Bruce Chalmers, Nicolas Cabrerra, Charles Frank, Dave Turnbull, Paul Flory, Joel Hildebrand, and I.N. Stranski at the Cooperstown Conference.

This is a veritable honor role of scientists who have contributed importantly to our field.

2. Interface instabilities

The original concepts for understanding interface instabilities are contained in the work of John Rutter and Bruce Chalmers [8]. They studied the formation

---

* Corresponding author. Tel.: +1-928-771-0211; fax: +1-928-771-0211.

E-mail address: kaj@aml.arizona.edu (K.A. Jackson).

0022-0248/$ – see front matter © 2004 Published by Elsevier B.V.
doi:10.1016/j.jcrysgro.2003.12.074

Originally published in
J. Crys. Growth 264 (4) (2004) 519–529.

Fig. 2. A decanted interface of tin showing cellular growth.

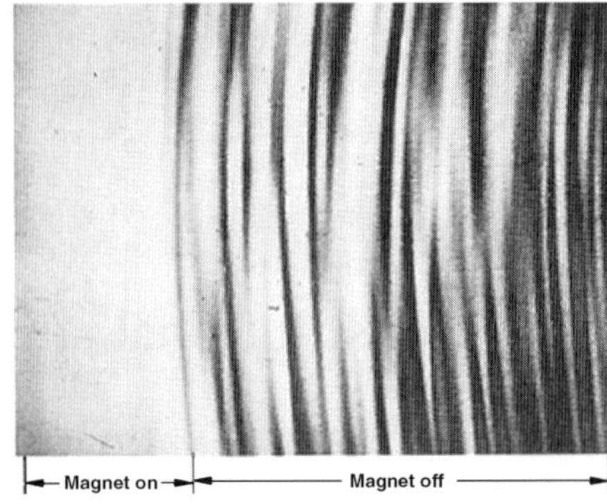

Fig. 3. Banding during crystal growth caused by convection in the melt.

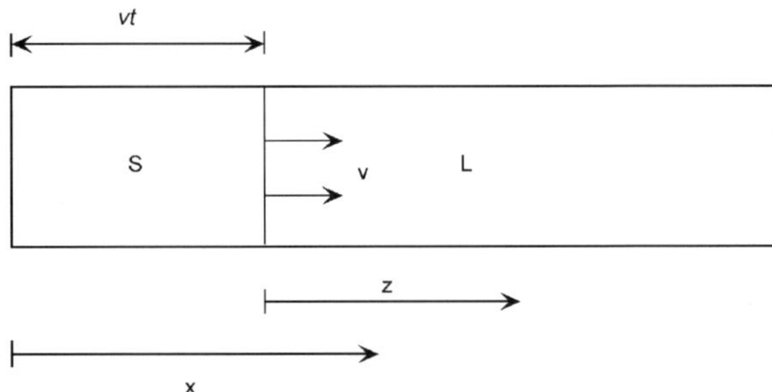

Fig. 4. Moving coordinate system for analyzing steady-state growth.

of the cellular substructure found in slightly impure metals. The cellular substructure consists of a honeycomb-like structure, elongated in the growth direction, with a more or less regular hexagonal pattern on the interface. The pattern can be revealed with standard etching. Rutter and Chalmers decanted the liquid from the growing interface, and revealed a cellular pattern as shown in Fig. 2, which established that the pattern was created during growth of the crystal.

They suggested that this structure resulted from an instability of the interface. They recognized that there was a boundary layer in the liquid, near the interface, which was rich in those components rejected by the growing crystal. This meant that the melting point of the liquid at the interface was depressed, and so the liquid ahead of growing interface could be supercooled even though it was at a higher temperature than the interface. They suggested that this condition caused the instability. They coined the term Constitutional Supercooling to describe this condition.

Bill Tiller and I joined Prof. Chalmer's research group soon after this work was published. Bill undertook a project studying the growth of lead crystals. He found that his crystals were banded, and he traced the banding structure to a periodic fluctuation in the growth rate caused by a slipping gear in the drive system for the furnace. Banding can be caused by convective fluctuations in the melt, as illustrated in Fig. 3.

The banding in Fig. 3 is suppressed on the left by a magnetic field, which dampens the liquid convection. Bill and professor Chalmers suggested that the banding was due to the presence of the same boundary layer which caused constitutional supercooling; that speeding up or slowing down of the growth rate would result in more or less of the boundary layer material being incorporated into the crystal.

Bill Tiller, John Rutter and I decided to try to analyze the mathematics of the diffusion process in the liquid ahead of the interface. Bill suggested that the problem should be treated in a coordinate system moving with the interface, as illustrated in Fig. 4.

The coordinate z is measured from the moving interface. The equation for steady-state diffusion in the coordinate system moving with the interface at a velocity v is

$$D\frac{d^2C}{dz^2} + v\frac{dC}{dz} = 0, \tag{1}$$

where C is the composition in the liquid, D is the diffusion coefficient, and z is the distance from the interface.

The concentration distribution which is a solution to this equation is

$$C = C_\infty + C_\infty\left(\frac{1}{k} - 1\right)\exp\left(-\frac{vz}{D}\right), \tag{2}$$

where C_∞ is the initial concentration of the liquid, and k is the equilibrium segregation coefficient, which is

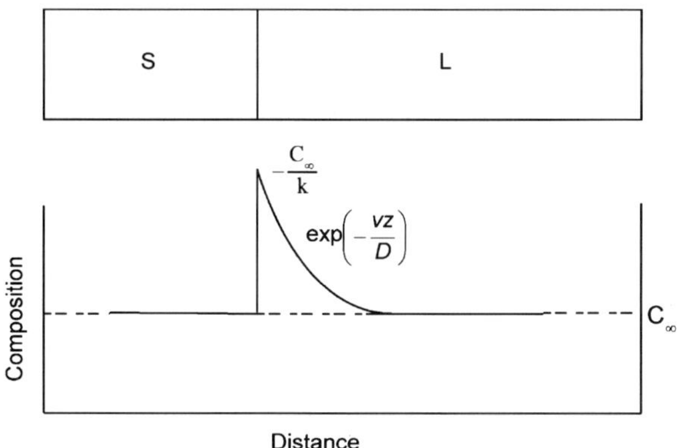

Fig. 5. The steady-state concentration distribution ahead of a moving interface.

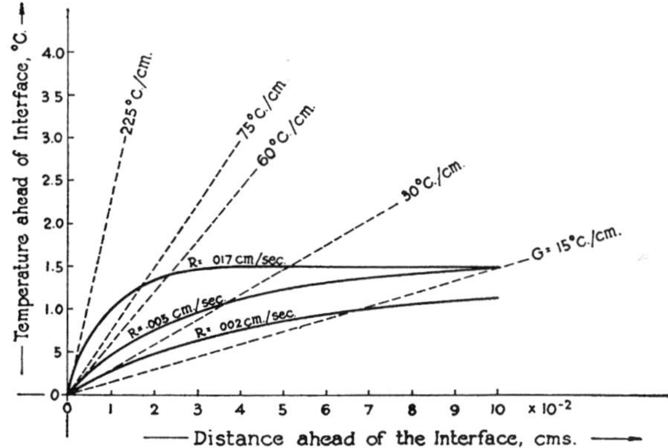

Fig. 6. The original illustration of constitutional supercooling.

the ratio of the composition in the solid to the composition of the liquid, ($k = C_S/C_L$) at the interface. This equation seems quite obvious now, but it seemed to be novel when we derived it. The concentration distribution given by Eq. (2) is illustrated in Fig. 5.

Next, we derived an approximate expression for how a step function change in the growth rate would result in a band of different composition in the crystal. Then we found approximate expressions for the composition profiles for the transient regions at the start of the growth where the boundary layer builds up, and at the end of the crystal, where the boundary layer finally crystallizes.

The next step, urged by Prof. Chalmers, was to find a mathematical expression for when constitutional supercooling occurred. We then derived the equation for constitutional supercooling, and drew Fig. 6 to illustrate it. This figure was copied directly for an illustration in the paper [9], and later appeared in Prof. Chalmers' book "Principles of Solidification" [10]. All this happened in about one week. The solid curves in the figure represent the melting point of the liquid as a function of distance ahead of the interface. Constitutional supercooling occurs when the slope of the melting point curve is greater than the slope of the temperature field as given by a dashed line.

The slope of the melting point curve at the interface can be determined from Eq. (2). The condition for constitutional supercooling is that the temperature gradient, G, should be less than the value given by

$$G < mC_\infty\left(\frac{1}{k} - 1\right)\frac{v}{D}. \tag{3}$$

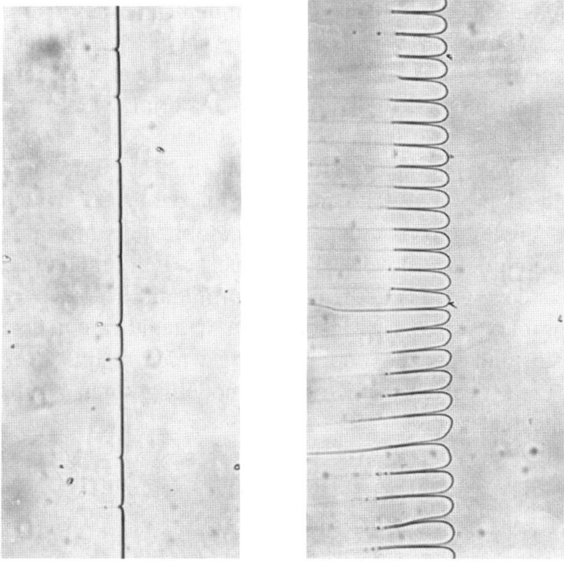

Fig. 7. On the left, a planar interface; on the right, a cellular interface.

Fig. 7 contrasts planar growth with cellular growth in a transparent alloy. An interface growing without constitutional supercooling is shown on left. The interface is planar, and very close to the melting point isotherm. On the right is a cellular interface which results from constitutional supercooling. The growth front can be switched back and forth between these two conditions by decreasing or increasing the growth rate, as predicted by Eq. (3).

The argument outlined above suggests that instability of the interface occurs when there is constitutional supercooling, but it does not demonstrate mathematically that there is an instability, or how the instability develops. That came a few years later, with the work of Mullins and Sekerka [11]. They performed a linear stability analysis, and derived a condition for the onset of an instability which is very similar to Eq. (3). Their mathematics described the early stages of the formation of the instability. Their work is classic, and a large literature on the mathematical modeling of the motion of interfaces has followed from their work.

The equations which describe the motion of an interface are very complex, because the boundary conditions for the diffusion field are applied at a moving interface, and the location and velocity of the interface depend on the diffusion field. These are nonlinear equations which can only be solved by computer simulations. The modeling of dendritic growth into an undercooled melt in a pure material, where the dendrites result from the thermal field, is reasonably well understood, although it is not a completely solved problem. The modeling of dendritic growth in an alloy is a much more complicated problem. A completely novel scheme, phase field modeling, was developed to address these problems. The level of complexity of these analyses is more than I care to deal with, and the new understanding which results from a great deal of mathematical modeling often seems meager.

3. Surface roughening

One of the aspects of crystal growth which intrigued me early on was why some crystals looked like crystals, whereas the metals, which are crystalline, "solidify", so that even single crystals of metals take up the shape of the container, with no apparent external evidence of crystallinity. This difference is illustrated in Fig. 8.

The crystal on the left is growing dendritically. The growing crystal is rejecting a red dye. The interface is rounded, showing no sign of facet formation. On the right, growing under identical conditions is a crystal of benzil, showing well-developed facets on the interface.

The idea was that perhaps there were many growth sites on the metal-like crystal, whereas there were few on the facet, so that nucleation of new layers was required there. A simple analysis of how many adatoms would be expected on a plane interface of a crystal was made. Fig. 9 illustrates the result of the analysis.

The parameter, α which is now known as the Jackson alpha factor [12], was identified by the analysis

$$\alpha = \frac{\eta}{Z} \frac{L}{kT_M}. \tag{4}$$

Here η is the number of nearest neighbor sites adjacent to an atom in the plane of the interface, Z the total

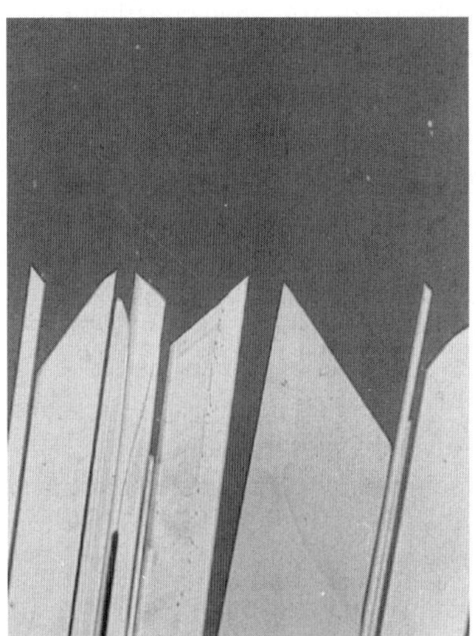

Fig. 8. On the left is a transparent metal analog; on the right, a crystal showing well-developed facets.

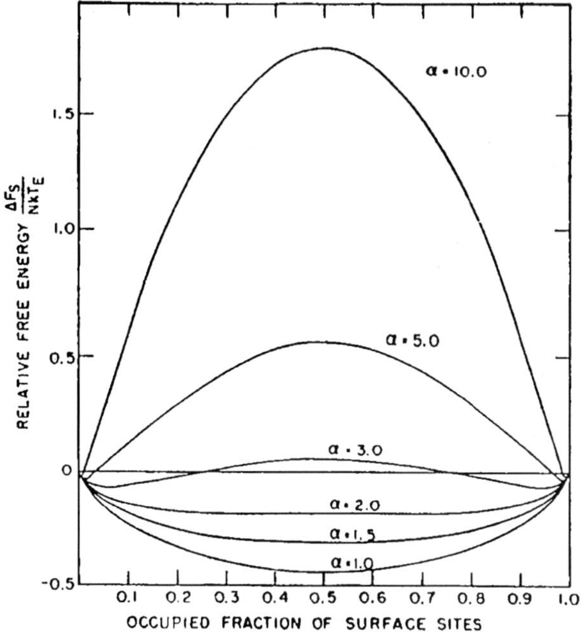

Fig. 9. Interfacial free energy as a function of surface coverage.

number of nearest neighbors of an atom in the crystal, L the latent heat of fusion, and T_M the melting point of the material. The equilibrium configuration is given by the lowest free energy on each curve. For large α, the lowest free energy configuration of the interface is with a few extra adatoms and a few missing atoms in the layer below. For small α, less than 2, the lowest free energy of the interface occurs when the interface is half covered with adatoms, that is, the surface is rough on the atomic scale.

This, in principle, solved the problem of why some crystals grow looking like crystals, and others do not. The factor η/Z is always less than 1, and L/kT_M for metals is typically 1, and so α is less than 2 for the metals, and their surfaces should be rough on an atomic scale. Many minerals and organic crystals have values of L/kT_M of 6 or so, and so they have some smooth faces which require surface nucleation. L/T_M is the entropy of fusion, and dividing by Boltzmann's constant, k, makes it dimensionless. This is a measure of the difference in the degree of order between the crystal and the melt. If the crystallizing entities are spherical atoms or molecules, then the difference is small, but the difference is much larger if a molecule must have the proper orientation to join the crystal.

I presented this story at the Cooperstown conference, unaware that Burton, Cabrerra, and Frank had presented a similar analysis a few years before [13]. Their analysis is contained in the last half of their famous paper on screw dislocation growth. The conclusion in that paper was that crystals grown by screw dislocations and that the surface roughening transition is irrelevant to crystal growth. And so their analysis of surface roughening was pretty much ignored. But they had approached the problem from different direction. They were concerned with why crystals grow at all.

Their story starts a few years earlier, when Becker applied the (then) new nucleation theory to crystal growth, and calculated that the nucleation of new layers would only occur at large very undercoolings. But crystals grow at much smaller undercoolings than the calculations indicated, so Charles Frank, who had some dealings with real crystals, suggested that new layers of a crystal could start at the defects which occur in natural crystals. And so the screw dislocation model for crystal growth was born. Frank had two bright young scientists working with him, Cabrerra and Burton, who helped with the analysis of screw dislocation growth, and who also applied the Onsager solution for the two-dimensional Ising model to obtain the roughness of a crystal surface. But they used the analysis to calculate *where* the surfaces of crystals should roughen, and concluded that it would be above the melting point, and so it was irrelevant to crystal growth. We know now that they should not have done that. The Ising model predicts the behavior of a system in the vicinity of a critical point, but not the location of the critical point. It does not predict the Curie temperature of iron or of nickel, but it does predict the change magnetization on approaching the transition. So their conclusion about the relevance of the surface roughening transition was wrong, even though their analysis was correct.

My concern was trying to understand why some crystals are faceted during growth, while others are not. Not having read the relevant parts of the BCF paper, I used the approximate Bragg–Williams model to describe surface roughening, rather than the exact Onsager solution for the two-dimensional Ising model, which BCF used. I made the same mistake that they did, by calculating the location of the roughening transition from the model, but luck was with me. The surface roughening transition is at a higher temperature in the Bragg–Williams model than in the Onsager solution, and happens to coincide with the experimental value for melt growth.

At the Cooperstown conference, after my presentation, Frank got up and said that it was wrong, then Cabrerra got up and said that they had done it before.

The bottom line is that dislocation free crystals of silicon are grown daily. They form facets on (1 1 1) interfaces during melt growth, and not on any of the other orientations, as correctly predicted by the α-factor criterion. So the α-factor explains the growth behavior not only of silicon, the most important man-made crystal, but also of melt-grown crystals in general.

A few years after Cooperstown, I went to visit Charles Frank in Bristol. We had a long discussion about surface roughening. I developed great respect for an eminent scientist who was eager to discuss how one of his famous contributions was only partly correct.

4. Transparent analogues

Soon after I joined Bell Labs, an outstanding young scientist, John Hunt, came to work with me as a post-doc. One day we decided to look and see if we could find any organic crystals with low α factors. We found several in the International Critical Tables. It turned out later that these materials were well-known to some chemists, who called them plastic crystals, because the also deform like metals. When materials in this class crystallize, their molecules are still free to rotate, and so they behave like the spherical atoms of metals. In his outstanding study of dendritic growth, Papetrou [14] used some of these materials. But John and I were looking for materials which solidify like metals in order to study phenomena found during the solidification of metals, and the α factor analysis gave us confidence that these materials would fit the bill [15].

John designed and built a temperature gradient microscope stage for observing the crystallization of transparent materials in a thin cell. It permitted independent control of the temperature gradient and the growth rate [16]. We studied many different materials, and after a time, we could estimate the entropy of fusion of a material from its crystallization

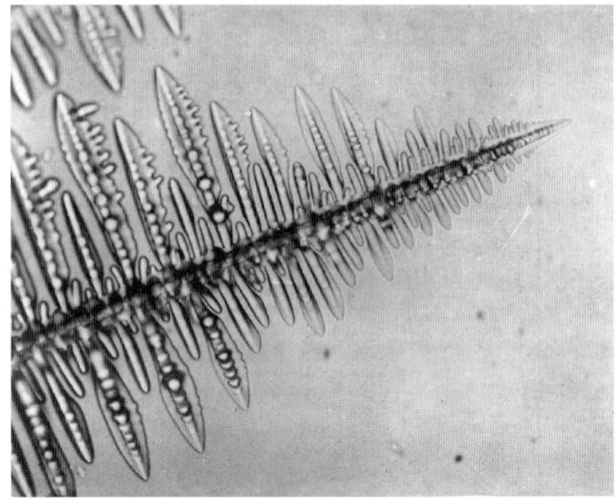

Fig. 10. Dendritic growth in succinonitrile.

morphology. We prepared a movie, which was widely distributed, to illustrate typical growth morphologies for materials with various entropies of fusion. One of the materials which we identified as a metal analog was succinonitrile, shown in Fig. 10, which Marty Glicksman [17] has used in extensive studies to obtain accurate data on dendritic growth for comparison with theoretical models.

John Hunt and I were aware of the old puzzle about where the nuclei come from to make the equiaxed zone in metal alloy castings [18]. The material system which we chose for a study castings was ammonium chloride–water solutions. We made movies of castings of this material, and studied its growth characteristics in the temperature gradient microscope stage. We discovered that the arms of dendrites would melt off the main stem during a temperature fluctuation. Qualitatively, this is because the dendrite side branches grow through the layer around the main stem which is enriched in the components which are rejected. So the parts of the side branches right near the main stem are less pure, and so have a lower melting point, than either the main stem or the parts of the side branches which are farther from the main stem. This is where the melting occurs during a temperature fluctuation, and the side branches separate from the main stem. In casting, these detached side branches can float away and be picked up by convection currents to make the new equiaxed grains in the center of the casting, as illustrated in Fig. 11.

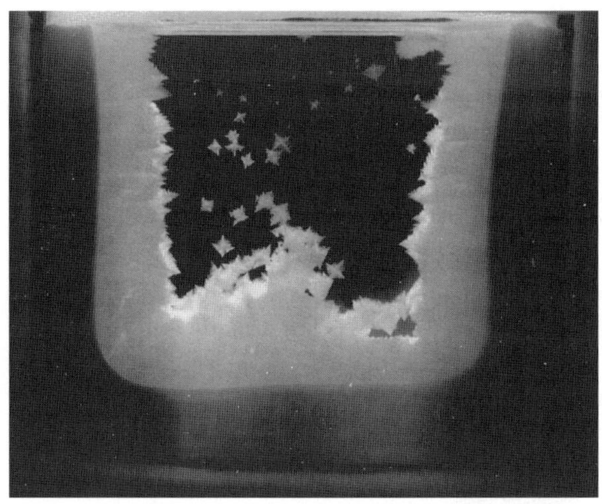

Fig. 11. Casting of an ammonium chloride–water solution.

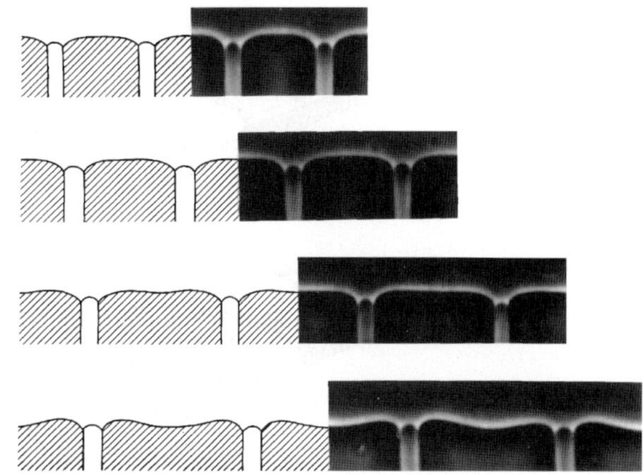

Fig. 12. Eutectic growth in a transparent metal analogue, compared with calculated interface shapes.

5. Modeling of crystal growth

When John and I presented this story at a TMS meeting, Cyril Stanley Smith got up and said that he had learned more about castings in the past 30 min than he had in the previous 20 years. The people at Pratt and Whitney were having a problem with a phenomenon known as "freckles" in the casting of turbine blades for jet engines. Soon after our publication [19] they made castings of ammonium chloride–water solutions, and observed the formation of freckles. They figured out how to mitigate the problem from their observations. The formation of freckles is still a problem that intrigues the people who model the solidification of castings. This is an important topic, since the design of the castings for new turbine blades is now based on this modeling. And ammonium chloride–water solutions are still used for modeling the macroscopic aspects of the structure of castings.

John Hunt and I next turned our attention to eutectics, which John had studied for his Ph.D., and on which I had done some modeling. We studied the solidification of many eutectic alloys in the temperature gradient stage, and observed the shape of a growing eutectic front for the first time, such as is illustrated in Fig. 12.

Together, we hammered out a theoretical paper on dendritic growth [20], and compared the calculated interface shapes with experiment. We defined the three classes of dendrite microstructures based on the entropies of fusion of the components [21].

The surface roughening transition as described by either the two-dimensional (single layer) Bragg–Williams model which I used, or the two-dimensional Onsager model which BCF used was clearly only an approximation to a multi-layer, three-dimensional interface. Harry Leamy and I tried various analytical models [22], but it was clear that computer modeling was necessary. George Gilmer joined our group, and he and Harry produced the widely reproduced images of surfaces below and above the surface roughening transition [23], as shown in Fig. 13.

They also determined the free energy of a step for various alpha factors [24]. The free energy of a step on the surface goes to zero at the surface roughening transition, as illustrated in Fig. 14, and so formally, there is no barrier to the formation of new layers of a crystal on an atomically rough surface.

The specific free energy of a step on a faceted interface is much lower than the specific free energy of the faceted surface. That is, why Becker's calculation gave the wrong answer.

For example, Kirk Beatty and I have recently shown that the edge free energy of a step on a (1 1 1) surface of silicon is only 10% of the specific surface free energy of the flat surface [25]. This decrease in free energy of the step is due to an increase the entropy of the step caused by the increased jaggedness of the step. The step disappears into the noise on a rough surface, as illustrated in Fig. 13.

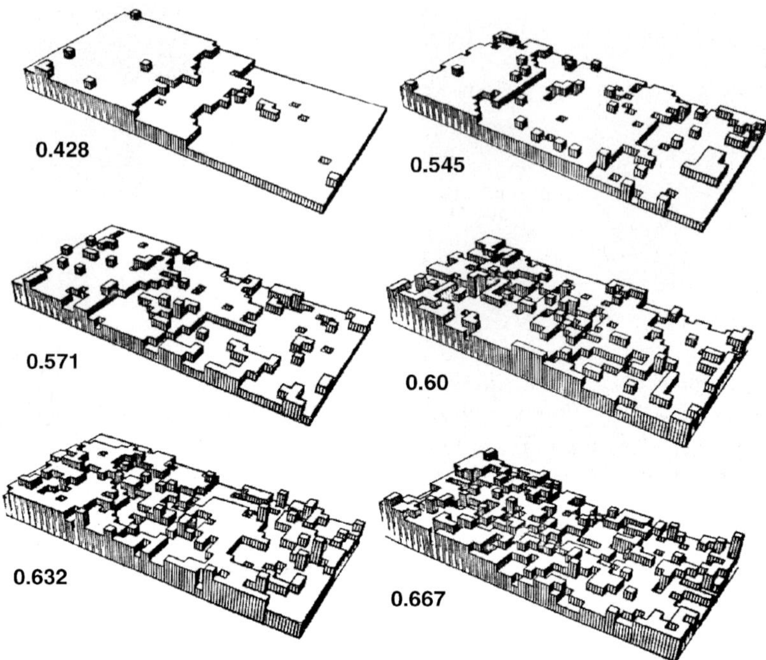

Fig. 13. Computer simulation of surfaces with various roughness.

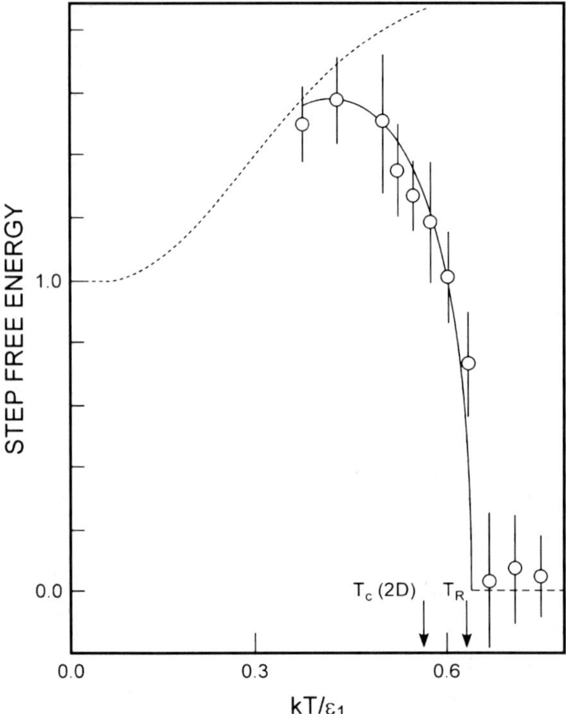

Fig. 14. The surface free energy of a step goes to zero at the surface roughening transition.

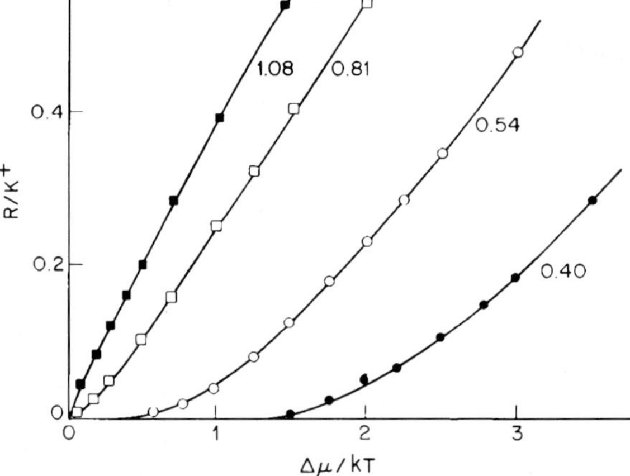

Fig. 15. Normalized growth rates, R/K^+, as a function of the chemical potential difference between the solid and the liquid at the interface, from computer simulations.

George Gilmer produced the growth rate curves for surfaces with various α factors [26]. The data are presented in Fig. 15.

The growth is linear with undercooling on rough surfaces, as for the curve labeled 1.08, which is just above the surface roughening transition at 1.0

in Fig. 15. The growth rate becomes increasingly slower on smoother interfaces below the roughening transition, as the growth rate becomes limited by the rate of nucleation of new layers. Chui and Weeks [27] performed a linear response analysis of the surface roughening transition. There is a very weak analytical singularity at the transition, but they proved that the growth rate curve comes in linearly to the origin on a rough surface, and comes into the origin horizontally on a smooth surface, as is suggested by Fig. 15. Not much happens to the equilibrium properties of an interface at the surface roughening transition. The surface continues to get rougher on going through the transition. But the free energy of a step goes to zero there, and the growth kinetics change abruptly from continuous growth to layer by layer growth.

Chui and Weeks also found that the partition function for a rough surface is the same as the partition function for a liquid-liquid interface. The rough interface does not know that the lattice is there.

6. The future

As Casey Stengel once said, predictions are very difficult to make, especially about the future.

Crystallization is a very complicated and complex topic and we have come along way in developing our understanding of the atomic or molecular scale processes involved. But there is still much to be done. I have been working over the past several years on what happens when an alloy crystallizes far from equilibrium. When the growth rate approaches the rate at which atoms can diffuse in the interface region, then the rules of equilibrium segregation as given by a phase diagram no longer apply. This is illustrated in Fig. 16, where the parameter β is a measure of how fast the interface is moving relative to how fast the atoms can move.

As the relative growth rate increases, the k-value increases from the equilibrium value on the left, towards one on the right. The open symbols are experimental data for dopants in silicon, the filled symbols are data from Monte Carlo computer simulations, and the curves are from analytical equations which I have developed to describe this phenomenon. This phenomenon has been called "solute trapping".

The non-equilibrium k-value, unlike the equilibrium value, depends on the orientation of the inter-

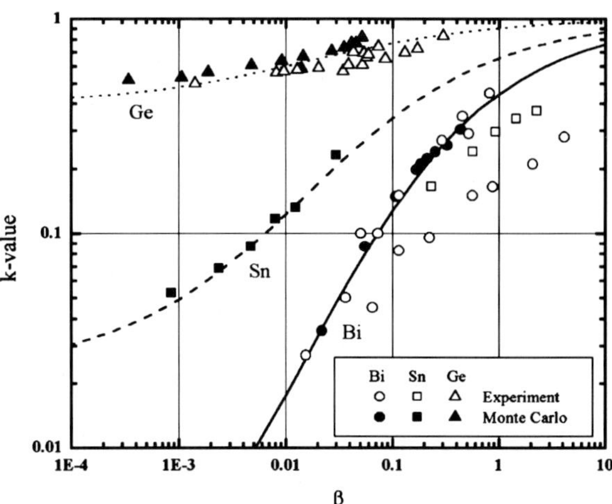

Fig. 16. The k-value increases with growth rate for dopants in silicon.

face. The k-value is larger on a faceted surface. The experimental data for this orientation dependence during very rapid solidification of silicon has been reproduced in Monte Carlo simulations. But orientation dependent segregation has been observed experimentally at slow growth rates in many semiconductors. This is called the "facet effect" because the k-value on a facet is larger than off the facet, just as in highspeed growth. But my equations, which describe the effect for high growth rates, do not predict a significant orientation dependence when extrapolated to slow speeds. The reason for this discrepancy is not clear. The Monte Carlo method works well to simulate growth in the high speed regime, but growth in the slow regime is too slow to model, and so computer modeling has not been able to resolve the issue.

Non-equilibrium segregation effects are very relevant to solid state phase transformations, where the diffusion rates for solutes are very slow. The understanding which has been developed for rapid crystallization should shed some light on what are known in metallurgy as massive transformations. There has been discussion in the literature about whether the driving force for these transformations should be measured from the solidus line or from the T_0 line (where the free energies are equal). The model developed for solute trapping suggests that it can be either, or something in between, depending on the parameter beta for the transformation conditions. The applicability of the solute trapping model to massive transformations has not been explored.

In general, the solid state phase transformations which are at the core of the discipline known as physical metallurgy are more complex than crystallization from a melt, because of the large role played by the stresses generated during the transformation. But the same atomic processes which are involved in crystallization also underlie these transformations. At present, the description of these transformations is largely qualitative, although there is a massive amount of empirical data available.

There are many aspects of alloy crystallization which are not understood in detail. One of these is the transition from alloy solidification to solution growth. Both occur in different parts of the same phase diagram. In solidification, the components joining the crystal are always present at the interface, and their relative rates of joining and leaving the crystal control the process. In solution growth, virtually all of the atoms at the interface are solvent atoms. The atoms which join the crystal diffuse through the solvent to the surface, so solution growth is largely controlled by diffusion. The transition between these two modes depends on the segregation coefficient, on the growth rate and on the diffusion process. Some aspects of this can be treated more or less simply by examining diffusion in the solvent, but there are effects due to surface interactions, boundary layers, and adsorption which complicate the transition.

Dendritic growth in pure materials seems to be fairly well-understood, but in alloys, there is a richer phenomenology. Phase field models hold promise for making advances in this area.

Modeling of castings has important commercial implications. Most of this involves the modeling of the fluid flow during crystallization, driven by buoyancy forces due to both composition variations and temperature variations. At present, the partially crystallized dendritic region is called the "mushy zone", and is characterized by an effective permeability and concentration of the liquid, rather than by the characteristics of dendritic solidification. The development of porosity, which depends on the dissolved gas content, and the segregation of those components during crystallization is also important.

Computer modeling of the crystallization processes are being pursued on many fronts. A coherent effort is being spearheaded by Alan Karma, bringing together molecular dynamics modeling to obtain accurate data for thermodynamic and kinetic properties of alloys, and coupling these with the results of kinetic Monte Carlo modeling and phase field modeling to generate critical tests of the equations which are used to describe crystallization processes, assumptions which go into the atomic level modeling, and the computer modeling methods which describe the resultant growth morphologies.

Most of the modeling studies to date have concentrated on modeling of metals and alloys, where the growth rate and other properties are reasonably isotropic, and the growth rate depends linearly on the undercooling. Modeling of layer-by-layer growth is significantly more complex mathematically, because the anisotropies are very large, the growth rate is not linear with undercooling, and the local growth rate depends not on the local conditions, but rather on the undercooling at the place on a facet where the new layers nucleate.

The details of how a crystal grower produces bigger and better crystals depends only in part on how well she/he understands the basic underlying processes, which are the primary subject of this paper. There are many other factors and talents involved in actually growing crystals.

References

[1] M. Knudsen, Annln. Phys. 29 (1909) 179.
[2] H.A. Wilson, Phil. Mag. 50 (1900) 238.
[3] J. Frenkel, Phys. Z. Sowjet Union 1 (1932) 498.
[4] I.N. Stranski, Z. Phys. Chem. (Leipzig) 136 (1928) 259.
[5] M. Volmer, Kinetik der Phasenbildung, Theodor Steinkopff, Dresden, 1939.
[6] G. Tammann, States of Aggregation (trans. R.F. Mehl), D. Van Nostrand, New York, 1925.
[7] R.H. Doremus, B.W. Roberts, D. Turnbull, Growth and Perfection of Crystals, Wiley, New York, 1958.
[8] J.W. Rutter, B. Chalmers, Can. J. Phys. 31 (1953) 15.
[9] W.A. Tiller, K.A. Jackson, J.W. Rutter, B. Chalmers, Acta Met. 1 (1953) 428.
[10] B. Chalmers, Principles of Solidification, Wiley, New York, 1964, p. 158.
[11] W.W. Mullins, R.F. Sekerka, J. Appl. Phys. 34 (1963) 444.
[12] K.A. Jackson, Liquid Metals and Solidification, ASM Cleveland, Ohio, 1958, p. 174.
[13] W.K. Burton, N. Cabrera, F.C. Frank, Phil. Trans. Roy. Soc. A 243 (1951) 299.
[14] A. Papapetrou, Zeit. Kristallogr. 92 (1935) 89.

[15] K.A. Jackson, J.D. Hunt, Acta Met. 13 (1965) 1212.

[16] J.D. Hunt, K.A. Jackson, H. Brown, Rev. Sci. Instrum. 37 (1966) 805.

[17] M.E. Glicksman, R.J. Schaefer, D. Ayers, Metall. Trans. A 7 (1976) 1747.

[18] F. Hensel, Trans. AIME 124 (1937) 300.

[19] K.A. Jackson, J.D. Hunt, D.R. Uhlmann, T.P. Seward III, Trans. Met. Soc. AIME 236 (1966) 149.

[20] K.A. Jackson, J.D. Hunt, Trans. Met. Soc. AIME 236 (1966) 1129.

[21] J.D. Hunt, K.A. Jackson, Trans. Met. Soc. AIME 236 (1966) 843.

[22] H.J. Leamy, K.A. Jackson, J. Appl. Phys. 42 (1971) 2121.

[23] H.J. Leamy, G.H. Gilmer, K.A. Jackson, in: J.M. Blakely (Ed.), Surface Physics on Materials, vol. 1, Academic Press, New York, 1975, p. 121.

[24] H.J. Leamy, G.H. Gilmer, J. Crystal Growth 24/25 (1974) 499.

[25] K.M. Beatty, K.A. Jackson, J. Crystal Growth 211 (2000) 13.

[26] G.H. Gilmer, J. Crystal Growth 36 (1976) 15.

[27] S.T. Chui, J.D. Weeks, Phys. Rev. B 14 (1976) 4978.

The theory of morphological stability

Robert F. Sekerka

Carnegie–Mellon University, Pittsburgh, PA 15213

'Twas the summer of '61 and I had just survived my first year of graduate studies in physics at Harvard. I was back in Pittsburgh, working at a summer job at what was then called Carnegie Institute of Technology (now Carnegie–Mellon University) under the direction of Bill Mullins, at that time Associate Professor of Metallurgical Engineering. I had worked the previous summer with Mullins at Carnegie, just after I had graduated from the University of Pittsburgh with a B.S. in Physics; several years prior to that, I had worked half-time as his technician at the Westinghouse Research Laboratories; the other half of my time was assigned to Bill Tiller who was also at Westinghouse in those days. Tiller was fresh out of graduate school at Toronto where he had worked with Bruce Chalmers, Ken Jackson and others on fundamental aspects of solidification having to do with solid liquid interface instability, cells and dendrites. There was a lot of excitement about these growth forms and about "constitutional supercooling" [1], a concept that I did not fully appreciate at the time.

Back to the summer of '61. Bill Mullins suggested that we might seek to quantify the phenomenon of interface instability by solving the appropriate field equations and examining the consequences of the resulting solutions. We turned our attention first to the shape-preserving solutions for phase transformations that were exhibited by Frank Ham [2] for precipitates of ellipsoidal shape.

We set out to test the stability of a growing spherical precipitate. This seemed to be a tractable problem and discussion led to our finding in the literature [3] an exact solution for the electrostatic potential about a solid conductor having the shape of two spheres that intersect one another everywhere at right angles. If one such sphere is considerably smaller than the other, the resulting body is essentially a sphere perturbed by the addition of a small hemispherical boss. We knew, from the work of Zener [4] and others, that the growth of a precipitate from a matrix having low supersaturation is governed approximately by Laplace's equation, so there was an analogue to the electrostatic case. It remained only to associate the equipotentials of the electrostatic problem with the isoconcentrates of the precipitation problem and to calculate the local concentration gradients, proportional to the growth rates in the vicinity of the large sphere and the hemispherical boss.

Not surprisingly, it was easy to see that the growth rate of the hemispherical boss was relatively faster than that of the main sphere; once having formed, such a boss would continue to grow at an accelerated rate, thus leading to instability of form or shape, which we now call morphological instability. We promptly wrote up our results and submitted them for publication.

Unfortunately, the report of the reviewer was not very sympathetic. The reviewer felt that the results, although interesting, were rather inconclusive because there was a sharp and unrealistic discontinuity in slope where the hemispherical boss met the large sphere at right angles. The reviewer proceeded to develop a solution to the problem in terms of an expansion

Previously published in AACG Newsletter Vol. 16(2) July 1986.

in Legendre polynomials, which are appropriate to the description of solutions to Laplace's equation with axial symmetry, and to attempt to show that the discontinuity might be a source of difficulty. We felt intuitively that our results were correct and proceeded to try to find ways to demonstrate this.

Perhaps inspired by the reviewer's expansions, or by Fermi's treatment of the fission of a nucleus according to the liquid drop model, or by some association with his work on grain boundary grooving or the smoothing of scratches—one wonders how the human mind works in such cases—Mullins decided to push through the calculation of the stability of a sphere perturbed by spherical harmonics. By some combination of inspiration and good fortune, he was successful. A key reason was the fact that the mean curvature, which plays a role in the Gibbs–Thompson equation that sets the local concentration or temperature at the interface, may be expressed (to first order) as the angular part of the Laplacian operator acting on the shape of the perturbed body. It turns out that the spherical harmonics are eigenfunctions of the angular part of the Laplacian (a fact well known in quantum mechanics, since the angular part of the Laplacian is essentially the operator for the square of the orbital angular momentum) and thus the effect of each harmonic may be treated independently.

The results of these calculations confirmed our earlier suspicions about the role of the gradient in the vicinity of a hemispherical boss (point effect of diffusion) as well as elucidating the effect of capillary forces (surface tension) as a stabilizing force. We learned that spherical harmonics of increasing index (and hence increasing number of nodes) became unstable at successively larger radii of the unperturbed sphere and that the important variable was not so much the size of the sphere itself but the interplay between the concentration gradient and the capillary effect. Similar results were obtained for the case of the solidification of a pure sphere, as opposed to the precipitation problem, and the results were published in our 1963 article [5] entitled "Morphological Stability of a Particle Growing by Diffusion or Heat Flow."

We were rather pleased with our new understanding of the morphological stability phenomenon and wanted to extend it to the case of the solidification of a dilute binary alloy, and hence to relate it to the constitutional supercooling principle. Since the problem of alloy solidification is governed by both compositional and thermal fields, we decided to treat the more tractable case of the stability of a planar interface during unidirectional solidification at constant velocity, as occurs in a typical crystal growth situation. Following our work on the sphere, we assumed that it would still be appropriate to use Laplace's equation to calculate the thermal and solutal fields for a perturbed interface shape of the form $z = A \cos(2\pi x/L)$. We obtained results in which the solutal gradients were destabilizing in opposition to the thermal gradients that were stabilizing (much as in constitutional supercooling) and capillarity appeared to be stabilizing for perturbations having short wavelengths, L. Our solutions were not very well behaved, however, and experienced some divergences for values of L near D/V, where D is the solute diffusivity and V is the unperturbed growth velocity of the planar interface.

I recall presenting these results to Bill Tiller and Fred Bolling at a sort of "chalk talk" at the Westinghouse R&D Center. It was almost prophetic that my chalk would not write the divergent terms on the greasy green chalkboard that hung in the seminar room and, although Tiller and Bolling were encouraging with respect to the progress we had made, they were quite critical of the work because they knew that D/V represented the solute boundary layer thickness for the unidirectional solidification problem and the divergences in our results were occurring right when L was comparable to this important length.

I returned to Harvard for the fall semester of '61 and, when I wasn't trying to keep my head above water with classwork, I worried a lot about the divergences in our calculation. Mullins had, by then, left for a year of academic leave in Paris, where he was working with Freidel, and we began to correspond. (Actually, a great deal of the work on the morphological stability of the sphere, discussed above, was done while Bill was on leave and later perfected via correspondence.) At any rate, I became convinced that the divergences in the binary alloy problem arose from using Laplace's equation, $\Delta c = 0$, for the solute problem and could be overcome by using the equation $\Delta c + (V/D)\partial c/\partial z = 0$, which is valid for a steady-state in a frame of reference that is *moving* with velocity V with respect to the solid. We could still get away with the use of Laplace's equation for the thermal fields because the corresponding thermal lengths were order K/V where

K is a thermal diffusivity and these thermal lengths are many orders of magnitude larger than D/V. I conveyed these thoughts by letter to Mullins in November of 1961 and he proceeded to work out the details of the analysis. By spring of '61, a new draft manuscript appeared and indeed, the divergences were gone! Several exchanges of manuscripts by mail over the next year or so were needed to eliminate all of the "bugs" in the development; the result was finally published in our 1964 paper [6] entitled "Morphological Stability During Directional Solidification of a Dilute Binary Alloy."

Two other pieces of work deserve mention in the historical context of morphological stability theory. The first was a paper published in 1956 by Wagner [7] in which he treated the instability, by means of a study of sinusoidal perturbations, of an interface during diffusion-controlled migration during an oxidation reaction. We did not become aware of this paper, however, and its possible connection to our work until our 1963 paper was in the page proof stage. The second was a paper published in the USSR by Voronkov, a student of Chernov, in which the conditions for the formation of a mosaic structure were derived; we only became aware of this work after its translation into English [8]. Apparently, the independent publication over a short span of time of such similar work was simply precipitated by an idea whose time had come; this is perhaps typical of progress in a number of areas of science.

One of the things left hanging in our '64 paper was a detailed evaluation of the stability criterion for the case in which capillarity played an important role. I attacked this problem one quiet Sunday afternoon and managed to whittle it down to the solution of a cubic equation for a quantity that is related to the critical wavelength at the onset of instability. While today's computers and software enable the numerical determination of roots of polynomials with ease, I recall searching through my father's old college algebra book for the "cubic formula" since I had heard somewhere that cubic and quartic equations could be solved exactly by algebraic methods, even though these were never taught in any of the courses that I had in school and have probably been expunged without a trace from the material taught to the present generation. At any rate, the cubic formula yielded nicely and I was able

to calculate a stability function [9] that exhibited in detail the stabilizing effect of capillarity.

By August of 1965 I had finished my doctoral thesis at Harvard under the direction of J.H. Van Vleck, whose work on the theory of magnetism eventually led to his receipt of a Nobel Prize in 1977. My thesis was entitled "The Theory of Magnetic Relaxation in Rare Earth Iron Garnets with Application to Europium Iron Garnet" and contains more quantum mechanics than I ever remember learning. I point this out because many people have the misconception that the work on morphological stability resulted from my doctoral dissertation and indeed, the timeframes were practically coincident. Nevertheless, it is fair to say that I never worked in the area of magnetism very seriously much thereafter. However, the morphological stability theory has permeated my work deeply over the last twenty years and, although I have become involved with many other things, it has continued to pose important and interesting problems that have now become an inseparable part of my work.

By 1965, I was employed by the Theoretical Physics Department of the Westinghouse R&D Center and was given quite free rein in the selection of a research topic. I was equipped with a number of powerful analytical tools that I learned from the Harvard applied math series, and proceeded to use simultaneous Fourier transforms (spatial variables) and a Laplace transform (time variable) to remove the steady-state approximations that were inherent in the 1964 paper. I presented the work at the International Conference on Crystal Growth held in Boston in 1966 and published it in the proceedings of that conference [10].

About that time, a number of other people began to appreciate the power of the stability analysis and made important contributions to its further development.

John Cahn [11] was the first to modify the original treatment to account for anisotropic surface tension and anisotropic interface attachment kinetics, and did so for the spherical geometry of the 1963 paper. In 1976, Coriell and I included the same anisotropies for the more tractable case of a planar interface [12]. Later, Chernov [13] gave an approximate treatment for crystals that grow with strong anisotropy.

In 1970, Wollkind and Segel [14] extended the stability theory into the nonlinear regime by means of an expansion to third order and showed that the amplitude of the interface shape satisfied a Landau equation of the type familiar in fluid dynamics. This sort of weak nonlinear stability analysis has been used by the Carolis [15] in France to treat a number of problems and has been extended by Wollkind and his co-workers [16] to the three-dimensional case in which the interface shape may be characterized by nodes or cells, as opposed to bands as in the two-dimensional case.

Don Hurle [17] and his co-workers in England were the first to modify the constitutional supercooling theory, and later the morphological stability theory, for the effect of fluid convection—but for the simple model of a stagnant fluid boundary layer near the interface. The boundary layer model really didn't work very well until modified later by Jacques Favier [18] from France in a clever way to allow it to deform along with the interface shape. The first really serious work to modify the morphological stability theory to account for convection was accomplished by Richard Delves [19] in England. In this work, Delves actually calculated the perturbed fluid flow fields by means of an approximate analytical technique and showed that perturbations with wave vector along the direction of flow could be strongly stabilized by the flow.

In the late 1960's, I began a close collaboration with Sam Coriell at the National Bureau of Standards that continues to this day. Early on, Coriell and Parker [20] had carried through a stability analysis for a cylindrical geometry and had accounted for isotropic interface attachment kinetics as well as interfacial diffusion. These calculations were compared with the careful experimental work of Steve Hardy [21] on single crystals of ice of cylindrical shape with the C-axis of the ice being the axis of cylindrical symmetry. Together [22] we modified the original theory to account for some mild nonlinearities and managed to obtain a value for the solid liquid surface tension of about 25 ergs/cm^2. This dynamical measurement of the surface tension was later found to be in good agreement with the value of 29 ergs/cm^2 measured by Hardy [23] by using the static grain boundary groove technique.

A major fraction of my work with Coriell [24] was in an effort to better understand the coupling between fluid dynamical instabilities and morphological instabilities. Although Delves had pioneered this area some years before by means of approximate analyti-

cal techniques, we had access to powerful numerical techniques and computers and were able to calculate accurately both fluid dynamical branches of instability that occur at long wavelengths and morphological branches of instability that occur at short wavelengths, as well as their coupling at intermediate wavelengths. In some of these cases, the onset of instability was found to be oscillatory, which, to the best of our knowledge, is never possible for purely morphological cases so long as there is local equilibrium at the crystal-melt interface. Coriell later teamed up with an applied mathematician, Geoff McFadden [25], whose facility with numerical computations has enabled some of these computations to be extended into the nonlinear regime.

In about the mid-1970's, Jim Langer became interested in the problem through our collaborative work in connection with Carnegie-Mellons' Materials Research Laboratory, then called The Center for the Joining of Materials and sponsored by the National Science Foundation. Jim brought some powerful analytical techniques to the table, including nonlinear expansion techniques, somewhat similar to those of Wollkind and Segel but involving the coupling of several modes. His article [26] in *Reviews of Modern Physics* served to bring the problem to the attention of a number of physicists who were interested in chaotic systems and pattern formation. Jim became very interested in the question of why dendrites grow at a definite speed determined by the melt undercooling. We all knew that simple needle crystal models, such as the Ivantsov paraboloid [27], admitted a family of solutions and that modifications to include capillarity (originally by Temkin [28], and later by Trivedi [29], Holtzmann [30], and Nash and Glicksman [31]), when augmented by a maximum velocity hypothesis, failed to agree with experiment. Langer became convinced that the problem was one of stability and he and H. Muller-Krumbhaar [32], who spent a few years as a visitor at CMU, carried through a stability analysis of essentially the Ivantsov paraboloid. Results agreed well with the careful experimental data of Glicksman on succinonitrile and my student, Fujioka, on ice [33]. A similar stability hypothesis had been made by Oldfield [34] some years before, but nobody was convinced that the results were significant because the grid size that was used for the computations was comparable to the wavelength of the instability and there was no convincing way to separate numerical instabilities from morphological instabilities.

Although the marginal stability hypothesis appeared to lead to agreement with experiment, theorists were still unsatisfied because they could not justify it on the basis of a nonlinear analysis. Consequently, they resorted to simplified but more tractable models—e.g., the string model of the Schlumberger group [35] and the boundary layer model of the Santa Barbara group [36]—in order to obtain better insight about the nonlinearities. These models allowed the computation of rather complex shapes and suggested that anisotropy plays a crucial role—at least in two dimensions—in enabling growth that would qualitatively be termed dendritic. It is not clear, however, whether these conclusions are valid in three dimensions or for the actual system in which the field equations lead to an effective repulsion of portions of crystal melt interfaces that closely approach one another. Currently, a controversy rages over whether or not the correct velocity for dendritic growth is given by a "microscopic solvability" condition, according to which one insists upon getting a strictly steady-state solution for a needle crystal model, at best a branchless dendrite. It is hard to refute the results of mathematical analysis of such specific models, but it is also hard to see how regions far from the dendrite tip could so strongly influence the tip behavior or, alternatively, how microscopic solvability can be reconciled with marginal stability that agrees so well with experiment, as recently reviewed by Glicksman [37]. Although tremendous progress has been made since the days of the Ivantsov solution, one suspects that the last chapter has not yet been written on this interesting but difficult problem.

Over the last five years or so, I have noticed an increasing interest in the morphological stability problem on the part of mathematicians and have frequently been invited to attend conferences where meaningful exchange of information with applied mathematicians could take place. Although the Stefan problem has been a long time favorite of applied mathematicians and has resulted in a number of important techniques and theorems, most solutions to it have been for one-dimensional problems. Two- and three-dimensional problems, although considerably more difficult, open up the world of shape of the free boundary, not just its location as in one dimension, as well as the 'exciting complication that the values of field quantities at the boundary depend on its local curvature, instead of being constant. It is my strong suspicion that this "modified Stefan problem" will keep applied mathematicians entertained for some years to come and that the results of this work will not only advance the frontiers of materials science but also a number of other areas—ranging from developmental biology to the spread of populations and languages.

Finally, I would like to take this opportunity to acknowledge the National Science Foundation, Division of Materials Research, for its continued support of my research on this problem under the auspices of a grant entitled "Toward a Unified Theory of Morphological Stability" (currently DMR8409397).

References

[1] W.A. Tiller, J.W. Rutter, K.A. Jackson, B. Chalmers, Acta Met. 1 (1963) 428.

[2] F.S. Ham, Quart. Appl. Math. 17 (1959) 137.

[3] J.J. Thomson, in: Elements of the Mathematical Theory of Electricity and Magnetism, fourth ed., Cambridge University Press, 1907, p. 162.

[4] C. Zener, J. Appl. Phys. 20 (1949) 950.

[5] W.W. Mullins, R.F. Sekerka, J. Appl. Phys. 34 (1963) 323.

[6] W.W. Mullins, R.F. Sekerka, J. Appl. Phys. 35 (1964) 444.

[7] C. Wagner, J. Electrochem. Soc. 103 (1956) 571.

[8] V.V. Voronkov, Soviet Physics Solid State 6 (1965) 2378.

[9] R.F. Sekerka, J. Appl. Phys. 36 (1965) 264.

[10] R.F. Sekerka, in: H.S. Peiser (Ed.), Crystal Growth, Pergamon Press, Oxford, 1967, p. 691.

[11] J.W. Cahn, in: H.S. Peiser (Ed.), Crystal Growth, Pergamon Press, Oxford, 1967, p. 681.

[12] S.R. Coriell, R.F. Sekerka, J. Crystal Growth 34 (1976) 157.

[13] A.A. Chernov, J. Crystal Growth 24/25 (1974) 11.

[14] D.J. Wollkind, L.A. Segel, Phil. Trans. Roy. Soc. London 268 (1970) 351.

[15] B. Caroli, C. Caroli, C. Misbah, B. Roulet, J. Physique 46 (1965) 1657.

[16] D.J. Wollkind, R. Siriranananthan, D.B. Oulton, Physica D 12 (1984) 215.

[17] D.T.J. Hurle, J. Crystal Growth 5 (1969) 162.

[18] J.J. Favier, A. Rouzaud, J. Crystal Growth 64 (1983) 367.

[19] R.T. Delves, J. Crystal Growth 3/4 (1968) 562, 8 (1971) 13.

[20] S.R. Coriell, R.L. Parker, J. Appl. Phys. 36 (1965) 632, 37 (1966) 1548.

[21] S.C. Hardy, S.R. Coriell, J. Crystal Growth 5 (1969) 329, 7 (1970) 147.

[22] S.R. Coriell, S.C. Hardy, R.F. Sekerka, J. Crystal Growth 11 (1971) 53.

[23] S.C. Hardy, Phil. Mag. 35 (1977) 471.

[24] S.R. Coriell, R.F. Sekerka, J. Physicochemical Hydrodynamics 2 (1981) 281.

[25] G.B. McFadden, R.G. Rehm, S.R. Coriell, W. Chuck, K.A. Moorish, Met. Trans. 15A (1984) 2125.

[26] J.S. Langer, Rev. Mod. Phys. 52 (1980) 1.

[27] G.P. Ivantsov, Doklady Akademii Nauk SSSR 58 (1974) 567.

[28] D.E. Temkin, Doklady Akademii Nauk SSSR 132 (1960) 1307.

[29] R. Trivedi, Acta Met. 18 (1970) 287.

[30] E. Holtzman, J. Appl. Phys. 41 (1970) 1460.

[31] G.E. Nash, M.E. Glicksman, Acta Met. 26 (1978) 1681, 1689, 1697.

[32] J.S. Langer, H. Müller-Krumbhaar, Acta Met. 26 (1978) 1681, 1689, 1697.

[33] J.S. Langer, T. Fujioka, R.F. Sekerka, J. Crystal Growth 44 (1978) 418.

[34] W. Oldfield, Matr. Sci. Engng. 11 (1973) 211.

[35] R.C. Brower, D.A. Kessler, J. Koplic, H. Levine, Phys. Rev. A 29 (1984) 1335, A 30 (1984) 3161, A 31 (1985) 1712.

[36] E. Ben-Jacob, Nigel Goldenfeld, J.S. Langer, Gerd Schon, Phys. Rev. A 29 (1984) 330;
E. Ben-Jacob, N.D. Goldenfeld, B.C. Kotliar, J.S. Langer, Phys. Rev. Lett. 53 (1984) 2110.

[37] M.E. Glicksman, J. Mat. Sci. Eng. 65 (1984) 45.

Morphology: from sharp interface to phase field models

Robert F. Sekerka *

Department of Physics, Carnegie Mellon University, 6416 Wean Hall, Pittsburgh, PA 15213, USA

Abstract

Over the last 50 years, there has been tremendous progress in the quantification of crystal growth morphology. In the 1950s, the dynamics of crystal growth from the melt was based on the sharp interface model (interface of zero thickness separating solid and liquid), often under the assumption of isotropy. Ivantsov had discovered analytical solutions to the Stefan problem for the special class of shapes known as quadric surfaces (ellipsoids, hyperboloids and paraboloids, including their special cases spheres, cylinders and planes). But in the 1960s, these solutions were shown to be morphologically unstable, resulting in cellular and dendritic growth forms that had long been known to exist from experimental work. Sharp interface models were used to model these growth forms, but it was necessary to include corrections of the interface temperature for capillarity and curvature (Gibbs–Thomson equation) in order to avoid instabilities at all wavelengths and to set the size scale of the resulting morphologies. Except for the case of total interface control, for which exact solutions even for facetted crystals had been provided by Frank using the method of characteristics, little could be done analytically to treat anisotropies. By the 1980s, our reliance on the sharp interface model began to change with the adaptation by Langer and others of diffuse interface models, of the Cahn–Hilliard type, to solve dynamical problems. This class of models, now known as phase field models, replaced the sharp interface model by the solution in the entire computational domain of coupled partial differential equations for thermal and compositional fields and for an auxiliary variable that keeps track of the phase. Moreover, the phase field equations incorporate automatically the Gibbs–Thomson equation, anisotropy and even departures from local equilibrium (interface kinetics) asymptotically for a sufficiently thin diffuse interface. But it has been only in about the last decade that massive improvements in computing power have rendered the numerical solution of the phase field model tractable. By means of this model, complex morphologies and related phenomena over a vast range of length scales can now be studied.
© 2003 Published by Elsevier B.V.

PACS: 81.10.Aj; 47.20.Hw; 07.05.Tp; 81.30.Fb

Keywords: A1. Crystal morphology; A1. Computer simulation; A1. Morphological stability; A1. Phase-field

0. Introduction

Over the last 50 years, there has been tremendous progress in our understanding of crystal growth morphology. Some of this has resulted from analytical solutions to simplified phenomenological models of the crystal growth process. Additional progress has been made by means of numerical solution of more elaborate but still phenomenological models. These models have benefited by incorporation of the results of modeling on the atomic scale, especially in the formulation of boundary conditions to be applied at the solid–liquid interface. Moreover, improvement in

* Tel.: +1-412-268-2362; fax: +1-412-681-0648.
E-mail address: rs07@andrew.cmu.edu (R.F. Sekerka).

0022-0248/$ – see front matter © 2003 Published by Elsevier B.V.
doi:10.1016/j.jcrysgro.2003.12.033

Table 1

Table illustrating disparity of length scales relevant to crystal growth morphology

Physical length	cm	Theoretical length
	10^2	
Sample size	10^1	
	10^0	Thermal length κ/V
	10^{-1}	
Cell size	10^{-2}	
Dendrite tip radius	10^{-3}	Diffusion length D/V
	10^{-4}	
Growth spirals	10^{-5}	
Terraces	10^{-6}	
	10^{-7}	Capillary length d_0
Interface thickness	10^{-8}	

The magnitudes are typical for metals but very approximate and may overlap several size ranges. Thermal lengths and diffusion lengths depend on the thermal diffusivity, κ, and the solute diffusivity, D. They depend inversely on the growth velocity, V, which can range over many orders of magnitude for dendritic growth. For nonmetals and slow diffusers, the thermal and diffusion lengths would be several orders of magnitude smaller for the same growth velocity.

experimental techniques, such as atomic resolution microscopy, has enabled the direct visualization of crystal surfaces and revealed such things as terraces, ledges, kinks, and growth spirals that had previously been postulated theoretically but only revealed indirectly.

Insofar as a quantitative understanding of crystal growth morphology is concerned, one must bear in mind the vast disparity of length scales, about 10 orders of magnitude, over which crystal growth phenomena take place. This is illustrated in Table 1 where the choice of length scales is typical although very approximate.

The main point to be appreciated from Table 1 is that a theoretical description that simultaneously incorporates phenomena over the entire range of length scales is currently intractable. The state of the art is rather to model over some narrower range of length scales and then "hand of" that information to an analysis on a neighboring length scale. For example, modeling on the atomic scale can be used to understand growth at steps and growth spirals, and this knowledge can be used to formulate boundary conditions for the solution of partial differential equations that describe transport in terms of solute or thermal diffusion on the scale of the crystal size.

This paper covers my personal interaction, over about the last 50 years, with modeling on the scale of transport in terms of solute or thermal diffusion. It begins in the 1950s with the "sharp interface" model of solidification (growth from the melt) for which some "exact solutions" of the governing partial differential equations are possible to obtain in limiting cases for oversimplified boundary conditions. It then progresses to the 1960s when these solutions were tested for morphological stability and found to be unstable. Understanding of this instability necessitated the use of more complicated boundary conditions that included the dependence of solid–liquid interface temperature on local curvature, the Gibbs–Thomson effect. During the decade of the 1970s, there was continued study, still with the sharp interface model, of more complex interface morphologies, such as cells and dendrites, that result subsequent to morphological instability. In the 1980s the phase field model, based on a diffuse interface, was developed to solve dynamical problems. This model incorporated interface boundary conditions along with diffusive transport, but remained rather intractable numerically until the 1990s when there were rapid advances in parallel supercomputing. Today, the phase field model is the model of choice for incorporating crystal growth phenomena over a vast range of length scales, but it is still phenomenological and remains to be integrated in the 21st century with simultaneous modeling on the atomic scale.

1. Sharp interface model

We begin by presenting a simple version of the sharp interface model for solidification of a single component crystal from its pure melt. For simplicity, it is assumed that the density is independent of phase and uniform throughout the system and that there is no fluid convection. Therefore, transport of heat is purely diffusive. The temperature T_L in the liquid (melt) is governed by the equation

$$\nabla^2 T_L = \frac{1}{\kappa_L} \frac{\partial T_L}{\partial t}, \tag{1}$$

where $\nabla^2 = \partial^2/\partial x^2 + \partial^2/\partial y^2 + \partial^2/\partial z^2$ is the Laplacian in Cartesian coordinates, t is the time and κ_L is the thermal diffusivity, assumed to be constant. Similarly in the solid (crystal) we have

$$\nabla^2 T_S = \frac{1}{\kappa_S} \frac{\partial T_S}{\partial t}. \tag{2}$$

In a crystal, the thermal diffusivity could be anisotropic, but for a cubic crystal it is still isotropic, so we treat this still realistic case of isotropy for simplicity.

For a solidification problem, solutions to Eqs. (1) and (2) must be joined at the solid–liquid interface, a free boundary of unknown shape and location that is assumed to be a surface of zero thickness where the temperature has the value T_I. One might first assume the temperature of the interface to be the thermodynamic melting point T_M, but a better approximation would be the equilibrium temperature T_E which corrects the melting temperature for the local interface mean[1] curvature K. For isotropic[2] solid–liquid surface tension γ, this correction for capillarity is given by the Gibbs–Thomson equation

$$T_E = T_M - T_M \frac{\gamma}{L_V} K, \tag{3}$$

where L_V is the latent heat of fusion per unit volume. But the interface temperature may differ from T_E due to interface motion. This effect can be represented by a kinetic law in which the normal growth speed U is the product of a kinetic coefficient μ and the interface undercooling $T_M - T_I$:

$$U = \mu(T_E - T_I). \tag{4}$$

In general, μ can depend strongly on crystallographic orientation as well as temperature, the latter leading to a nonlinear dependence on interface undercooling $T_E - T_I$. By combining Eqs. (3) and (4) we obtain

$$T_I = T_M - T_M \frac{\gamma}{L_V} K - \frac{U}{\mu}. \tag{5}$$

For very rapid interface kinetics, $\mu \to \infty$ and $T_I \to T_E$, a condition known as local equilibrium. At the moving solid–liquid interface, energy must be conserved, which leads to the additional boundary condition

$$L_V U = (k_S \nabla T_S - k_L \nabla T_L) \cdot \hat{\boldsymbol{n}}, \tag{6}$$

where k_S and k_L are thermal conductivities of solid and liquid and $\hat{\boldsymbol{n}}$ is the unit outward normal to the crystal.

Eqs. (1) and (2), together with initial conditions, far-field boundary conditions, and the interfacial boundary conditions Eqs. (5) and (6), constitute a free boundary problem for the shape and location of the crystal–melt interface, and hence for crystal morphology. This is obviously a difficult problem, so we turn next to some special cases that allow for exact solutions in order to gain further insight.

1.1. Shape preserving solutions

For the special case of local equilibrium with negligible correction for surface tension, Eq. (5) becomes

$$T_I = T_M, \quad \mu \to \infty; \ \gamma \to 0 \tag{7}$$

so the solid–liquid interface temperature is a constant, the thermodynamic melting temperature. This results in the so-called Stefan problem. With this simplification, and special initial and far-field boundary conditions, a class of exact shape preserving solutions was discovered by Ivantsov [3,4] and elaborated by others [5–8]. Ivantsov considered problems for which the solid was isothermal at its melting point. Thus, the solution to Eq. (2) is just $T_S = T_M$ and the term involving ∇T_S on the right-hand side of Eq. (6) is zero. He then generalized the remaining Eq. (6) into a first-order partial differential equation valid in all space, found solutions to it by envelope formation, and then found compatible solutions to Eq. (1). By this ingenious method, he found exact solutions for solids having the shapes of quadric surfaces, which are surfaces described in space by polynomials of degree two. Such surfaces are ellipsoids, paraboloids and hyperboloids, and all of their special cases such as prolate and oblate spheroids, spheres, circular paraboloids, parabolic cylinders, elliptic cylinders, circular cylinders and planes. In all of these cases, the isotherms are members of the same shape class. These forms evolve so that they can change size but not shape. The following cases result:

Ellipsoids These have the form

$$\frac{x^2}{\xi_0^2 - a^2} + \frac{y^2}{\xi_0^2 - b^2} + \frac{x^2}{\xi_0^2} = 4\kappa_L t, \tag{8}$$

where ξ_0 is a generalized coordinate that is constant on the solid surface and $\xi_0 \geqslant b \geqslant a$, where a and b

---

[1] The mean curvature $K = 1/R_1 + 1/R_2$ where R_1 and R_2 are the principal radii of curvature, signed positive for a spherical crystal.

[2] For anisotropic surface tension, the result is more complicated and involves derivatives of γ, the so-called Herring equation [1,2].

are shape parameters. Any linear dimension of such an ellipsoid increases in proportion to $t^{1/2}$.

Paraboloids These have the form

$$\frac{x^2}{\xi_0} + \frac{y^2}{\xi_0 + B} = \frac{4\kappa_L^2}{V^2}\xi_0 - \frac{4\kappa_L}{V}(z - Vt), \tag{9}$$

where B is a shape parameter. They translate in the z direction with constant velocity V while thickening in the x or y direction, at constant $z = \xi_0\kappa_L/V$, in proportion to $t^{1/2}$.

Hyperboloids These have the form

$$-\frac{x^2}{a^2 - \xi_0^2} - \frac{y^2}{b^2 - \xi_0^2} + \frac{x^2}{\xi_0^2} = 4\kappa_L t, \tag{10}$$

where $\xi_0 < a \leqslant b$. In this case, hyperboloids of two sheets melt away from the midplane that separates them in proportion to $t^{1/2}$.

Since Eqs. (1) and (2) are parabolic, they are not invariant under the transformation $t \to -t$. For example, one cannot just reverse the sign of t in Eq. (10) for the melting hyperboloids so get a solution for growing hyperboloids. Therefore, growth and melting are intrinsically different problems. By using separation of variables, Ham [6] showed that some of these shapes could be made to move in the reverse direction, but only for a finite time.[3] A detailed description of all of these solutions, including equations for determination of growth or melting rates as a function of supercooling or superheating, can be found in Ref. [8]. As far as we know, these are the only exact analytical solutions to the solidification problem with an isothermal interface.

1.2. Anisotropic interface control

Another class of exact solutions to the sharp interface problem can be obtained in essentially the opposite limit of that considered for shape preserving solutions. The appropriate limit is one in which heat flow is so fast that the entire system, both solid and liquid, is practically at a uniform temperature T_I. In this case, Eqs. (1), (2) and (6) can be ignored[4] and growth is

governed by Eq. (5). If the capillary term in γ can also be neglected, we have

$$U = \mu(\hat{\mathbf{n}})\Delta T, \tag{11}$$

where the interface undercooling $\Delta T := T_M - T_I$ is a constant and we have exhibited the dependence of the kinetic coefficient on interface orientation. It turns out that the growth law represented by Eq. (11) leads to an exact solution for any initial shape. The trick is to update the shape in time by following trajectories of constant orientation, rather than following the movement of each element along its local normal. This is based on the method of characteristic curves [9] of partial differential equations, but has also been related to a physical model by Frank [10]. In terms of the unit vectors $\hat{\mathbf{n}}$, $\hat{\boldsymbol{\theta}}$ and $\hat{\boldsymbol{\phi}}$ of a spherical coordinate system, for which $U(\theta, \varphi) = \mu(\theta, \varphi)\Delta T$, the velocity along a trajectory of constant orientation is given by

$$\mathbf{V} = U\hat{\mathbf{n}} + \frac{\partial U}{\partial \theta}\hat{\boldsymbol{\theta}} + \frac{1}{\sin\theta}\frac{\partial U}{\partial \varphi}\hat{\boldsymbol{\varphi}}. \tag{12}$$

But since Eq. (12) is independent of time, these trajectories are straight lines! Moreover, the directions of these straight lines can be determined geometrically as follows: a polar plot of the reciprocal of U can be represented in spherical coordinates by the equation

$$f(r, \theta, \varphi) := r - \frac{1}{U(\theta, \varphi)} = 0. \tag{13}$$

Its normal is therefore along

$$\nabla f = \hat{\mathbf{r}} + \frac{1}{r}\frac{1}{U^2}\frac{\partial U}{\partial \theta}\hat{\boldsymbol{\theta}} + \frac{1}{r\sin\theta}\frac{1}{U^2}\frac{\partial U}{\partial \varphi}\hat{\boldsymbol{\varphi}} = \frac{\mathbf{V}}{U}, \tag{14}$$

where we have identified $\hat{\mathbf{r}} = \hat{\mathbf{n}}$ and used Eq. (13) to eliminate r.

Examination of Eqs. (12)–(14) shows that they bear an analogy to the problem of finding the equilibrium shape of a particle of fixed volume for anisotropic surface tension $\gamma(\hat{\mathbf{n}}) = \gamma(\theta, \varphi)$. In this case, one minimizes the surface energy $\int \gamma(\hat{\mathbf{n}})\,dA$ of the particle, subject to the constraint of constant volume. The result is the so-called Wulff shape which is self similar to the convex hull of a polar plot of the $\boldsymbol{\xi}$-vector of Hoffman and Cahn [11,12] which in spherical coordinates is given by

$$\boldsymbol{\xi} = \gamma\hat{\mathbf{n}} + \frac{\partial\gamma}{\partial\theta}\hat{\boldsymbol{\theta}} + \frac{1}{\sin\theta}\frac{\partial\gamma}{\partial\varphi}\hat{\boldsymbol{\varphi}}. \tag{15}$$

---

[3] The resulting solutions depend on $\sqrt{\tau - t}$ where τ is a constant.

[4] In regard to Eq. (6), the growth must be so slow that the difference in temperature needed to carry off the latent heat is negligible.

Furthermore, it is well known that the direction of $\boldsymbol{\xi}$, as a function of θ and φ, is along the normal of a polar plot of $1/\gamma(\theta, \varphi)$. We therefore see that U is analogous to γ and $\mathbf{V}$ is analogous to $\boldsymbol{\xi}$. Thus, when Eq. (11) applies, a crystal grows such that trajectories of surface elements having constant orientation will move along straight lines until a shape, self-similar to the Wulff shape for its normal growth speed $U(\theta, \varphi)$, is approached asymptotically. This asymptotic shape is often referred to as the "kinetic Wulff shape", since it is also the Wulff shape for the kinetic coefficient $\mu(\theta, \varphi)$ which is proportional to $U(\theta, \varphi)$. As is well known, the Wulff shape can have missing orientations, which occur where the $1/\gamma(\theta, \varphi)$ plot is concave. So for the "kinetic Wulff shape", the faster growing orientations "grow out" and eventually cease to exist, leaving the crystal shape to be bounded by its more slowly growing orientations. If the kinetic Wulff shape is facetted, the resulting crystal shape is facetted, in agreement with our common impression of a crystal morphology.

If the temperature throughout solid and liquid is uniform and equal to T_I but the curvature term in Eq. (5) is retained, it is still possible [13] to formulate a differential equation for the evolution of the crystal shape. In two dimensions, detailed calculations [14] have demonstrated that the kinetic Wulff shape is nearly approached asymptotically, except that capillarity comes into play in regions where the curvature is large. Thus, sharp corners that result from missing orientations on the kinetic Wulff shape are replaced locally by rounded corners, so no orientations are actually missing.

2. Morphological instability

In principle, it would seem possible for crystals to have morphologies in accord with the shape preserving solutions discussed in Section 1.1. Nevertheless, one must ask whether such solutions are stable, given the possibility that a small perturbation of such a shape might actually grow. Such a possibility is suggested by directional solidification experiments in which the solid–liquid interface is observed sometimes to be cellular, rather than planar, as well as free growth into supercooled liquid that can result in dendritic forms.

This possibility of morphological instability was studied theoretically by Mullins and Sekerka [15] by considering the growth of a sphere perturbed by spherical harmonics. In polar coordinates, the perturbed shape was represented in the form

$$r = R + \delta Y_{\ell m}(\theta, \varphi), \tag{16}$$

where $R(t)$ is the time-dependent radius of the unperturbed sphere, $\delta(t)$ is the time-dependent amplitude of a perturbation, and $Y_{\ell m}(\theta, \varphi)$ is a spherical harmonic. Local equilibrium ($\mu \to \infty$) at the solid–liquid interface is assumed, so Eq. (5) takes the form

$$\begin{aligned} T_I &= T_M[1 - \Gamma K] \\ &= T_M\left[1 - \frac{2\Gamma}{R} - \frac{\Gamma \delta}{R^2}(\ell - 1)(\ell + 2)Y_{\ell m}(\theta, \varphi)\right], \end{aligned} \tag{17}$$

where $\Gamma = \gamma/L_V$ is a capillary length and higher order terms in δ/R have been neglected. In Eq. (17), the term $2\Gamma/R$ comes from the unperturbed sphere and the term in $Y_{\ell m}(\theta, \varphi)$ comes from the perturbation.

The analysis proceeds by solving for the temperature fields in solid and liquid to first order in δ. If the interface were an isotherm, the isotherms of these fields would become distorted near the perturbation into shapes that resemble the perturbation. This would tend to enhance the growth of the perturbation. But this distortion of the isotherms is mitigated by the fact that the interface is not an isotherm, as represented by the term containing $Y_{\ell m}(\theta, \varphi)$ in Eq. (17), and this leads to stabilization. Detailed analysis of the sign of the quantity $(1/\delta)\,\mathrm{d}\delta/\mathrm{d}t$ leads to the conclusion that the sphere is unstable whenever $\ell > 1$ and

$$R > \frac{R^*}{2}\left[(\ell + 1)(\ell + 2) + \ell(\ell + 2)\frac{k_S}{k_L} + 2\right], \tag{18}$$

where $R^* := 2\Gamma T_M/(T_M - T_\infty)$ is the nucleation radius, in which T_∞ is the far-field temperature. Thus, the sphere becomes unstable to an ellipsoidal shape ($\ell = 2$) whenever $R/R^* > 7 + 4k_S/k_L$ and to more undulating shapes at larger values of R, corresponding to larger values of ℓ. If it were not for capillarity, i.e., if $\Gamma = 0$, the sphere would be unstable at all sizes to perturbations of all wavelengths.

The criterion for instability can be written in an alternative way in terms of the magnitude $-G_L =$

$[T_{\mathrm{M}}(1 - R^*/R) - T_\infty]/R$ of the (negative) temperature gradient at the solid–liquid interface of the unperturbed sphere. Thus, instability occurs whenever

$$-G_{\mathrm{L}} > \frac{T_{\mathrm{M}}\Gamma}{R^2}\left[(\ell+1)(\ell+2) + \ell(\ell+2)\frac{k_{\mathrm{S}}}{k_{\mathrm{L}}}\right]. \qquad (19)$$

Eq. (19) supports the interpretation that growth into a supercooled melt $(-G_{\mathrm{L}} > 0)$ is destabilizing while capillarity (term in Γ) is stabilizing. For large ℓ, one can interpret $\lambda \sim 2\pi R/\ell$ as the wavelength of a perturbation, in which case Eq. (19) at marginal stability (replace $>$ by $=$) and for $k_{\mathrm{S}} = k_{\mathrm{L}}$ yields

$$\lambda \sim 2\pi\left[\frac{2T_{\mathrm{M}}\Gamma}{|G_{\mathrm{L}}|}\right]^{1/2}. \qquad (20)$$

Thus the scale of the instability is the geometric mean of a capillary length Γ and a diffusion length $T_{\mathrm{M}}/|G_{\mathrm{L}}|$. Another form of Eq. (20) can be obtained in terms of the growth velocity $V = |G_{\mathrm{L}}|k_{\mathrm{L}}/L_{\mathrm{V}}$ and the capillary length used in Table 1, namely $d_0 = \Gamma T_{\mathrm{M}}c_{\mathrm{V}}/L_{\mathrm{V}}$ where c_{V} is the heat capacity per unit volume. The result is

$$\lambda \sim 2\pi\left[2d_0\frac{\kappa_{\mathrm{L}}}{V}\right]^{1/2}, \qquad (21)$$

which is essentially the geometric mean of the capillary length d_0 and the thermal length κ_{L}/V. This is a general characteristic of morphological instability phenomena, independent of the shape of the unperturbed body. It was on this basis that Langer and Müller-Krumbhaar [16] first proposed that a dendrite tip radius ρ should be about equal to λ, which leads to

$$\sigma^* := \frac{2d_0\kappa_{\mathrm{L}}}{\rho^2 V} = \frac{1}{(2\pi)^2} \approx 0.025. \qquad (22)$$

It is amazing that Eq. (22) is in pretty good agreement [17] with experiment, although the value of the numerical constant is surely fortuitous. It turns out that the scaling suggested by Eq. (22) is essentially correct, but the value of σ^* depends on a more delicate analysis, such as that provided by microscopic solvability theory [18–20].

Morphological stability theory is very general, and applies to growth by solute diffusion [15] as well as solidification of alloys [21]. It can also be extended to include departures from local equilibrium (interface kinetics) as well as anisotropy. For a comprehensive review, see Coriell and McFadden [22].

3. Phase field model

The main lesson to be learned from morphological stability analysis is that computations of crystal morphology require the solution of a more complex free boundary problem in which the effects of capillarity must be included. Neglecting these effects gives rise to solutions for idealized shapes that are unstable on all length scales of continuum models. Adding to this the fact that the surface tension is actually anisotropic and that anisotropic interface kinetics (see Section 1.2) can give rise to shapes related to this anisotropy as well, we were faced with a formidable free boundary problem. This provided motivation for the phase field model in which all of these effects could be incorporated in a more tractable way.

In the phase field model [23–25], the sharp interface is replaced by a diffuse interface and an auxiliary parameter φ, the phase field, is introduced to indicate the phase. The quantity φ is a continuous variable that takes on constant values in the bulk phases, say 0 in the solid and 1 in the liquid, and increases from 0 to 1 over a thin layer, the diffuse interface. A partial differential equation is formulated to govern the time evolution of φ. It incorporates the interfacial physics of the problem in such a way that the diffuse interface has an excess energy, which gives rise, for a sufficiently thin interface, to a surface tension γ. Bending of the diffuse interface automatically introduces capillarity, Eq. (3). A diffusivity related to the time evolution of φ gives rise to a linear kinetic law, Eq. (4). Both the surface tension and the kinetic coefficient can be made to be anisotropic. The partial differential equation for φ is coupled to other equations that determine the relevant fields that govern transport, temperature in the case of energy transport and composition in the case of solute transport.

We indicate briefly the general procedure for constructing the phase field equations for solidification of a single component from its pure melt. For simplicity of presentation, we assume that all quantities are isotropic, that the density is uniform in solid and liquid, and that there is no convection in the liquid. We postulate that the internal energy $\mathcal{U}$ and the entropy $\mathcal{S}$ in any subvolume $\mathcal{V}$ of our system are given by

$$\mathcal{U} := \int_{\mathcal{V}}\left[u + \frac{1}{2}\varepsilon_u^2|\nabla\varphi|^2\right]\mathrm{d}^3x \qquad (23)$$

and

$$\mathscr{S} := \int_{\mathscr{V}} \left[s(u, \varphi) - \frac{1}{2} \varepsilon_s^2 |\nabla \varphi|^2 \right] d^3 x, \tag{24}$$

where $u(\mathbf{r}, t)$ is the local density of internal energy, $\varphi(\mathbf{r}, t)$ is the phase field, $\mathbf{r}$ is the position vector, t is time, and ε_u^2 and ε_s^2 are constants. We regard these expressions to be *functionals* of $u(\mathbf{r}, t)$ and $\varphi(\mathbf{r}, t)$; in other words, $\mathscr{U}$ and $\mathscr{S}$ depend on functions, rather than just variables. The quantities u and s are internal energy and entropy densities that pertain to a homogeneous phase having a uniform value of φ. The terms involving $|\nabla \varphi|^2$ are corrections that are only important in the diffuse interface where φ changes from its value $\varphi = 0$ in bulk solid to its value $\varphi = 1$ in bulk liquid. The term $\frac{1}{2}\varepsilon_u^2 |\nabla \varphi|^2$ is called a gradient energy and $-\frac{1}{2}\varepsilon_s^2 |\nabla \varphi|^2$ is called a gradient entropy. Together they give rise to a gradient free energy, such as used in Cahn–Hilliard theory [26].

Dynamical equations are based on the functionals given by Eqs. (23) and (24) and the concepts of local energy conservation and local entropy production. Since energy is conserved,

$$\dot{U}_{\mathrm{prod}} := \frac{d}{dt}\mathscr{U} + \int_{\mathscr{A}} \left[\mathbf{q} - \varepsilon_u^2 \dot{\varphi} \nabla \varphi \cdot \hat{\mathbf{n}} \right] d^2 x = 0. \tag{25}$$

The rate of entropy production is

$$\dot{S}_{\mathrm{prod}} := \frac{d}{dt}\mathscr{S} + \int_{\mathscr{A}} \left[\frac{\mathbf{q}}{T} + \varepsilon_s^2 \dot{\varphi} \nabla \varphi \cdot \hat{\mathbf{n}} \right] d^2 x \geqslant 0. \tag{26}$$

Here, $\mathscr{A}$ is the area surrounding the arbitrary subvolume $\mathscr{V}$, $\hat{\mathbf{n}}$ is its unit outward normal, and a dot above a variable denotes partial differentiation with respect to time. The vector $\mathbf{q}$ is the classical heat flux and $\mathbf{q}/T$ is the classical entropy flux. The additional fluxes in the area integrals are nonclassical fluxes associated with the gradient energy and gradient entropy corrections. These nonclassical fluxes arise whenever elements of the diffuse interface enter or leave a control volume, as discussed by Wang et al. [27]. From Eq. (25) we obtain

$$\int_{\mathscr{V}} \left[\dot{u} + \nabla \cdot \mathbf{q} - \varepsilon_u^2 \dot{\varphi} \nabla^2 \varphi \right] d^3 x = 0. \tag{27}$$

Since Eq. (27) holds in every arbitrary subvolume, the integrand itself must vanish and we obtain

$$\dot{u} + \nabla \cdot \mathbf{q} - \varepsilon_u^2 \dot{\varphi} \nabla^2 \varphi = 0. \tag{28}$$

Fro m Eq. (26) we obtain

$$\int_{\mathscr{V}} \left[\mathbf{q} \cdot \nabla \left(\frac{1}{T} \right) \right] d^3 x$$
$$+ \int_{\mathscr{V}} \left[\left(\frac{\partial s}{\partial \varphi} \right)_u + \frac{\varepsilon_f^2}{T} \nabla^2 \varphi \right] \dot{\varphi} \, d^3 x \geqslant 0, \tag{29}$$

where $\varepsilon_f^2 = \varepsilon_e^2 + T\varepsilon_s^2$. Eq. (29) can be satisfied for every subvolume $\mathscr{V}$ by assuming linear constitutive laws of the form

$$\mathbf{q} = M_u \nabla \left(\frac{1}{T} \right) = -k \nabla T, \tag{30}$$

where $M_u > 0$ and $k = M_u / T^2$ is the thermal conductivity, and

$$\tau \dot{\varphi} = \left(\frac{\partial s}{\partial \varphi} \right)_u + \frac{\varepsilon_f^2}{T} \nabla^2 \varphi, \tag{31}$$

where $\tau > 0$.

Eq. (31) is the equation for the time evolution of the phase field. Substitution of Eq. (30) into Eq. (28) leads to a compatible energy equation, essentially an equation for time evolution of the temperature. This becomes clear once explicit functions for u and s in terms of independent variables T, φ are specified. For example, we could take an internal energy density of the form

$$u(T, \varphi) = u_0 + c_{\mathrm{V}}(T - T_{\mathrm{M}}) + L_0 p(\varphi) + \frac{W_u}{2} g(\varphi), \tag{32}$$

where u_0 is a constant, c_{V} is a constant heat capacity per unit volume, L_0 is a constant latent heat per unit volume, $g(\varphi) = \varphi^2 (1 - \varphi)^2$ is a double well potential, W_u is a constant strength parameter for the double well, and $p(\varphi) = \varphi^3 (10 - 15\varphi + 6\varphi^2)$ is a smooth function of φ that increases monotonically from $p(0) = 0$ to $p(1) = 1$. We could also take a Helmholtz free energy density $f = u - Ts$ of the form

$$f(T, \varphi) = u_0 - Ts_0 + c_{\mathrm{V}}(T - T_{\mathrm{M}})$$
$$- c_{\mathrm{V}} T \ln(T / T_{\mathrm{M}})$$
$$+ L_0 (1 - T/T_{\mathrm{M}}) p(\varphi) + \frac{W_f}{2} g(\varphi), \tag{33}$$

where s_0 is a constant and $W_f = W_u + T W_s$, where W_s is a constant. Then, by means of a thermodynamic equation, we deduce that

$$\left(\frac{\partial s}{\partial \varphi}\right)_u = -\frac{1}{T}\left(\frac{\partial f}{\partial \varphi}\right)_T$$

$$= L_0\left(\frac{1}{T_{\mathrm{M}}} - \frac{1}{T}\right)p'(\varphi) - \frac{W_f}{2}g'(\varphi), \qquad (34)$$

where the primes on p and g denote differentiation. We therefore obtain the phase field equations

$$c_V\dot{T} + L_0\dot{p}(\varphi) = k\nabla^2 T + \varepsilon_u^2\dot{\varphi}\nabla^2\varphi - \frac{W_u}{2}\dot{g}(\varphi) \quad (35)$$

and

$$\tau\dot{\varphi} = \frac{\varepsilon_f^2}{T}\nabla^2\varphi + L_0\left(\frac{1}{T_{\mathrm{M}}} - \frac{1}{T}\right)p'(\varphi)$$

$$- \frac{W_f}{2}g'(\varphi). \qquad (36)$$

The term containing L_0 in Eq. (35) gives rise to latent heat evolution at the interface and incorporates the boundary condition Eq. (6) of the sharp interface model. The term containing L_0 in Eq. (36) provides a bias to the double well potential represented by $g(\varphi)$ and this causes the crystal to melt or grow, depending on the sign of $T - T_{\mathrm{M}}$. By means of asymptotic analysis [28] in the limit of a very thin interface, one can show that the interface thickness (as φ varies from 0.05 to 0.95) is about 6ℓ where

$$\ell = \frac{\varepsilon_f}{\sqrt{W_f}}. \qquad (37)$$

The surface tension is given by

$$\gamma = \frac{W_f\ell}{6} = \frac{\varepsilon_f\sqrt{W_f}}{6} \qquad (38)$$

and the kinetic coefficient is given by

$$\mu = \frac{6L_0\ell}{T_{\mathrm{M}}^2\tau}. \qquad (39)$$

Eqs. (37)–(39) can be used to relate the parameters of the model to physical properties and to the thickness of the diffuse interface, a computational parameter. Asymptotic analysis for a thicker interface [29] leads to a somewhat different relationship of model parameters to physical properties. Anisotropy in γ and μ can be introduced by allowing ε_f and τ to depend on $\hat{\mathbf{N}} = \nabla\varphi/|\nabla\varphi|$, which plays the role of a unit normal vector in the interfacial region. In this case, Eq. (36) must be replaced by a more complicated equation that contains derivatives of ε_f. Details and examples are presented in Refs. [30–33], where these anisotropies are related to the $\boldsymbol{\xi}$-vector formalism of Hoffman and Cahn.

4. Conclusion

We conclude by mentioning briefly some generalizations and recent developments.

In addition to solidification of a pure material, the phase field model applies equally well to the analogous problem of isothermal precipitation. In addition, it has been generalized by Wheeler and Boettinger [34–37] and others [38–40] to apply to the solidification of alloys, in which case there are coupled partial differential equations for time evolution of the phase field, the temperature and the composition. Computations based on the alloy phase field model have led to a much better understanding of solute segregation and pattern selection at cellular interfaces and during dendritic growth [41–45].

The phase field model has also been extended to include hydrodynamics, both for pure materials and for alloys [46–50]. Computations including hydrodynamics are difficult, but results are beginning to emerge [51,52]. Hydrodynamics has also been added to the phase field model by means of hybrid methods by Tönhardt and Amberg [53–55] and Beckermann et al. [56] and has led to somewhat more tractable models [57] for computing solidification microstructures.

Solution adaptive grids have been used to facilitate phase field modeling in two dimensions [58,59]. They are especially useful in helping to resolve dendrite sidebranching with computational efficiency. Solution adaptive grids are practically mandatory for modeling dendrites in three dimensions and were first used for this purpose for the sharp interface model [60]. Computations using the phase field model have also been extended to three dimensions, both by using adaptive grids [61,62] and by means of hybrid methods involving random walk algorithms [63]. Computations based on these models have allowed for simulations that can be used to study dendrite sidebranching and to test three-dimensional predictions of microscopic solvability theory [64].

It has been demonstrated that the phase field model gives rise to phenomena such as solute trapping and solute drag [34,65–70] as well as other effects related to departures from local equilibrium at a sharp solid–liquid interface. In order to eliminate these nonequilibrium phenomena and compare with well-known results for the case of local equilibrium, "corrections" based on thin interface asymptotics have been formulated [71,72].

Recently, Kobayashi et al. [73] have generalized the model by including a complex order parameter that enables the orientation of a grain to be tracked. This allows one to model grain growth and grain rotation in solid–solid transformations. Moreover, Kassner et al. [74] have included the effects of strain for a solid in contact with a melt. Finally, Gránásy et al. [75] have used a diffuse interface model to compute nucleation and have incorporated this with the phase field model for growth to simulate nucleation and growth of polycrystalline aggregates.

Today, the phase field model is the model of choice for computation of complex interface morphologies that result subsequent to morphological instability. Already it has enhanced our theoretical understanding of the origin and complexity of these morphologies. But it is also beginning to be incorporated in codes to enable realistic simulations of crystal growth processes that can serve as guidance for the design of engineering systems to improve crystal yield and quality.

References

[1] W.E. Kingston (Ed.), Conyers Herring, Surface Tension as a Motivation for Sintering, The Physics of Powder Metallurgy, a Symposium held at Bayside, L.I. New York, 24–26 August, 1949, McGraw-Hill, New York, 1951, p. 143.

[2] R. Gomer, C.S. Smith (Eds.), Conyers Herring, The Use of Classical Macroscopic Concepts in Surface-Energy Problems, Structure and Properties of Solid Surfaces, a Conference Arranged by the National Research Council, Lake Geneva, WI, September 1952, University of Chicago Press, Chicago, 1953, p. 5.

[3] G.P. Ivantsov, Dokl. Akad. Nauk SSSR 58 (1947) 567; G. Horvay (English Transl.), General Electric Research Report No. 60-RL-(2511M), Schenectady, NY (1960).

[4] G.P. Ivantsov, Dokl. Akad. Nauk SSSR 83 (1952) 573.

[5] G. Horvay, J.W. Cahn, Acta Metall. 9 (1961) 695.

[6] F.S. Ham, Quart. J. Appl. Math. 17 (1959) 137.

[7] D. Canright, S.H. Davis, Metal. Trans. 20A (1989) 225.

[8] R.F. Sekerka, Shun-Lien Wang, Moving phase boundary problems, in: H.I. Aaronson (Ed.), Lectures on the Theory of Phase Transformations, second ed., TMS, Warrendale, 1999, p. 231.

[9] A.A. Chernov, Sov. Phys. Crystallogr. 8 (1964) 401.

[10] F.C. Frank, On the kinematic theory of crystal growth and dissolution processes, in: R.H. Doremus, B.W. Roberts, D. Turnbull (Eds.), Growth and Perfection in Crystals, Wiley, New York, 1958, p. 411.

[11] D.W. Hoffman, J.W. Cahn, Surf. Sci. 31 (1972) 368.

[12] J.W. Cahn, D.W. Hoffman, Acta. Metall. 22 (1974) 1205.

[13] S. Angenent, M.E. Gurtin, Arch. Ration. Mech. Anal. 108 (1989) 323.

[14] E. Yokoyama, R.F. Sekerka, J. Crystal Growth 125 (1992) 389.

[15] W.W. Mullins, R.F. Sekerka, J. Appl. Phys. 34 (1963) 323.

[16] J.S. Langer, H. Müller-Krumbhaar, J. Crystal Growth 42 (1977) 11.

[17] J.S. Langer, R.F. Sekerka, T. Fujioka, J. Crystal Growth 44 (1978) 414.

[18] A. Barbieri, J.S. Langer, Phys. Rev. A 39 (1989) 5314.

[19] M. Ben Amar, P. Pelcé, Phys. Rev. A 39 (1989) 4263.

[20] Y. Pomeau, M. Ben Amar, Dendritic growth and related topics, in: C. Godrèche (Ed.), Solids Far from Equilibrium, Cambridge University Press, Cambridge, 1992, p. 365.

[21] W.W. Mullins, R.F. Sekerka, J. Appl. Phys. 35 (1964) 444.

[22] S.R. Coriell, G.B. McFadden, Morphological stability, in: D.T.J. Hurle (Ed.), Handbook of Crystal Growth 1A, Fundamentals, Transport and Stability, North-Holland, Amsterdam, 1993, p. 785.

[23] J.S. Langer, August 1978, unpublished notes.

[24] J.B. Collins, H. Levine, Phys. Rev. B 31 (1985) 6119.

[25] J.S. Langer, in: G. Grinstein, G. Mazenko (Eds.), Directions in Condensed Matter Physics, World Scientific, Singapore, 1986, p. 165.

[26] J.W. Cahn, J.E. Hilliard, J. Chem. Phys. 28 (1958) 258.

[27] S.-L. Wang, R.F. Sekerka, A.A. Wheeler, B.T. Murray, S.R. Coriell, R.J. Braun, G.B. McFadden, Physica D 69 (1993) 189.

[28] G.B. McFadden, A.A. Wheeler, D.M. Anderson, Physica D 144 (2000) 154.

[29] A. Karma, W.-J. Rappel, Phys. Rev. E 53 (1996) 3017.

[30] G.B. McFadden, A.A. Wheeler, R.J. Braun, S.R. Coriell, R.F. Sekerka, Phys. Rev. E 48 (1993) 2016.

[31] A.A. Wheeler, G.B. McFadden, Eur. J. Appl. Math. 7 (1996) 367.

[32] A.A. Wheeler, G.B. McFadden, Proc. Roy. Soc. London A 453 (1997) 1611.

[33] R.F. Sekerka, Fundamentals of phase field theory, in: K. Sato, Y. Furukawa, K. Nakajima (Eds.), Advances in Crystal Growth Research, Elsevier, Amsterdam, 2001, p. 21.

[34] A.A. Wheeler, W.J. Boettinger, G.B. McFadden, Phys. Rev. A 45 (1992) 7424.

[35] A.A. Wheeler, W.J. Boettinger, in: M.E. Gurtin, G.B. McFadden (Eds.), On the Evolution of Phase Boundaries, in: The IMA Volumes in Mathematics and its Applications, vol. 43, Springer, Berlin, 1992, p. 127.

[36] A.A. Wheeler, W.J. Boettinger, G.B. McFadden, Phys. Rev. E 47 (1993) 1893.

[37] W.J. Boettinger, A.A. Wheeler, G.B. McFadden, R. Kobayashi, in: S.P. Marsh, C.S. Pande (Eds.), Modeling of Coarsening and Grain Growth, TMS, Warrendale, 1993, p. 45.

[38] Z. Bi, R.F. Sekerka, Physica A 261 (1998) 95.

[39] Ch. Charach, P.C. Fife, SIAM, J. Appl. Math. 58 (1998) 1826.

[40] Ch. Charach, P.C. Fife, Open Syst. Inform. Dyn. 5 (1998) 99.

[41] J.A. Warren, W.J. Boettinger, Acta Metall. Mater. 43 (1995) 689.

[42] M. Conti, Phys. Rev. E 56 (1997) 3197.

[43] J.A. Warren, W.J. Boettinger, in: M. Cross, J. Campbell (Eds.), Modeling of Casting, Welding and Advanced Solidification Processes, vol. VII, TMS, Warrendale, 1995, p. 601.

[44] J.A. Warren, W.J. Boettinger, in: J. Beach, H. Jones (Eds.), Solidification Processing, Department of Engineering Materials, University of Sheffeld, UK, 1997, p. 422.

[45] Zhiqiang Bi, R.F. Sekerka, J. Crystal Growth 237–239 (2002) 138.

[46] M.E. Gurtin, D. Polignone, J. Viñals, Math. Models Meth. Appl. Sci. 6 (1996) 815.

[47] D.M. Anderson, G.B. McFadden, Phys. Fluids 9 (1997) 1870.

[48] D.M. Anderson, G.B. McFadden, A.A. Wheeler, Annu. Rev. Fluid Mech. 30 (1998) 139.

[49] D.M. Anderson, G.B. McFadden, A.A. Wheeler, Physica D 135 (2000) 175.

[50] R.F. Sekerka, Z. Bi, Phase field model of multicomponent alloy with hydrodynamics, in: M.K. Smith, M.J. Mixis, G.B. McFadden, G.P. Neitzel, D.R. Canright (Eds.), Interfaces for the Twenty-First Century, Imperial College Press, London, 2001, p. 147.

[51] D.M. Anderson, G.B. McFadden, A.A. Wheeler, Physica D 2711 (2001) 1.

[52] D.M. Anderson, G.B. McFadden, A.A. Wheeler, A phase-field model of solidification with convection: numerical simulations, in: M.K. Smith, M.J. Mixis, G.B. McFadden, G.P. Neitzel, D.R. Canright (Eds.), Interfaces for the Twenty-First Century, Imperial College Press, London, 2001, p. 131.

[53] R. Tönhardt, G. Amberg, J. Crystal Growth 194 (1998) 406.

[54] R. Tönhardt, G. Amberg, J. Crystal Growth 213 (2000) 161.

[55] R. Tönhardt, G. Amberg, Phys. Rev. E 62 (2000) 828.

[56] C. Beckermann, H.J. Diepers, I. Steinbach, A. Karma, X. Tong, J. Comput. Phys. 154 (1999) 468.

[57] W.J. Boettinger, J.A. Warren, C. Beckermann, A. Karma, Annu. Rev. Mater. Res. 32 (2002) 163.

[58] R.A. Braun, B.T. Murray, J. Crystal Growth 174 (1997) 41.

[59] N. Provatas, N. Goldenfeld, J. Dantzig, J. Comput. Phys. 148 (1999) 265.

[60] A. Schmidt, J. Comput. Phys. 125 (1996) 293.

[61] N. Provatas, N. Goldenfeld, J. Dantzig, Phys. Rev. Lett. 80 (1998) 3308.

[62] Jun-Ho Jeong, N. Goldenfeld, J.A. Dantzig, Phys. Rev. E 64 (2001) 41602.

[63] A. Karma, M. Plapp, Phys. Rev. Lett. 84 (2000) 1740.

[64] A. Karma, Youngyih H. Lee, M. Plapp, Phys. Rev. E 61 (2000) 3996.

[65] W.J. Boettinger, A.A. Wheeler, B.T. Murray, G.B. McFadden, Mater. Sci. Eng. A 178 (1994) 217.

[66] M. Conti, Phys. Rev. E 55 (1997) 701.

[67] M. Conti, Phys. Rev. E 55 (1997) 765.

[68] N.A. Ahmad, A.A. Wheeler, B.J. Boettinger, G.B. McFadden, Phys. Rev. E 58 (1998) 3436.

[69] Ch. Charach, C.K. Chen, P.C. Fife, J. Stat. Phys. 95 (1999) 1141.

[70] Ch. Charach, P.C. Fife, J. Crystal Growth 198/199 (1999) 1267.

[71] G.B. McFadden, A.A. Wheeler, D.M. Anderson, Physica D 144 (2000) 154.

[72] A. Karma, Phys. Rev. Lett. 87 (2001) 115701.

[73] R. Kobayashi, J.A. Warren, W.C. Carter, Physics D 140 (2000) 141.

[74] K. Kassner, C. Misbah, J. Mülller, J. Kappey, P. Kohlert, J. Crystal Growth 225 (2001) 289.

[75] L. Gránásy, T. Börzsönyi, T. Pusztai, J. Crystal Growth 237–239 (2002) 1813.

Dendritic crystal growth in pure materials

M.E. Glicksman*, A.O. Lupulescu

Materials Science and Engineering Department, Rensselaer Polytechnic Institute, 110 8th Street, CII 9111, Troy, NY 12180-3590, USA

Abstract

Dendritic growth is a fundamental crystal growth phenomenon accompanying most casting and solidification processes, and, occasionally, occurring during the growth of single crystals, where it is detrimental to crystalline quality. Dendrites are the ubiquitous crystal form in freezing alloys and supercooled melts, because their shapes are most suited for efficient heat and mass transfer at small scales. Dendritic scales are typically the smallest length scales of interest in ingots and castings, typically associated with: (1) chemical processes, such as microsegregation, (2) thermal processes, for example, latent heat release, and (3) mechanical processes, for instance, the volume change during phase transformation. All of these processes operate at the dendritic solid–melt interface. Understanding dendritic growth is therefore considered essential for controlling basic solidification and crystal growth processes. A brief history of dendrites will be sketched, showing how the subject of dendritic solidification evolved to its present status as a modern sub-field of general crystal growth. The comprehensive understanding of dendrites and developing a predictive capability of practical utility to the crystal grower, however, remain as works in progress. The subject of dendritic growth will be presented on the basis of heat and mass transfer, capillarity effects at the solid–melt interface, and interfacial dynamics, including morphological stability, and side-branching dynamics. Experimental verification of dendritic scaling laws using microgravity experimentation is included as a brief attempt to encapsulate this important subject within crystal growth science.
© 2003 Published by Elsevier B.V.

PACS: 44.25; 51.10; 68.00; 89.02

Keywords: Dendritic growth; Diffusion transport; Segregation; Crystal patterns

1. History and background

1.1. Approach

Websters International Dictionary defines the word "dendrite" as (1) a branching figure resembling a tree produced on or in a mineral; (2) a crystallizing arborescent form. Indeed, in keeping with these definitions, snow flakes and frost patterns are among the

most obvious examples of naturally formed dendritic crystals, the occurrence of which is ubiquitous. The fundamental solidification process to be considered here is dendritic growth. Dendrites (tree-like crystals, from the Greek word $\delta\varepsilon\nu\delta\rho\sigma\nu$) are now known to represent the evolved microstructure of an unstable solid–melt interface [13]. Dendrites, in fact, are the most common form of crystal growth encountered when metals and alloys solidify under low thermal gradients. From a commercial standpoint, dendrites also invariably constitute the crystal form encountered in the manufacture of alloy castings, primary metal in-

---

* Corresponding author. Tel.: + 1-518-276-6721.
  E-mail address: glickm@rpi.edu (M.E. Glicksman).

0022-0248/$ – see front matter © 2003 Published by Elsevier B.V.
doi:10.1016/j.jcrysgro.2003.12.034

gots, and industrial weldments. Except in the restrictive cases of controlled growth of bulk single crystals, where dendritic morphologies are often purposely avoided, dendrites virtually always appear when supercooled melts and solutions are solidified.

Aside from its underlying technical importance and applications in engineering, geology, and biology, dendritic growth also represents a fascinating category of self-organizing pattern formation phenomena, which in recent years has become a deeply researched subject within the broader field of non-linear dynamics. In fact, the current interest among condensed matter physicists in studying dendritic growth is directly attributable to the apparent simplicity of dendritic growth as an important non-linear phase transformation process, and to the extraordinary richness of dendritic behavior observed in experiments and predicted from relatively simple theory.

The purpose of this paper is to develop a contemporary view of dendritic crystal growth from a chronological perspective, and thereby provide a guide to the reader of the rudimentary concepts underlying dendritic crystal growth. We will provide neither an exhaustive annotated bibliography nor a complete history of the subject, but rather a focussed overview of the major phenomena that have been fused together over the past five decades to form the framework of our current understanding of dendritic crystal growth.

1.2. Characteristics of dendritic crystals

In most cases, the formation of dendritic crystals involves the coupling of two different processes: (1) the steady-state propagation of the tip region, accounting for the formation of the main, or primary stem, and (2) the time-dependent crystallization of the secondary and tertiary side branches. These processes establish the most obvious length scales of a dendrite. Fig. 1 shows two typical dendrites of pure materials: on the left, is a dendrite of a body-centered cubic crystal, succinonitrile (SCN) [CN–(CH$_2$)$_2$–CN]; on the right is a dendrite of a face-centered cubic crystal, pivalic acid (PVA), [(CH$_3$)$_3$–C–COOH]. A fully developed dendrite consists of a smooth paraboloidal shaped tip region, behind which occurs a trailing periodic "wake" of branches that spreads away from the primary stem with an opening angle from about 30° to 60° depending on the crystal. In a yet larger field of view (not

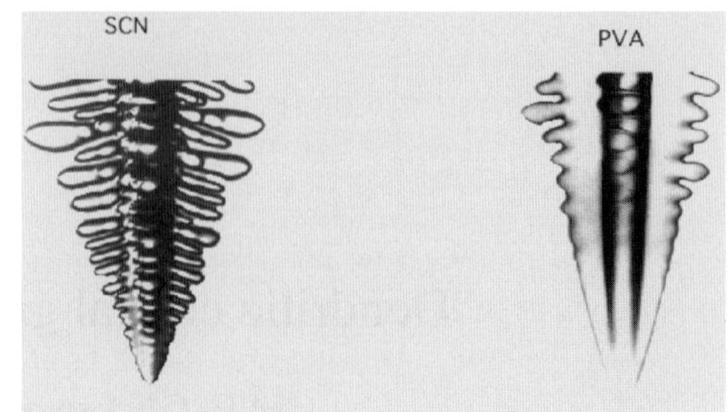

Fig. 1. Steady-state dendritic crystals growing in high-purity supercooled melts: (left) succinonitrile (SCN) a body-centered cubic crystal; (right) pivalic anhydride (PVA) a face-centered cubic crystal.

shown), the open angle of the side branches for these materials will eventually approach 90° because of its underlying cubic crystallographic symmetry.

Curiously, until recently, the time-dependent aspects of dendritic growth were ignored. Most theories were limited to mathematical descriptions of branchless, or so-called "needle crystals" growing at steady state. Secondary and tertiary branches are important insofar as they establish the length scales and pattern over which chemical impurities or alloy components would be concentrated. This microsegregation scale is of immense practical importance in determining the engineering properties of materials that solidify dendritically. In a pure material, on the other hand, such as the crystals illustrated in Fig. 1, the patterns observed during crystal growth are best thought of as revealing "microsegregation" of the enthalpy. Such spatial "enthalpy distributions", in contrast to chemical segregation patterns, disappear entirely when the dendritic crystal growth process is completed.

1.3. Physico-chemical basis for dendritic growth

Dendritic crystal growth is generally acknowledged to be controlled by a diffusion-limited process. For example, in pure materials, the growth rate of a dendrite is controlled by the diffusion of latent heat away from the advancing crystal–melt interface. In addition, one must recognize that molecular or atomic transfer across the crystal–melt interface, as well as the spontaneous creation of the interface itself, requires expenditure of the free energy available for the dendritic transformation. These energetic processes encompass the basic thermophysical phenomena that lead to the formation of dendritic patterns.

Unlike dendrites in pure materials, alloy dendrites propagate as the crystalline solid grows and rejects its excess solute, which flows away from the interface by chemical diffusion through the surrounding melt or solution. In addition to the solute rejection process occurring at the crystal–melt interface and the related chemical diffusion in the melt, the latent heat of fusion is also released from the creation of an orderly crystalline phase from its more random parent phase. Latent heat also must flow away from the dendritic interface by transport processes such as thermal conduction, convection, or radiation. Chemical diffusion, which is generally slower than thermal transport, is often the rate-controlling process in alloy dendrites. In this case, the solute transport equations can be scaled to a form equivalent to that for enthalpy diffusion in pure-material dendrites. However, the presence of two coupled transport fields in alloy melts, as well as the temperature dependence of the equilibrium phase concentrations, complicates the analysis of the process somewhat. Therefore, in reviewing the salient features and theoretical approaches toward modeling of dendritic growth, we focus in this paper on the subject of solidification of pure materials from their melt.

1.4. Thermodynamics and kinetics of dendritic crystal growth

The thermodynamic driving forces and the kinetic resistances encountered in dendritic growth were first clearly described about 30 years ago by Temkin [1] and by Bolling and Tiller [2]. These investigators considered dendritic crystal growth of a pure substance and identified three coupled *kinetic* effects: (1) transport (conduction) of latent heat; (2) molecular attachment at the crystal–melt interface; and (3) creation of interfacial area. Temkin, Bolling, and Tiller associated each of these kinetic effects with the dissipation or consumption of a fraction of the total free energy available for dendritic crystal growth. This total free energy can be expressed as a quantity that is related to the supercooling, $\Delta T = T_{\mathrm{m}} - T_{\infty}$, where T_{m} is the bulk melting point, i.e., the equilibrium temperature at a stationary, planar crystal–melt interface, and T_{∞} is the temperature of the supercooled melt far from the interface. The relationship between the

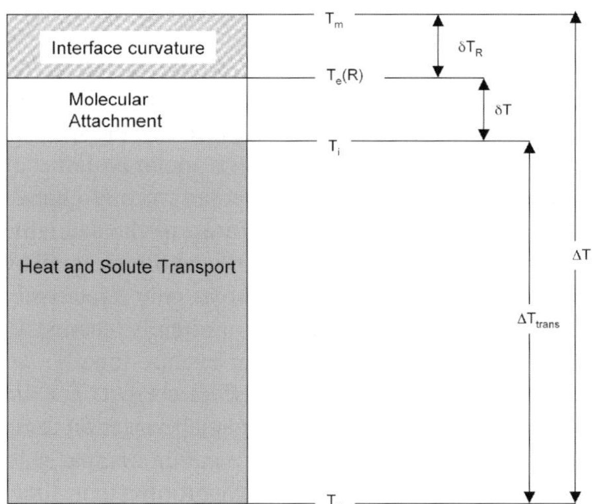

Fig. 2. Distribution of supercooling during steady-state dendritic crystal growth. The total (applied) supercooling is ΔT, which is the difference between the bulk melting temperature, T_{m} and the supercooling temperature, T_{∞}, of the melt. The greatest portion of ΔT, needed for thermal transport, is $\Delta T_{\mathrm{trans}}$. The temperature drop caused by kinetic molecular attachment at the crystal–melt interface is δT. The shift in the equilibrium melting point caused by curvature of the crystal–melt interface is δT_R. Thus, the equilibrium temperature of a curved interface, such as a dendrite tip, is $T_{\mathrm{e}}(R)$. The overall crystal–melt interface temperature is T_{i}.

available free energy and the supercooling is linear in ΔT provided that $\Delta T / T_{\mathrm{m}} \ll 1$—a condition that is well satisfied in general practice. The total free energy is then subdivided among the coupled physico-chemical processes identified above, as is also shown schematically in Fig. 2. In respective declining order of magnitude, the undercoolings used to drive these processes are: (1) $\Delta T_{\mathrm{trans}}$, the temperature drop associated with thermal transport; (2) δT, the supercooling required to "drive" the net atomic or molecular transfer across the crystal–melt interface from the fluid phase to the solid and, thereby, effect attachment to the crystal lattice; and (3) δT_R, the geometrically induced small temperature depression, often called the Gibbs–Thomson capillary temperature shift, which yields the interfacial equilibrium temperature, $T_{\mathrm{e}}(R)$, [3–5,29]. This depression of the equilibrium interface temperature from the presence of excess interfacial free energy at a curved interface may be expressed as

$$\delta T_R = T_{\mathrm{m}} - T_{\mathrm{e}}(R) = \frac{\gamma \Omega}{\Delta S_{\mathrm{f}}} \kappa. \tag{1}$$

Here we define the interfacial curvature, κ, as twice the mean curvature, $\kappa = \kappa_1 + \kappa_2$, where κ_1 and κ_2 are the principal curvatures at a point on the crystal–melt interface; γ, Ω, and ΔS_f are, respectively, the specific excess interfacial free energy, the molar volume of the crystalline phase, and the molar entropy change for melting. As Fig. 2 suggests, most of the available free energy is dissipated by latent heat diffusing away from the dendritic interface, whereas only a relatively small amount of free energy is normally needed to activate the interfacial molecular events. Finally, we note that the Gibbs–Thomson effect corrects for the fact that a curved interface has a small variation in its thermodynamic equilibrium temperature compared to that of a planar interface that is proportional to its local mean curvature.

The description of free energy dissipation just described shows that the process of dendritic growth, even in pure crystals, is complex. For example, the interface temperature, T_i, depends on *both* geometrical effects (through the Gibbs–Thomson relation) and certain additional kinetic details, including the functional relationships among the interface supercooling, δT, velocity, V, orientation, θ, and mobility, M. These kinetic factors are usually expressed as a combined interface kinetic term of the form $\delta T = K(V, \theta, M)$. Every material has an unique interfacial kinetic relationship, but, fortunately, most fall into just a few broad categories. Jackson [6,7] has shown that metals, some ionic compounds, and a few organic materials, such as SCN and PVA shown in Fig. 1, tend to form crystal–melt interfaces that are "rough" on an atomic scale. The so-called rough interfaces easily accommodate atomic or molecular transfer and attachment from *all* interfacial orientations with respect to the principal crystal axes, so δT tends to be extremely small (high molecular mobility, M) and only weakly dependent on orientation, θ. Such materials virtually always crystallize as dendrites. Semiconductors and most covalently bonded materials, on the other hand, display much greater directionality in bonding, and therefore tend to exhibit "smooth," atomically faceted, interfaces. Covalent materials often have a δT that is small in their "rough" orientations and large in the "smooth" or faceted ones. Such materials tend to form faceted dendrites containing internal twinning defects. Finally, polymers and complex network-forming silicate materials have low mobilities so that δT is almost as large as the total supercooling, ΔT, so that transport of heat and species become relatively unimportant components of the overall crystallization process. As a consequence, dendrites seldom ever form in these materials. Although some thermal or constitutional supercooling is *always* required to form dendrites, polymers and complex oxide and sulfide melts can crystallize under extraordinarily large supercoolings in a nearly isothermal, non-dendritic manner.

2. Steady-state dendritic growth

Numerous theoretical [8–10] and quantitative experimental dendritic crystal growth studies have been reported over the past 25 years [11,12,14–19]. Dendritic solidification requires the coupling of two independent growth processes: (1) the steady-state evolution of the dendrite tip, and (2) the non-steady-state development of dendrite branches. The free energy of any system decreases as a crystal freezes from its supersaturated melt. For this to happen, the latent heat generated during crystallization must be carried away from the crystal–melt front by thermal transport. Not surprisingly, "thermal" dendrites are simplest to describe as the crystallization of a pure, supercooled molten phase. By contrast, when alloy dendrites grow from a supersaturated melt, both thermal and solutal boundary layers are involved. Mathematically, however, the dendritic growth problem for pure and alloy melts are essentially identical, consisting of solving: (1) the diffusion equation, (2) boundary conditions of heat and mass conservation at the moving front, and (3) capillary effects introduced at the curved crystal–melt interface.

2.1. Transport theory

The classical theory of "diffusion-limited" dendritic growth is attributed to Ivantsov. The Ivantsov transport solution remains valid for any solidifying system, and may be applied provided that the diffusivity for heat (or solute) is known. Ivantsov modeled the steady-state growth of a dendrite as a smooth paraboloidal body of revolution. Dendrites, of course are not smooth, but contain side branches. Time dependent features, such as side branches, are ignored in Ivantsov's theory [20]. His theory predicts a mathematical relation between a dendrite's tip velocity, V, and its radius of curvature R, as functions of the supercooling (or supersaturation). Furthermore, dendrites are assumed to grow at a steady speed, V, into a pure melt with uniform supercooling, $\Delta T = T_{\mathrm{m}} - T_{\infty}$. A steady-state shape was found, and the crystal–melt interface was assumed to remain at its melting temperature, T_{m}. Ivantsov's transport solution predicts only the growth Péclet number, Pe, as a function of dimensionless supercooling. Ivantsov's solution is usually given as

$$\Delta \vartheta = Pe\,\mathrm{Ei}(Pe)e^{Pe}. \tag{2}$$

In Eq. (2), $\mathrm{Ei}(Pe)$ is the first exponential integral, a tabulated function, with its argument, the growth Péclet number, Pe, defined as

$$Pe \equiv \frac{VR}{2\alpha}. \tag{3}$$

The dimensionless supercooling, or supersaturation, driving the crystallization process is defined as

$$\Delta \vartheta \equiv \Delta T \frac{C_p}{\Delta H_{\mathrm{f}}}. \tag{4}$$

In Eqs. (3) and (4), the materials parameter α represents the thermal diffusivity of the molten phase, C_p is the molar specific heat of the melt, and ΔH_{f} is the molar latent heat of fusion. Each value of the growth Péclet number uniquely corresponds to a given supercooling. However, for a fixed $\Delta \vartheta$, Eq. (2) yields only a *single* relationship between the two independent unknowns: the speed, V, and tip radii, R. This transport relationship is of the form $VR = \mathrm{const}$. Unique values for V and R are, however, not predicted by transport theory alone. Thus, for any supercooling, or supersaturation, an infinite range of velocity and radii combinations satisfy Eq. (2), which is therefore incapable of predicting the specific operating states of a dendrite as observed in the laboratory for various supercoolings.

Theoretically predicted Péclet numbers can be verified experimentally by measuring V and R simultaneously at a sequence of known supercoolings. The normal presence of gravitationally induced convective heat transfer alters the diffusion-limited conditions under which Ivantsov's prediction of the growth Péclet number holds true. The approach taken by the authors to provide a diffusion-controlled environment to mea-

sure dendritic growth will be discussed later in this review.

2.2. Interfacial physics

The growth Péclet number predictions from transport theory can be decomposed into unique speed and tip radius predictions if one introduces an additional equation that provides an independent, second length scale to the problem. The additional length scale combined with the Ivantsov transport solution selects the unique dendritic operating state. Although the physical mechanisms invoked to provide this extra length scale differ in detail in each interfacial theory, their effect may be expressed in terms of a "scaling factor", $\sigma^{\star}$, defined as

$$\sigma^{\star} = \frac{2\alpha d_0}{VR^2}. \tag{5}$$

In Eq. (5) d_0 is the "capillary length", a microscopic quantity equal to circa 10^{-7} cm, which is conventionally defined as

$$d_0 \equiv \frac{2\gamma \Omega C_p}{\Delta H_{\mathrm{f}} \Delta S_{\mathrm{f}}}. \tag{6}$$

In Eq. (6), γ is the crystal–melt specific interfacial energy. It is now accepted that d_0 is the required second length scale in the dendrite growth rate problem. This important discovery was originally proposed by Nash [22,23] in 1974, who developed the modern scaled form of the "dendrite equation".

Remarkably, $\sigma^{\star}$, defined in Eq. (5), is found experimentally to fall in a narrow range for different materials over wide ranges of supercoolings. Thus, Eq. (5) may be written in the form of the dendritic "scaling law",

$$VR^2 \approx \mathrm{const}. \tag{7}$$

Laboratory experiments, to be discussed in Section 3 of this review, verify the unique values of V and R observed experimentally over a range of supercoolings and substances. The explicit relationships for V and R are derived in Ref. [13], and given as

$$V = \frac{2\alpha \Delta H_{\mathrm{f}} \Delta S_{\mathrm{f}}}{\gamma \Omega C_p}(Pe)^2 \sigma^{\star} \tag{8}$$

and

$$R = \frac{d_0}{\sigma^{\star} Pe}. \tag{9}$$

The validity of this overall theoretical approach has a number of interesting implications for applying dendritic growth theory to practical situations, by providing microstructure rules appropriate to dendritic crystal growth processes. In fact, all the stability-based theories, e.g., Oldfield's [21] numerical model, the spherical stability model [17], Langer and Müller-Krumbhaar's paraboloidal model [8,24] eventually reduce to the identical dendritic scaling law, Eq. (7), and to the specific expressions, Eqs. (8) and (9). Theoretical estimates for the interfacial scaling factor, $\sigma^\star$, may be found straightforwardly using marginal stability theory or "microscopic solvability" [25–27]. These approaches all yield similar predictions to the original marginal stability theory, but also take into account the anisotropy of the crystal–melt interfacial energy, and therefore include some information about the crystalline atomic bonds and their symmetry. The predictions may be summarized for materials with nearly isotropic γ-values as

$$\sigma^\star = \frac{1}{4\pi^2} \approx 0.025. \tag{10}$$

In summary, Eqs. (8)–(10) collectively permit quantitative estimation of dendritic growth kinetics in a wide variety of nearly isotropic materials. It is especially interesting to point out that this estimate for $\sigma^\star$ (linear morphological stability theory for a planar interface) is both virtually independent of the assumed geometry of the crystal–melt interface and the physical properties of the system [9].

3. Experimental verification

3.1. Model test systems

Quantitative descriptions of dendritic microstructures, suitable for critically testing theories and hypotheses, have been reported over the past 25 years. These data and observations come almost exclusively from a few critical experiments carried out on well-characterized model transparent systems [11,17]. Under terrestrial conditions, dendrites invariably interact with buoyancy-induced hydrodynamic flows in the melt. Consequently, the basic theory of dendritic growth, i.e., the combination of Eqs. (8)–(10), are of necessity best tested under strictly diffusion-controlled conditions, where gravitational acceleration is reduced to almost zero.

The *isothermal dendritic growth experiment* (IDGE) is a microgravity materials science space flight experiment that was designed to provide terrestrial and microgravity measurements on the kinetics, morphology, and dynamics of dendritic solidification under pure diffusion control [14,17,18,28]. Before the advent of IDGE, it was not possible to test separately and quantitatively the Ivantsov transport solution and the interface scaling factor, $\sigma^\star$. The IDGE instrument was flown three times aboard the space-shuttle orbiter *Columbia,* as part of NASAs periodic USMP-2, -3, and -4 shuttle missions. Flights were carried out in 1994, 1996, and 1997. The IDGE space flight data provide the first solid evidence that Ivantsov's solution describes heat transport during dendritic growth. It has now been tested for dendrites in two, cubic, transparent materials: (1) ultrapure (6–9's) succinonitrile (SCN), and (2) in pure (5–9's) pivalic anhydride (PVA). The two test materials differ markedly in their anisotropy of the crystal–melt energy. Specifically, SCN is nearly isotropic, with about 0.5% anisotropy of γ around its fourfold [1 0 0] zone axis, whereas PVA is extremely anisotropic with almost 10 times as much anisotropy (5%). This difference in crystal–melt anisotropy accounts for their significantly different dendritic morphologies, as shown in Fig. 1. The IDGE flight instruments provided electronic CCD images (as in-flight data), 35 mm films (as post-flight data), and then for the first time on USMP-4, near-real-time full gray-scale video data were streamed to Earth at 30 frames/s.

3.2. Verification of transport theory

On the space flights USMP-2 and USMP-3, data were gathered on dendritic growth speed and tip radius as functions of the supercooling [28,29]. Over 200 experiments were conducted on ultra-pure SCN. These microgravity data were assembled as growth Péclet numbers, $Pe = VR/2\alpha$ (see again Section 2.1) along with comparable speed and radius data taken under ordinary terrestrial conditions (i.e., at unit gravity, $g = 9.8$ m/s^2). The measurements taken both under microgravity conditions and under terrestrial conditions are compared with Ivantsov's theory in Fig. 3. It is clear that under microgravity conditions the ob-

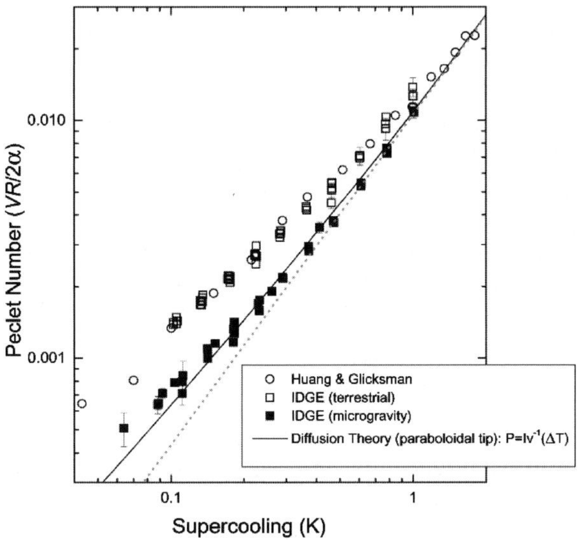

Fig. 3. Growth Péclet number versus supercooling. The Péclet number is calculated from measurements of dendritic tip speed, V, and tip radius, R, knowing the thermal diffusivity, α. The solid line is predicted from Ivantsov's transport theory for paraboloidal dendrites. Open data symbols were collected under terrestrial conditions, with convection present in the melt. Solid data symbols were obtained under convection-free microgravity conditions.

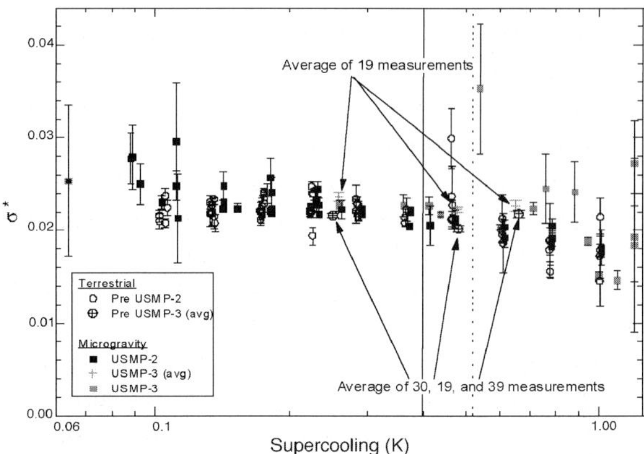

Fig. 4. Plot of the scaling factor, $\sigma^{\star} \propto 1/V R^2$ versus supercooling. These data show that the scaling law, $V R^2 \approx$ const, holds over a wide range of supercoolings, under both terrestrial and microgravity conditions.

served Péclet numbers are in basic agreement with theory. It is important to note that there are not any adjustable parameters in this comparison. The terrestrial data, taken for dendrites growing parallel to gravity, remain well above those measured in microgravity, where convective effects in the melt are eliminated. This data set constitutes the most exacting test of thermal transport during dendritic growth carried out to date.

3.3. Verification of interfacial physics

Similar data sets were then assembled to calculate $\sigma^{\star} \propto 1/V R^2$. As explained in Section 2.2, contemporary theories of dendritic growth [8,9,25] show that $\sigma^{\star}$ should be invariant with the supercooling. All the applicable dendritic growth data—including speed and radii measured under terrestrial and microgravity growth conditions—are shown as values of the dendritic scaling factor, $\sigma^{\star}$ in Fig. 4. These data prove conclusively that the dendritic scaling law, $V R^2 =$ const is valid *both* under microgravity and terrestrial crystal growth conditions. It is interesting that the hy-

drodynamic state of the melt has negligible influence on this scaling law. This behavior stands in sharp contrast with the Péclet number behavior discussed in the previous section. Except for a slight downward drift with increased supercooling, the values of $\sigma^{\star}$ are virtually constant, and independent of both the supercooling and the convection state of the melt. These experiments serve to verify the robustness of the dendritic scaling law, allowing its application to a wider variety of materials and crystal growth processes.

4. Summary and conclusions

1. The basic theory of dendritic growth is shown to consist of two components: (a) diffusive transport theory that describes how heat and (if present) solute are redistributed around a growing dendrite; (b) interfacial stability theory that leads to the scaling relationship that $\sigma^{\star} \propto 1/V R^2 \approx$ const.

2. When combined, the theory components labelled (a) and (b), above, yield quantitative predictions for the steady-state dendritic speed, V, and tip radius, R, as functions of the supercooling. Such estimates are in reasonable agreement with experiments conducted under microgravity conditions, where convection is absent.

3. The interfacial stability theory prediction that $V R^2 \approx$ const is confirmed quantitatively, insofar as the value of $\sigma^{\star}$ is almost independent of

the supercooling, gravitational level, and material. Robustness of this crucial scaling law allows reliable estimates to be made of the response during crystal growth.

4. Microgravity experiments have directly verified Ivantsov's thermal transport solution for paraboloidal dendrites. Péclet number data for SCN are in agreement with predictions based on Ivantsov's theory.

5. The detailed shapes of PVA dendrites, and their micromorphologies are different from those for SCN dendrites, due primarily to their differing anisotropies of the interfacial energy, γ. Both test materials differ from the ideal, smooth, paraboloids of revolution assumed in the transport theory, the differences being primarily in their tip shapes. Although tip shape effects do indeed impose quantitative influences on the transport behavior they do not change the basic agreement with transport theory.

6. Fifty years of research into dendritic growth has transformed this topic in crystal growth and pattern formation from a purely qualitative, descriptive one to a highly predictive and quantitative one. Dendritic scaling laws, which derive from fundamental theory, are being used for engineering applications in the field of casting and welding as well as for further scientific enquiry. As modeling efforts continue, further progress will unfold concerning our understanding and control of this ubiquitous form of crystallization.

Acknowledgements

The authors are pleased to acknowledge financial support received from the National Aeronautics and Space Administration, Code U, Office of Biological and Physical Sciences, Washington, DC, under the Rensselaer Isothermal Dendritic Growth Project, NAG8-1696.

References

[1] D.E. Temkin, Dokl. Akad. Nauk SSSR 132 (1960) 1307.
[2] G.F. Bolling, W.A. Tiller, J. Appl. Phys. 32 (1961) 2587.
[3] J.W. Cahn, D.W. Hoffman, Acta Metall. 22 (1974) 1205.
[4] D.W. Hoffman, J.W. Cahn, Surf. Sci. 31 (1972) 368.
[5] M.E. Glicksman, J. Crystal Growth 42 (1977) 347.
[6] K.A. Jackson, B. Chalmers, Can. J. Phys. 34 (1956) 473.
[7] K.A. Jackson, in: R.H. Doremus (Ed.), Growth and Perfection of Crystals, Wiley, New York, 1958, p. 319.
[8] J.S. Langer, H. Muller-Krumbhaar, Acta Metall. 26 (1978) 1681.
[9] J.S. Langer, in: Critical Problems in Physics, in: Princeton Series in Physics, Princeton University Press, Princeton University, 1996, Chapter 2.
[10] W.W. Mullins, S.F. Sekerka, J. Appl. Phys. 34 (1963) 323.
[11] M.E. Glicksman, R.J. Schaefer, J.D. Ayers, Metall. Trans. A 7A (1976) 1747.
[12] M.E. Glicksman, in: E. Kaldis (Ed.), Crystal Growth of Electronic Materials, Elsevier, Amsterdam, 1985, pp. 57–69 (Chapter 5).
[13] M.E. Glicksman, S.P. Marsh, in: D.T.J. Hurle (Ed.), Handbook of Crystal Growth, Elsevier, Amsterdam, 1993, pp. 1075–1122.
[14] M.E. Glicksman, M.B. Koss, V.E. Fradkov, R.E. Rettenmayr, S.S. Mani, J. Crystal Growth 137 (1994) 1.
[15] M.E. Glicksman, M.B. Koss, L.T. Bushnell, J.C. LaCombe, E.A. Winsa, ISIJ Int. 35 (6) (1995) 604.
[16] M.E. Glicksman, M.B. Koss, L.T. Bushnell, J.C. LaCombe, E.A. Winsa, Mater. Sci. Forum 215–216 (1996) 179.
[17] S.C. Huang, M.E. Glicksman, Acta Metall. 29 (1981) 701.
[18] J.C. LaCombe, M.B. Koss, V.E. Fradkov, M.E. Glicksman, Phys. Rev. E 52 (3) (1995) 2778.
[19] R. Trivedi, J.T. Mason, Metall. Trans. A 22 (1991) 235.
[20] G.P. Ivantsov, Dokl. Akad. Nauk USSR 58 (1947) 56.
[21] W. Oldfield, Mater. Sci. Eng. 11 (1973) 211.
[22] G.E. Nash, M.E. Glicksman, Acta Metall. 22 (1974) 1283.
[23] G.E. Nash, M.E. Glicksman, Acta Metall. 22 (1974) 1291.
[24] R.D. Doherty, B. Cantor, S. Fairs, Metall. Trans. A 9 (1978) 621.
[25] D. Kessler, H. Levine, Phys. Rev. B 33 (1986) 7867.
[26] D. Kessler, H. Levine, Phys. Rev. Lett. 57 (1986) 3069.
[27] D. Kessler, H. Levine, Acta Metall. 36 (1987) 2693.
[28] M.B. Koss, L.A. Tennenhouse, J.C. LaCombe, M.E. Glicksman, E.A. Winsa, Metall. Mater. Trans. A 30 (1999) 3177.
[29] M.E. Glicksman, A. Lupulescu, Dendritic Growth, in: D.M. Stephanescu, R. Ruxanda, M. Tierean, C. Serban (Eds.), Proceedings of the International Conference on the Science of Casting and Solidification, Editura Lux Libris Brasov, Romania, 2001, pp. 15–21.

The first Czochralski silicon

Ernie Buehler

Bell Laboratories, Murray Hill, NJ

The first Czochralski growth of silicon is generally credited to G.K. Teal and E. Buehler of Bell Laboratories. Their short paper describing these early experiments was published in Phys. Rev. 87 (1952) 190. The paper reported that the pulling technique was generally similar to that used previously for germanium and was preferred over solidification in a container since it resulted in more crack- and twin-free crystals. Crystals up to 5″ in length and 1″ in diameter had been grown at the time of publication. Carrier lifetimes up to 200 μs and carrier mobilities of 1200 cm² / Vs, four times higher than in previously available materials, were reported.

The first silicon single crystal ingot was pulled from the melt at Bell Labs in 1949 by the Czochralski method. In retrospect the technique had a tremendous impact in making it possible to provide reproducible single crystal raw material to a rapidly growing semiconductor solid state device industry. What can be said about how such a technique becomes a major milestone in crystal growing endeavors? How much credit would one attribute to the *individuals* who were responsible for generating the ideas in design, implementation and growth of silicon crystals? How much influence did an *ideal working environment* have on motivating people? Was it essential that the involved researchers be endowed with a "*common cause*" attitude? While all three factors were important in the origin of single crystal silicon perhaps the most significant was the "*common cause.*" In simplest terms the basic driving force ("a common cause") was a goal set by management many years before to obtain a *working solid state amplifier.*

As early as 1936 Bell Labs Management set as one of its research goals the development of a solid state amplifier (M.J. Kelly). Oxide semiconductor research yielded such devices as thermistors and limited performance rectifiers. (Pearson, Green, Becker, Brattain, Storks, The Christensens). Polycrystalline ingots of germanium and silicon grown by the Bridgman method evolved during World War II resulting in high frequency solid state rectifiers and early transistors. (Ohl, Scaff, Theurer, Pfann, Brattain, Bardeen and Shockley). The demand for higher purity, more uniform large grain single crystal germanium and silicon ingots snowballed. Gordon Teal then successfully applied the Czochralski pulling method to germanium. (Teal, Little, Gomperez and Zinc). Thereafter the same procedure with extensive thermal modifications was used with equal success on silicon.

The modifications required extensive changes to the apparatus but pressure to attempt the first pulling experiments on silicon had been mounting. On the day the first crystal was grown, Gordon Teal was adamant, "We're going to do it today" (even though he was going to a cocktail party that evening). The apparatus was not ready until about 4:00 p.m., and although I wasn't feeling well at the time, I carried on with the run and the first sample was actually finished around midnight. It is now in the Smithsonian Museum.

Previously published in AACG Newsletter Vol. 13(2) July 1983.

Incidentally, although the very first silicon ingot pulled out of a melt was polycrystalline (Teal, Buehler; 5/16″ in diameter × 5/8″ long), it revealed that this procedure would yield single crystals with very little additional extension. When crystals were grown up to 3″ in length, self-selection would occur among the grains of the polycrystalline seed, and single crystal regions could be grown.

References to publications relevant to the large semiconductor research and development effort in progress at that time are too numerous to be included in this short account. These are available in library computer bibliographies. More importantly silicon single crystal ingots came about through the efforts of dedicated individuals, a good working environment, and a "common cause."

How zone melting was invented

W.G. Pfann

Bell Laboratories, Murray Hill, NJ

Zone melting was conceived not once, but twice. In the late 30's I was a laboratory assistant in a small group of physical metallurgists in the Research Department of Bell Labs. My main job was preparing metals and alloys for microscopic examination. Evenings I attended New York's Cooper Union, working toward a B.Ch.E. degree, but I was already leaning strongly toward research, and away from engineering. Even as a lowly lab assistant I was thrilled by the idea of doing RESEARCH.

In 1939, Earle Schumacher, director of the group, did an amazing thing. He said, "Bill, take half your time, and do research on anything you want." Hard to believe. I grew some foot-long lead single crystals by the Bridgman method. I wanted to deform them, and study the slip bands. I wanted to add antimony, and knew it would segregate, so I thought to zone level it, although I did not know those words "zone level" then. I would just melt a short length, add antimony and pass it along the crystal, so as to spread out the antimony uniformly. We then were plunged into urgent defense activity, and my short research holiday was over. I forgot about the zone melting, as it seemed so obvious that I thought everyone knew it.

During WWII I became a full-fledged member of this same Metallurgy Research Department, in the group headed by Jack Scaff, which played a key role in providing ingots of germanium and silicon, and in developing point contact rectifiers for use in radar receivers. Late in 1947 the first transistor was made, and we became much in demand. Henry Theuerer provided directionally frozen, polycrystalline ingots of germanium, and worked to improve them. I devised a housing for the point-contact transistor, and we made hundreds of them.

Bill Shockley put intense pressure on materials people to provide single crystals of germanium of unheard-of purity. Gordon Teal and John Little provided the single crystals by their well known pulling method, but the segregation and purity problems remained. I had no responsibility for the material, but I worked very closely with Henry and was keenly aware of the severe segregation problem. The 1939 zone leveling idea reoccurred to me, and I began to build apparatus for zone leveling germanium, still unaware that it was a unique idea.

Just then another colleague spoke to me about his idea for applying fractional crystallization to germanium; that is, directionally freeze, pour off last 5 or 10% of the liquid, remelt, and repeat 4 or 5 times without removing crucible from furnace. I told him it sounded like a lot of work. That lunch hour I leaned back in my chair, head on windowsill, and began my daily 10 or 15 minute catnap, in which I never actually slept, but just relaxed, and I suddenly saw the answer to this fractional crystallization problem. Pass a long ingot through a series of heating coils! (I remember the feet of the chair hitting the floor with a loud clack as I jumped up.) I ran next door and told Henry about it. He thought it over for about 5 seconds and said, "That's it! You've done it, Bill." (Other coworkers had serious doubts for months, believe it or not.)

Previously published in AACG Newsletter Vol. 12(3) November 1982.

Success in the zone refining of germanium was immediate and spectacular. Success with zone leveling followed shortly as we showed, using radioactive antimony as the desired solute. (A low-k solute such as antimony segregates the most in directional freezing, but the least in zone-leveling.) Once the lid was off, other variations came rapidly, e.g. temperature-gradient zone melting, zone remelting, continuous zone refining. An unexpected dividend came from the zone leveling work. It was like falling into the lake and coming up with fish in your boots. I casually asked my assistant, Dan Dorsi, to use a seed crystal to grow a zone-leveled *single crystal* of germanium. I thought it would be easy, but it took a year and a half of hard work. Low-angle boundaries, which reduced the minority carrier lifetime were very persistent. To sum up, we showed that the low angle boundaries were tilt boundaries comprised of edge dislocations which were revealed as etch pits, thereby setting off a large activity of fundamental research on dislocations, and also providing a tool for the quality control of single crystals.

Meanwhile Henry Theuerer had turned his attention to silicon, which could not be zone refined in a container. He conceived the floating zone technique, reduced it to practice, and was awarded the basic patent. This important invention not only extended the range of zone melting applications, but also became a valuable crystal growth method in itself.

Analysis: I was trying to solve segregation using the zone idea (zone leveling), and by chance was made to think of the purification idea (fractional crystallization). In a semiconscious moment I saw how to use the zone idea to solve the purification problem. That was the origin of zone refining.

A lot has been said about creativity. Some think it can be taught. I think not. The creative act is an ephemeral thing. It requires first of all hard work and total absorption in the subject. Well, many qualify for this. But it also requires an uncanny ability to connect disparate ideas. I say uncanny because the process certainly does not appear logical. And I say ephemeral because there's just that fleeting moment, triggered by accident, in which the connection is made. In the above account, the trigger was the suggestion, by a colleague, of his idea for fractional crystallization of germanium, of which I disapproved, and which left me with a troubled, incomplete feeling. It is interesting to think that this whole chain of events began when an astute director trusted his instincts, and gave a young, unlettered lab assistant time to do research on whatever he wanted.

A brief history of defect formation, segregation, faceting, and twinning in melt-grown semiconductors

D.T.J. Hurle [a,*], P. Rudolph [b]

[a] *University of Bristol, Scotscraig House, Storridge, Malvern, Worcs WR 13 5EY, UK*
[b] *Institut für Kristallzüchtung, Max-Born-Str. 2, Berlin 12489, Germany*

Abstract

A historical review of the development of knowledge of defect formation in semiconductor crystals is given. The treatment starts with zero-dimensional defect types, especially native point defects in Si and GaAs. One-dimensional structural disturbances—dislocations and their patterning—are discussed next. Whereas in Si the total elimination of extended dislocations is well established, in semiconductor compounds, like III–Vs with low critical resolved shear stress, this seems to be impossible. In a further section micro- and macro-segregation phenomena—striations and the effects of constitutional supercooling—are reviewed. Finally, two-dimensional features are discussed. First the interplay between facets and inhomogeneous dopant incorporation is described. Then the problem of twinning, especially in InP, is outlined. The paper is focused on the grassroots from the beginning of the 1950s—the birth of semiconductor melt growth. For each defect type the current state of knowledge and methods of control are indicated. Problems remaining to be solved in the future are summarised.
© 2004 Elsevier B.V. All rights reserved.

PACS: 81.10F; 61.66; 61.72; 61.72J; 61.72F; 81.05D; 81.10A

Keywords: A1. Characterization; A1. Defects; B2. Semiconducting gallium arsenide; B2. Semiconducting silicon

1. Introduction

The development of the transistor in 1948 [1] and, later, the integrated circuit placed unprecedented demands for the enhancement and control of the perfection of semiconductor crystals. In this review we explore the history of the development of understanding and control of chemical and structural inhomogeneities in melt-grown bulk crystals. Strong emphasis is given to semiconductor crystal growth since it is from this class of materials that most has been first learned, the resulting knowledge then having been applied to other classes of materials, notably to refractory oxides and other inorganic materials used in optics and electro-optics.

2. Native point defects

About 50 years ago the understanding of intrinsic point defects and their role in crystals progressed markedly. After Huntington and Seitz [2] had calculated for the first time the defect formation energy in copper, Zener [3] and Le Claire [4] estimated the va-

---

* Corresponding author. Tel.: +44-1886-880348.
E-mail address: donhurle@iname.com (D.T.J. Hurle).

0022-0248/$ – see front matter © 2004 Elsevier B.V. All rights reserved.
doi:10.1016/j.jcrysgro.2003.12.035

cancy density in FCC crystals. Shockley compiled [5] a pioneering review of semiconductor defects. Later Kröger [6] published his fundamental compendium on defects in semiconductor compounds which remains to today one of the basic guides for the crystal grower. It was shown that, at all temperatures above absolute zero, equilibrium concentrations of vacancies, self-interstitials and, in the case of compound semiconductors, anti-site defects will exist. This is because point defects increase the configurational entropy leading to a decrease in free energy of a crystal. Hence, a crystallising system is composed of two opposite processes—regular (enthalpy part), and defective incorporation (entropy part) of the "growth units". The entropic contribution to the defect concentration varies exponentially with temperature. Considering this combination of ordering and disordering forces we understood that it is not possible to grow an absolutely perfect crystal. In reality no "ideal" but only an "optimal" crystalline state can be obtained. In other words, in thermodynamic equilibrium and for uncharged native point defects, the crystal perfection is limited by incorporation of a given vacancy and/or interstitial concentration n, having the form:

$$n = N \exp(-E_d/kT), \tag{1}$$

where E_d is the defect formation energy, N is the total number of possible sites, k is the Boltzmann constant.

In the case of formation of vacancy-interstitial complexes, i.e. Frenkel defects, n is somewhat modified to:

$$\sqrt{N_i N} \exp(-E_d/kT), \tag{2}$$

where N_i is the total number of interstitial positions depending on the given crystal structure.

Setting $N \approx N_A$ (Avagadro's constant) $= 6 \times 10^{23}$ mol^{-1} and $E_d = 1$ eV in Eq. (1) the equilibrium (i.e. minimum) native point defect concentrations at 1000 and 300 K are about 6×10^{18} and 6×10^7 mol^{-1}, respectively. Hence, a quite low equilibrium point defect content is estimated for room temperature. In reality, however, the limitation of finite defect diffusion rate leads to the freezing-in of a considerable fraction of the high-temperature defect concentration as the crystal cools down. As a result the existing native point defect content exceeds markedly the equilibrium concentration at room temperature.

As we can see, the equilibrium concentrations of each point defect are dependent on the crystal growth conditions. Hence, the crystal grower must obtain an understanding of native point defect equilibrium in order to carry out accurate defect engineering.

In recent years interest has been increasingly focussed on these native point defects in silicon because of their role in the formation of unwanted oxidation-induced stacking faults (OSF). A good understanding of this problem has now been obtained and described by fairly sophisticated modelling [7,8]. Generally, at the melting point the concentration of vacancies and interstitials in silicon is about 10^{14}–10^{15} cm^{-3}. At high pulling rates vacancies are incorporated in excess and condense during cooling down to form octahedral voids of ~ 100 nm in size which were first observed by Itsumi et al. [9] using transmission electron microscopy. At low pulling rates interstitials are in excess forming a network of dislocation loops. In between, a defect-free region is obtained which is bounded by the OSF ring (Fig. 1). The balance between the number of vacancies and interstitials is the controlling factor.

There are three ways to obtain micro-defect-free silicon:

(i) the growth of defect-free crystals by keeping the growth conditions within the defect-free regime which is approximately $\pm 10\%$ around the critical ratio $R/G = 1.34 \times 10^{-3}$ cm^2 K^{-1} min^{-1}, where R is the growth rate and G is the temperature gradient at the interface. But such a small tolerance permits only very low pulling velocities of about 0.5 mm min^{-1}. Falster and Voronkov [10] described the in situ out-diffusion of interstitials. In this case, the crystals are pulled under interstitial-rich conditions and maintained at high temperatures for extended times thus utilising the very high migration speed of interstitials. However, extended cooling times with very low cooling rates are required;

(ii) keeping a maximum pulling rate with fast cooling followed by a wafer annealing process to reduce the grown-in defect sizes;

(iii) using a new cost-optimised approach, a so-called "flash wafer" step [8], where only a thin Si layer of 0.5 μm is deposited onto the wafer surfaces. That combines maximum pull rate and fast cooling with low-cost treatment.

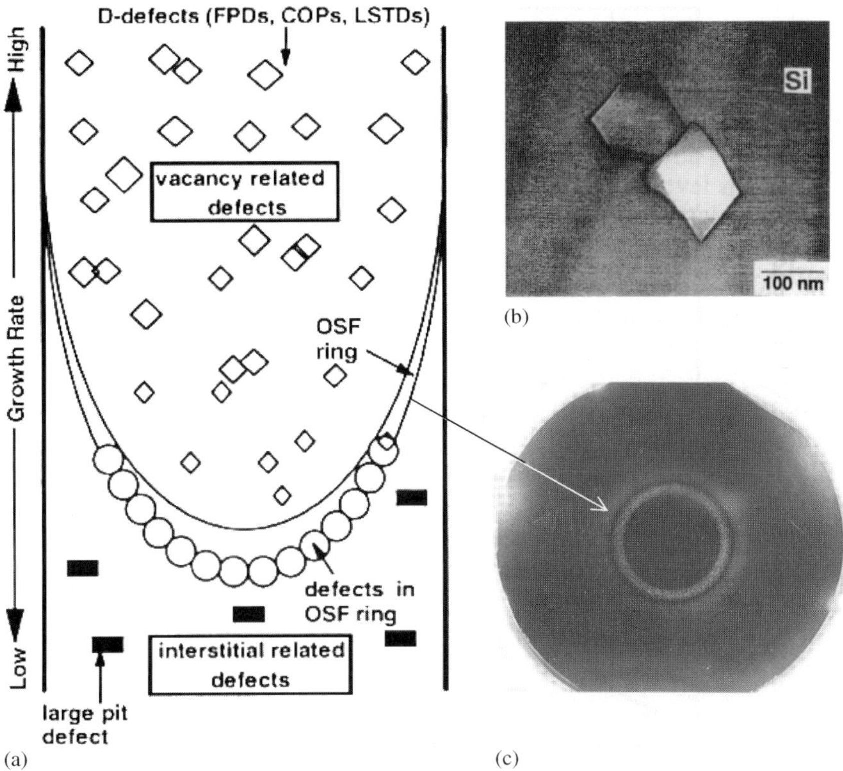

Fig. 1. Voids and interstitials in Si: (a) Sketch of the defect behavior in silicon crystals as a function of pulling rate [8], (b) TEM micrograph of two overlapping voids [9], (c) OSF ring (E. Dornberger, Thesis, University of Catholique de Louvain, 1999).

More complicated and less studied is the situation in the III–V compounds [11] and even less in the II–VI ones [12]. The equilibrium point defect concentrations at the melting point tend to be much higher in the compound semiconductors than in Si and Ge. At such high temperatures the point defects are isolated and usually electrically charged influencing the Fermi level position. (It is usually and, in our view, often erroneously assumed that electron–hole pair formation by bond breaking determines the Fermi level [11]).

The density of the ionised fraction depends on the type of charged carrier (n or p) and on the carrier concentration. This correlation is often referred to as the Fermi-level-effect [13]. For instance, the fraction of vacancies V^{z-} in the charge state z^- is then

$$\frac{V^{z-}}{V_0} = \exp\frac{zE_F - \sum E_i}{kT}, \qquad (3)$$

where V_0 is the concentration in the uncharged state, z is the charge, E_F the Fermi energy, E_i $(i = 1 \ldots z)$

are the ionisation levels of the vacancy and T is the absolute temperature.

One of the biggest experimental challenges is the analysis of point defect types and concentrations. Possible methods include precise measurement of density and lattice constant [14], vapour pressure scanning [15], coulometric titration [16] and positron annihilation [17]. Bublik et al. [14] compared the actual mass per unit cell, calculated from measurement of crystal density and lattice parameter, with that of an ideal stoichiometric crystal. Oda et al. [18] used coulometric titration which directly determines the deviation from stoichiometry. The data are compared in Fig. 2. A good agreement can be observed. As can be seen from this curve, stoichiometric GaAs will be obtained only from a markedly Ga-rich melt. At mole fractions of As in the melt higher than 0.47 arsenic interstitials are in excess.

However, for defect concentrations below $\sim 10^{17}$ cm^{-3} these measurement techniques begin to fail.

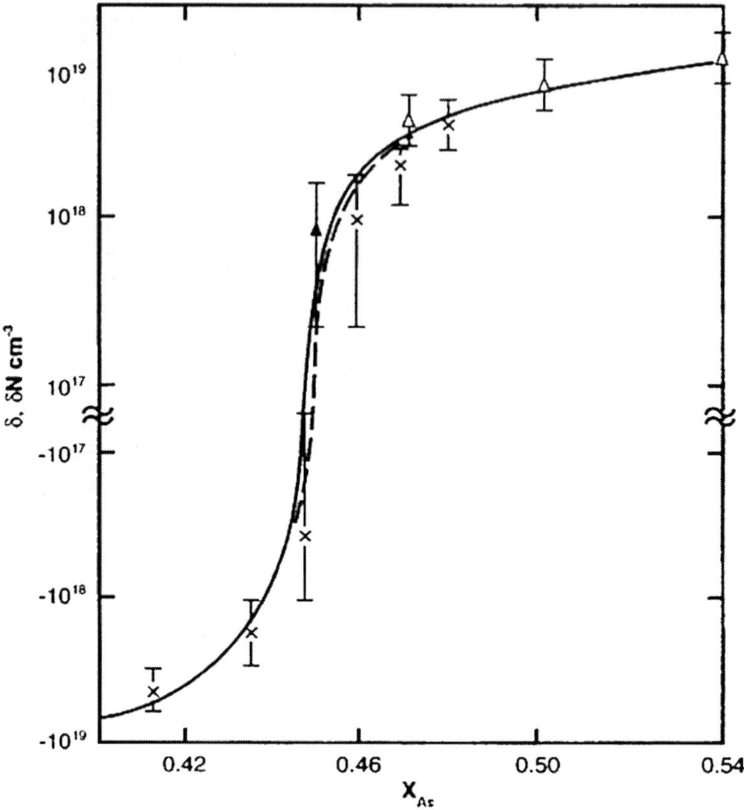

Fig. 2. Bublik's data on density and lattice parameter (crosses) and Oda's titration data (triangles) showing the excess concentration of As-vacancies/interstitials vs. melt composition [11].

Hurle [11,19] has pioneered an additional approach—the use of equilibrium thermodynamics to fit chemical dopant solubility data. Since the Fermi level position influences and is influenced by the concentrations of both electrically active dopants and charged native point defects, by fitting to room temperature carrier concentration data it is possible to calculate the concentrations of charged native point defects grown into crystals containing known concentrations of dopant atoms provided that charge states are known (e.g. from positron annihilation studies [17]).

The behaviour of doped material in compound semiconductors is particularly complicated by the fact that some dopants are amphoteric. Not only can they substitute on both sub-lattices of a binary compound, they can also form complexes with charged vacancies to produce compensating deep levels [11]. The interaction of native point defects with dislocations can

have a dramatic effect on the lifetime of lasers in some III–V compounds [20].

Ab initio theoretical treatments by computer modelling have come to the fore in recent years (see for example Ref. [21]). These usually seek to calculate system total energy using a pseudopotential description and a local density functional approximation. This enables one to calculate the energies of the different possible configurations that a defect (such as an interstitial for example) can take up. However, since entropic contributions are not included, results strictly apply only at the absolute zero of temperature. In fact good agreement with experimental data obtained from high temperature growth processes is not generally obtained [11] possibly as the result of the neglect of entropy.

One of the most serious consequences of compound crystal growth under conditions of native point defect formation is their condensation in precipitates and mi-

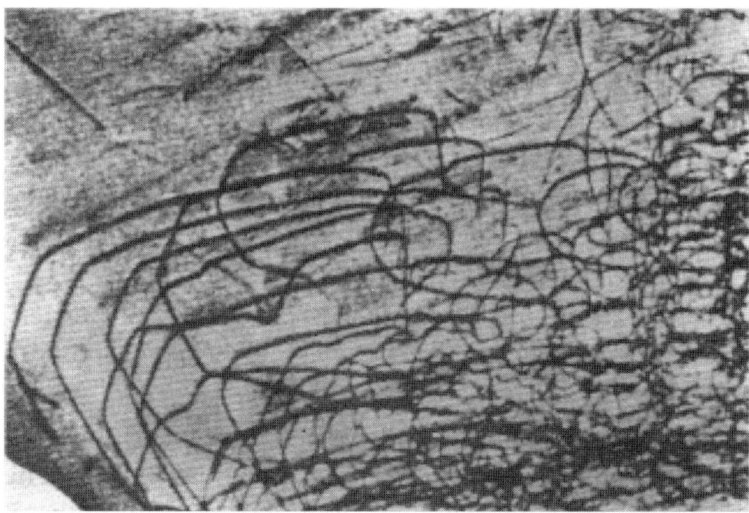

Fig. 3. X-ray topography of dislocations in a silicon crystal analysed by Lang in 1959 (courtesy of A.R. Lang).

crovoids affecting the crystal quality. This phenomenon is due to the retrograde behaviour of the boundary of the compound existence region and, therefore, related to non-stoichiometry. Practical measures of point defect engineering during melt growth by in situ control of stoichiometry were carefully analysed for GaAs, e.g. by Nishizawa [22] and for CdTe and ZnSe e.g. by Rudolph and co-workers [23,24]. Oda et al. [18] proposed post-growth wafer annealing as the most effective controlling step for producing stoichiometric and, hence precipitate-free, GaAs substrates.

3. Dislocations

The growth of Ge and Si single crystals in 1950 by Teal and Little [25] and in 1952 by Teal and Buehler [26], respectively, using a development of the pulling technique devised by Czochralski [27], produced the first melt-grown semiconductor crystals having a sufficiently low dislocation density for it to be possible to image individual dislocations. The concept of the dislocation was well established by then [28] and the role of dislocations on the kinetics of growth was just being expounded by Burton et al. [29]. The emergence of individual dislocations at crystal surfaces were revealed by chemical etching, first in 1953 on Ge [30], and 1956 on Si [31]. Imaging dislocations internally became possible shortly afterwards with the invention

by Lang of X-ray projection topography [32]. (Fig. 3 shows an image by Lang from 1959 with dislocation patterns in a silicon crystal.) This enabled individual dislocations to be imaged in thin wafers where the dislocation density was sufficiently low. Using the criterion of image disappearance [33] Burgers vector analysis became possible:

$$\vec{g} \cdot \vec{b} = 0, \tag{4}$$

where $\vec{g}$ is the diffraction vector and $\vec{b}$ is the Burgers vector.

Application of this technique to Ge and more especially to Si wafers revealed that the dominant dislocation formation mechanism was slip produced by the thermal stresses generated during cooling of the grown crystal to room temperature.

Dash [34] showed that all the dislocations propagated from the seed crystal could be removed by rapid initial growth of a tapered-down extension of the seed (so-called "Dash necking"). In the absence of any dislocations, the crystal lattice became much stronger and the crystal diameter could be extended to large values without the introduction of any new dislocations. Fig. 4 shows the necking effect in an early Si floating zone crystal.

So successful is this technique that freedom from dislocations has been part of the Si wafer specification for many years now. It is attainable right up to current production of dislocation-free 300-mm and development of even 400 mm diameter silicon crystals [35].

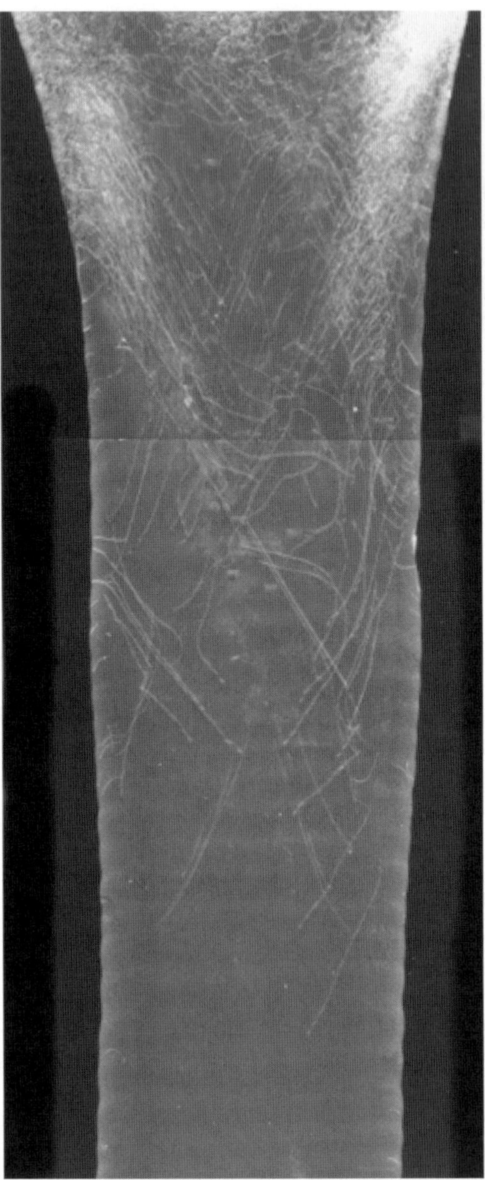

Fig. 4. The neck region with out-growing dislocations in an early FZ Si crystal taken by X-ray topography (courtesy of W. Schröder, H. Riemann, A. Alex, IKZ, Berlin).

Again, the situation in semiconductor compounds is more complicated due to yielding as a result of the much lower critical resolved shear stress (CRSS) near the melting point for GaAs and CdTe of only 0.5 and 0.2 MPa, respectively. Nevertheless, using Dash necking with the hot-wall pulling technique without boric oxide encapsulant, Steinemann and Zimmerli [36] succeeded in 1967 to grow small dislocation-free undoped GaAs crystals of diameters below 20 mm. However, due to the proportionality between diameter and acting shear stress the production of bigger dislocation-free compound semiconductors has not yet been possible. Combining the thermal stress in a cylindrical growing crystal with the basic $\langle 1\,1\,0 \rangle$ $\{1\,1\,1\}$ glide system of the zinc-blende structure the Schmidt contour can be theoretically calculated [37] which shows that the acting stress relaxes by radial glide along the $\langle 1\,1\,0 \rangle$ directions. This was very clearly demonstrated theoretically and experimentally for GaAs by Jordan et al. [38].

It was Billig [39] in 1956 who discovered that the dislocation density of Ge crystals was correlated with the imposed temperature gradient. Indenbom [40] demonstrated that thermally induced stresses arise from temperature non-linearity, i.e. divergence of the isotherm from an idealised planarity. Theoretically, that implies the simplified, but for the crystal grower quite useful, formula

$$\sigma = \alpha_T E L^2 \left(\partial^2 T / \partial z^2 \right) \approx \alpha_T E \delta T^{\mathrm{max}}, \qquad (5)$$

where σ is the thermal stress, α_T the coefficient of thermal expansion, E Young's modulus, L characteristic length (approximately the crystal diameter), T the temperature, z the given coordinate (pulling axis), and δT^{max} the maximum deviation of the isotherm from linearity. From this formula one can see that only a very small isotherm deviations from planarity δT^{max} of 1–2 K is needed to exceed the CRSS for dislocation multiplication in III–V and II–VI compounds.

Hence knowledge and control of the temperature field at all process stages are of essential significance. Due to the practical difficulties of experimental measurement, numerical simulations are becoming more and more important for heat flow analysis and "tailoring" of the growth process. The excellent analysis of dynamic deformation behaviour of diamond-like crystals made by Alexander and Haasen [41] in 1968 is used to estimate the local dislocation density from the constitutive law linking plastic shear rate and dislocation density with the stress generated in the course of the crystal cooling procedure. A profound review has been given by Völkl [42].

One of the most interesting structural features in semiconductor compounds, which is not yet com-

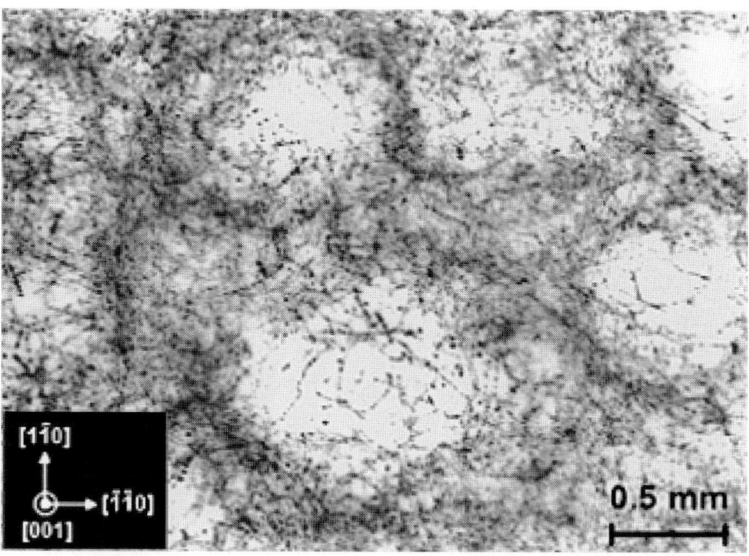

Fig. 5. Dislocation cells in a GaAs crystal decorated by As precipitates. The image was taken by laser scattering tomography [43].

pletely clarified, is the patterning of dislocations into a cellular network. Fig. 5 shows such cells in a GaAs crystal, decorated by As precipitates and revealed by laser scattering tomography [43]. The cells are of globular-like shape. Their size decreases with increasing average dislocation density yielding diameters of 1–2 mm at a dislocation density of $\leqslant 10^4$ cm^{-2} and of < 500 μm at a density of 10^5 cm^{-2}. Characteristic features of classic polygonized networks [44] are observed only in the cell walls. Probably, a superposition of a macroscopic self-organised cell formation and a mesoscopic substructuring of the cell walls by dynamic polygonization has to be considered as discussed in newer concepts of dynamic interaction between dislocations in ensembles (e.g. Ref. [45]). Maybe this will provide an explanation of the strange observation of cell absence in InP.

4. Micro- and macro-segregation

The phenomena of micro- and macro-segregation are both based on the effect of different solubilities of foreign atoms (impurities) in the solid (crystal) and mother phases (melt). Pfann did pioneering work to find out experimentally the segregation coefficients and characteristic impurity distributions along crystals by applying his zone melting technique. In 1952 he purified numerous metals and, for the first time, germanium crystals by this method [46]. The technique was later further developed to industrial multi-zone arrangements and today is one of the most effective methods for obtaining highly purified starting materials for compound semiconductor synthesis.

Non-uniform segregation of solutes on the microscale can occur by a variety of mechanisms. Early methods of forming p–n junctions in Ge, developed in Nobel-prize winning work at Bell Laboratories, relied on co-doping the crystal-growing melt with both donor and acceptor dopants. Because the diffusion rate in the melt of each chemical species was a little different, p–n junctions could be formed by modulating the growth rate and/or crystal rotation rate [47]. Needless to say this did not result in very sharp p–n junctions and solid state in-diffusion of dopant soon supplanted the technique as the principal method for forming transistor junctions.

However this early need to understand dopant incorporation led to the seminal studies by Burton, Prim and Slichter (EPS) in 1953 [48] on the dependence of effective distribution coefficient k_e on growth rate R and crystal rotation rate ω:

$$k_e = \frac{k_0}{k_0 + (1 - k_0)\exp(-R\delta_s/D)}, \tag{6}$$

Fig. 6. X-ray diffraction topograph of a Czochralski silicon wafer showing the strain distribution due to microsegregation of carbon (courtesy of M.J. Hill).

where k_0 is the equilibrium segregation coefficient, D is the melt diffusion coefficient and δ_s is the solute boundary layer, the thickness of which was solved by Levich [49]. By applying the Cochran flow solution at the surface of an infinite rotating disk he obtained the expression $\delta_s \sim 1.6 D^{1/3} \nu^{1/6} \omega^{-1/2}$ (ν is the kinematic viscosity of the melt). Later Ostrogorski and Müller [50] quantified δ_s more physically for situations where natural convection rather than rotating disc flow dominated by considering its dependence on lateral convection velocity and length of the interface.

This famous BPS equation probably represents the single most useful piece of theory ever produced for practical crystal growers.

Chemical etchants that reveal dislocations on $\{1\,1\,1\}$ surfaces of Ge and Si also reveal micro-scale variations in the concentrations of electrically active dopants (n or p) on non-$\{1\,1\,1\}$ surfaces. Etched longitudinal sections of the small Ge crystals grown at Bell Labs in the 1950s showed that the dopant concentration down the crystal had a helical modulation with a helix pitch equal to the amount of crystal grown per revolution of the crystal [47]. A longitudinal section cut slightly away from the axis of the helix thus reveals a striated dopant distribution. If the interface is curved the helix is 'dished' and a transverse section shows a single start spiral pattern (Fig. 6). The cause of the dopant striations was a lack of thermal symmetry in the melt. This caused the local crystal growth rate to vary as the crystal rotated in a thermally asymmetric melt. A precise geometric analysis of such rotationally induced striations was given in Ref. [51].

Later, as crystals got larger and materials of higher melting points were grown, other, non-periodic, dopant striations were observed. These were present also in crystal growth configurations (such as Bridgman growth) where there was no rotation and were shown by Müller and Wilhelm [52], Witt and Gatos [53] and Hurle [54] to have their origin in convective temperature fluctuations in the melt. Chedzey and Hurle [55] and, independently, Utech and Flemings [56] demonstrated in 1966 that these temperature fluctuations could be suppressed by application of a steady magnetic field. Hurle and co-workers [57] made a detailed study of the onset of this non-steady flow in metallic melts and showed that it was due to bifurcation from steady flow produced by buoyancy forces.

Incidentally, this all occurred at about the time that NASA was seeking funding for the Space Shuttle and was looking for experimentation that uniquely required a near-zero gravity environment. The simplistic argument was made that convection was 'bad' and therefore growth in space, where there should be no buoyancy-driven convection, must be 'good'. From this emerged the unrealistic expectation of growing 'perfect' crystals in space. However, initially, Marangoni convection [58,59] (which is present in zero-gravity), was overlooked. Also it turned out that residual gravity levels in space were sufficient to give rise to solutally driven buoyant convection [60].

Today ground-based experiments have gained increased importance. For instance it is well known that alternating magnetic fields can also help very effectively to damp convection and, hence, temperature oscillations also. This requires only a low magnetic field strength of some millitesla [61]. Rotating fields were first used by Hulme and Mullin [62] in 1959 to homogenise an InSb melt. Subsequently other techniques have been introduced to achieve homogenisation and to prevent temperature oscillations. These include the accelerated crucible rotation technique (ACRT) [63] and vibration stirring [64]. From these experiments we experienced practically that the growing interface acts as a low-pass filter, as predicted theoretically by Hurle and Jakeman [65], so that high rotational or vibrational frequencies do not produce dangerous disturbances and can smooth out composition fluctuations very effectively.

Segregation phenomena can also occur on a more macro-scale. In the presence of solute mixing in the melt due either to convection or to forced flow driven, for example, by crystal and/or crucible rotation, axial segregation of all solutes occurs. The degree of that axial segregation depends on the extent to which the solute segregation coefficient (k_0) differs from unity. For unidirectional solidification in a completely mixed melt this is described by the well-known Scheil equation [66] being valid for a stable planar crystal–melt interface.

However, under certain conditions, especially if the melt is not mixed by convection or stirring, (i.e. if the solute boundary layer is well developed), the interface can become morphologically unstable. In the late 1950s Esaki [67] devised a novel device structure which involved the tunnelling of charged carriers through a heavily doped p–n junction. For the first time this required that the semiconductor be doped to very high levels. This necessitated dopant concentrations in the melt of up to several atomic per cent and produced equilibrium liquidus temperature reductions of up to several tens of degrees. This tended to give rise to a condition known as constitutional supercooling of the melt, a phenomenon which had been discovered already in 1953 by the group of Professor Chalmers at Toronto in metal alloy systems [68].

Both an enriched ($k_0 < 1$) or depleted ($k_0 > 1$) solute boundary layer δ_s, having a diffusion determined (i.e. exponential) concentration distribution ahead of the growing interface, can give rise to constitutional instability. A zone of constitutional supercooling exists if the equilibrium liquidus temperature gradient exceeds the actual temperature gradient in the melt at the crystal/melt interface. In the presence of a zone of constitutional supercooling random formation of a projection (A) on the interface advances that portion of the interface into a region of increased supercooling (Fig. 7a). Here it can grow more rapidly. This results in a lateral segregation of solute that suppresses growth in the neighbouring region (B). A close packed array of such projections form having as length scale the lateral diffusion distance D/R. This is exactly what happens in most metal alloy systems.

In semiconductors however, it was found that once the amplitude of the projections grew to the point that their interface with the melt became tangential to a {1 1 1} faceting direction (see Section 5), micro-facets formed on the interface (Fig. 7b). These facets exhibited the facet effect, trapping excess dopant (Fig. 7c) [69]. Indeed this occurs even if the dopant is not the cause of the constitutional supercooling. A non-congruent melt produces rejection of the component in excess and this lowers the liquidus temperature in exactly the same way as a solute (but one having a near-zero segregation coefficient). Because of the development of the micro-facets the morphology of the resulting cellular structure is orientation dependent (*note*: not to be confused with the polygonized cell

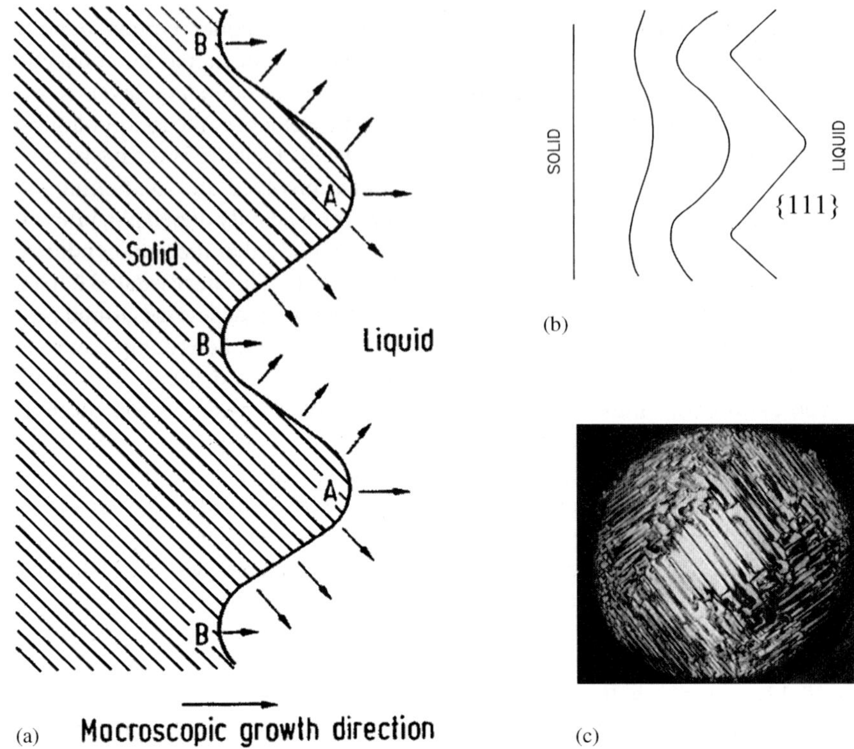

Fig. 7. Morphological instability of a growing melt–solid interface: (a) lateral segregation; (b) formation of micro-facets in semiconductors; and (c) autoradiograph showing excess Te incorporation on {1 1 1} facets in an InSb crystal.

structure described in Section 2). The microstructures obtained and the resulting micro-segregation and dislocation substructures were studied by Bardsley and co-workers [70,71].

Tiller et al. [72] deduced theoretically the condition for the prevention of constitutional supercooling (and hence, approximately at least, the condition for the preservation of morphological stability of the interface). They showed that the ratio of the temperature gradient in the melt at the interface G to the growth rate R must exceed a critical value given by:

$$\frac{G}{R} \geqslant \frac{mC_0(1 - k_0)}{k_0 D}, \tag{7}$$

where C_0 is the starting solute concentration and m is the slope of the liquidus from the T–x-phase diagram projection.

A linear stability analysis predicting the exact conditions of onset of the morphological instability appeared a few years later in the now-famous paper by Mullins and Sekerka [73]. In the following years

the theory has been widely extended to include higher order bifurcations and additional physical effects such as melt flow, atomic kinetics, Soret diffusion, applied electric fields, etc. An excellent review appears in Ref. [74].

5. Faceting and twinning

When Hulme and Mullin [75] looked in 1957 at the segregation of a number of solutes in InSb they found that the segregation coefficient was different for crystals grown in a ⟨1 1 1⟩ direction as compared with growth in any other direction. Using radio-tracer techniques, they demonstrated that a non-equilibrium concentration was incorporated into those parts of the crystal that had been grown on a faceted interface (Fig. 8).

Facets form normal to crystal directions for which 2-D nucleation is required in order to initiate the growth of a new layer. On non-facetted ('rough')

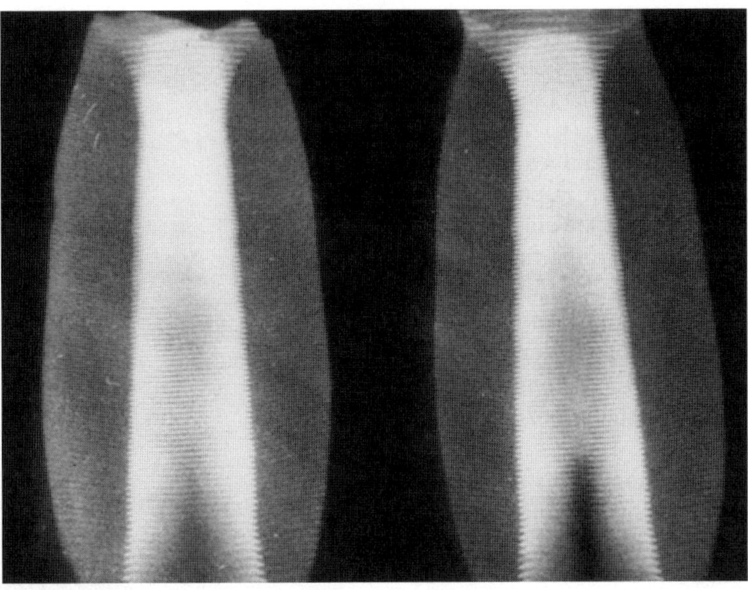

Fig. 8. Audio-radiographs of a longitudinally sectioned $\langle 1\,1\,1\rangle$-oriented InSb crystals doped with Te. The bright central column in each section is a region of enhanced Te concentration where the crystal grew with a faceted interface. The fine horizontal lines are growth striations produced by crystal rotation (courtesy of J.B. Mullin).

surfaces atoms can be added singly without the need for nucleation. At a given growth temperature, all crystals will have some surfaces which are rough. However most crystals will have one or more surfaces which are 'smooth' requiring nucleation. Early on Jackson [76] provided a simple thermodynamic model which indicated that the magnitude of the entropy of fusion of a material was a guide to its likelihood of forming facets during growth, materials having a low entropy of fusion (such as metals) having lowest probability. The common semiconductor materials, with their covalent bonding, tend to form facets during melt growth only on their most close-packed (i.e. $\{1\,1\,1\}$) planes. A further, geometric, requirement for facet formation during constrained crystal growth (such as by Bridgman or Czochralski methods) is that the radial temperature gradient be such that the freezing point isotherm is convex when viewed from the melt (Fig. 9). This ensures that, if the crystal starts to lag behind the isotherm, it experiences an increased supercooling which ultimately promotes the nucleation of a new layer. The lateral extension of the facet d is proportional to the supercooling ΔT^* as [77]:

$$d = 2\sqrt{2r\,\Delta T^*/G_{\mathrm{r}}},\qquad(8)$$

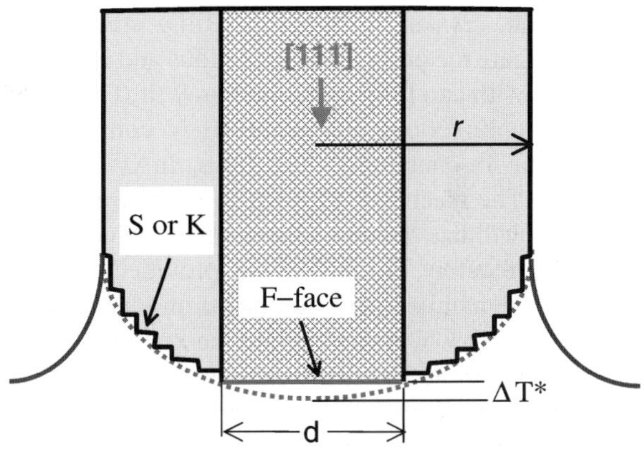

Fig. 9. Sketch of Czochralski growth in $\langle 1\,1\,1\rangle$ direction with convex interface where a $\{1\,1\,1\}$ core facet is formed.

where G_{r} is the radial temperature gradient and r is the radius of curvature of the interface. As can be seen d increases with decreasing convexity of the interface, i.e. reducing radial temperature gradient. The rapid lateral growth tends to trap in the surface adsorbed (equilibrium) solute concentration thereby increasing the effective segregation coefficient of solutes that are preferentially adsorbed at the interface. The most dramatic effect occurs with Te-doping of InSb. Here the equilibrium segregation coefficient of Te is ~ 0.5 whereas the effective segregation coefficient on the $\{1\,1\,1\}$ facets is ~ 4.0 giving the remarkable ratio of 8:1 (Fig. 8) [78].

If growth is promoted at a temperature below the melting point (i.e. from solution or vapour phase) then, as the growth temperature is lowered progressively, more crystal orientations develop facets. Theoretical studies of this 'roughening transition' following on from the original BCF theory [79] have been extensive and a rather complete understanding now exists. Bauser [80] studied $\{1\,1\,1\}$ faceting in LPE-grown semiconductor layers and showed excellent experimentally obtained images.

Twinning in the diamond-cubic and zinc-blende lattices, which is closely related to facet formation, is specified by a rotation of the lattice by $60°$ about a $\langle 1\,1\,1\rangle$ axis, the twin lying on the orthogonal $\{111\}$ plane. It can occur during the melt growth of such semiconductors and it was earlier recognised by Billig [81] that such twinning occurred principally on $\{1\,1\,1\}$ facets which form adjacent to the three-phase boundary of melt, crystal and ambient. Billig studied Ge crystals where the problem is not serious and is totally avoidable with carefully controlled growth. The problem is a more serious one in the III–V compounds, notably the In-containing ones InSb, InAs and InP (Fig. 10). The mechanism by which such twins form during growth has defied explanation for many years but, fairly recently, Hurle [82] has provided a possible thermodynamic description based on ideas due to Voronkov [83], which can explain the key features of the process. The model demonstrates that, because of the orientation dependence of interfacial energies in the presence of facets, there is a configuration of the three-phase boundary for which, for sufficiently large supercooling, the free energy of formation of a critical nucleus is actually lowered by forming that nucleus at the three-phase boundary in twinned orientation. This will occur only if a critical angle of conical growth presenting a portion of crystal surface normal to $\langle 1\,1\,1\rangle$ is sampled during the growth.

Such a twinned nucleus is thermodynamically favoured if the supercooling exceeds the critical value

$$\delta T^* = (\sigma T_\mathrm{m}/h\Delta H)A^*, \qquad (9)$$

where σ is the twin plane energy, T_m is the melting temperature, h is the nucleus height, ΔH is the

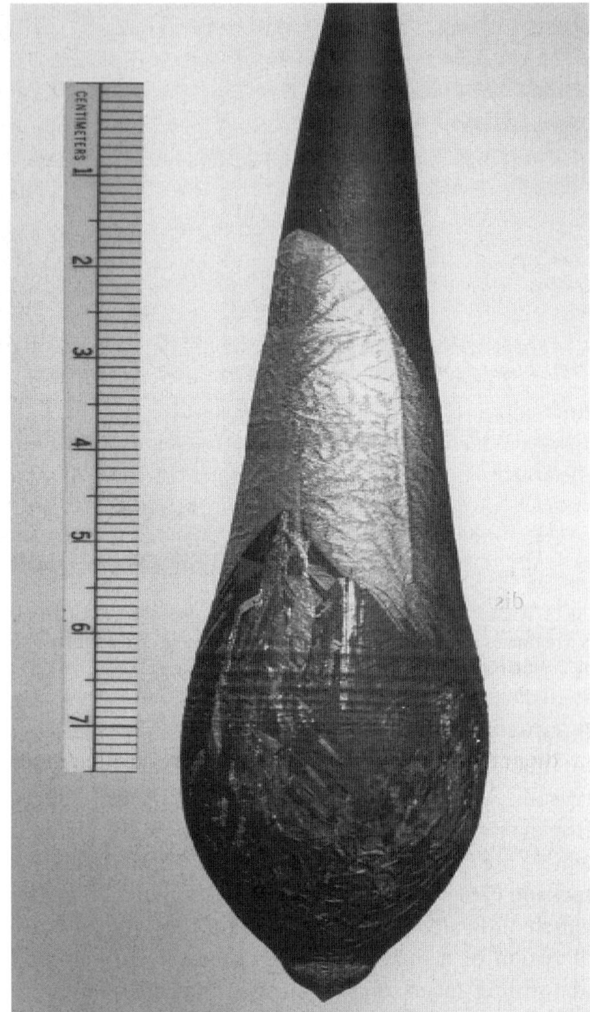

Fig. 10. $\{1\,1\,1\}$ twinned InP crystal (courtesy of W. Bonner).

latent heat of fusion, and A^* is the reduced work of formation of a nucleus intersecting the three-phase boundary. Experimental test of this model had been provided by the groups of Müller at Erlangen [84] and Dudley at the State University of NY at Stonybrook [85,86].

Many LEC and Bridgman experiments have demonstrated that the twin probability is reduced markedly if the temperature oscillations of the growth system, and therefore excursions of the angle of the contacting meniscus, are minimised. This is due to the reduced probability of encountering the critical angle described above when the meniscus angle fluctuations are reduced. In fact Japanese producers succeeded recently in twin-free InP crystal growth with diameters up to 100 (150) mm by careful maintenance of thermal stability during growth [87] which was achieved by applying damping magnetic fields around the melt [88].

6. Summary and outlook

Over the half-century of the development of semiconductor technology, most of the important defect-forming mechanisms have become well understood. That is not to say that all defects can be avoided. For instance, the relatively poor thermal conductivity and low yield stresses of III–V and II–VI compounds as compared with Ge and Si mean that it is not possible to reduce the thermal stresses to a sufficiently low level to avoid dislocation multiplication. However, recent developments in vertical Bridgman and gradient freeze techniques have produced a marked reduction in dislocation density in undoped GaAs and InP [89–92]. In LEC growth the reduction of the axial temperature gradient by using the vapour pressure controlled modification (VCz) [92,93] has led to dislocation densities reduced to below 10^4 cm^{-2} in 4 and 6 inch diameter GaAs crystals.

Again, although we understand fully the conditions under which morphological instability occurs, it is still not possible to obtain conditions of growth that permit the production of large, homogeneous alloy (mixed) single crystals which would be invaluable as tailored substrates (e.g. $In_{1-x}Ga_xAs$).

Finally twinning remains a serious limiter of yield in the growth of InP single crystals. However, promising reports have been published recently which show that VCz arrangements in combination with marked damping of temperature fluctuations by magnetic fields are able to yield excellent twin-free 4-inch InP crystals [88].

So what of the future? We need to obtain a much better understanding of the thermodynamics and kinetics of native point defects and their interactions with chemical dopants in both Si and the common compound semiconductors. We need to understand their behaviour both during growth and in any post-growth annealing schedules that are employed to optimise the material. Such knowledge offers the potential to increase both performance and yield of electronic and optoelectronic devices.

Scaling up to achieve cost reduction is an ever-present pressure. Increasing crystal diameter increases the thermal stresses experienced during cooling. Avoiding increased dislocation density requires ever-more refinement of furnace design and here computer modelling plays a key role. An additional problem posed by scaling-up is the increased turbulence which occurs in the melt. This produces violent fluctuations in growth rate which probably also increases the incidence of twinning (at least in the case of III–V compounds). The use of both static and variable magnetic fields is being exploited to achieve damping of this turbulence.

Perhaps the biggest prize remaining to be grasped is the overcoming of morphological instability associated with the phenomenon of constitutional supercooling. The authors believe that if this could be achieved for the growth of solid solution crystals of Si–Ge and of a number of pseudo-binary semiconductor III–V and II–VI alloys, like $In_{1-x}Ga_xAs$, $Cd_{1-x}Zn_xTe$, this would open a range of new opportunities for device design.

References

[1] W. Shockley, Bell Syst. Tech. 28 (1949) 435.

[2] H.B. Huntington, F. Seitz, Phys. Rev. 76 (1949) 1728.

[3] C. Zener, J. Appl. Phys. 22 (1951) 372.

[4] A.D. Le Claire, Acta Metall. 1 (1953) 438.

[5] W. Shockley (Ed.), Imperfections in Nearly Perfect Crystals, Wiley, New York, Chapman & Hall, London, 1952.

[6] F.A. Kröger, The Chemistry of Imperfect Crystals, North-Holland, Amsterdam, 1964, 1973.

[7] V. Voronkov, J. Crystal Growth 59 (1982) 625.

[8] E. Dornberger, J. Virbulis, B. Hanna, R. Hoelzl, E. Daub, W. von Ammon, J. Crystal Growth 229 (2001) 11.

[9] M. Itsumi, H. Akiya, T. Ueki, M. Tomita, M. Yamawaki, J. Appl. Phys. 78 (1995) 10.

[10] R. Falster, V. Voronkov, Mater. Sci. Eng. B 73 (2000) 87.

[11] D.T.J. Hurle, Appl. Phys. Rev. 85 (1999) 6957.

[12] T.E. Schlesinger, J.E. Toney, H. Yoon, E.Y. Lee, B.A. Brunett, L. Franks, R.B. James, Mater. Sci. Eng. R 32 (2001) 103.

[13] T.Y. Tan, H.M. Yu, U. Gösele, Appl. Phys. A 56 (1993) 249.

[14] V.T. Bublik, V.V. Kartataev, R.S. Kulagin, M.G. Milvidskii, V.B. Osvenski, O.G. Stolyarov, L.P. Kholodnyi, Sov. Phys. Crystallogr. 18 (1973) 218.

[15] J. Greenberg, Thermodynamic Basis of Crystal Growth, Phase P–T–X Equilibrium and Non-Stoichiometry, Springer, Berlin, 2002.

[16] K. Terashima, S. Washizuka, J. Nishio, A. Okada, S. Yasuami, in: Institute of Physics Conference Series, vol. 79, 1985, p. 37.

[17] R. Krause-Rehberg, H.S. Leipner, Positron Annihilation in Semiconductors, Springer, Berlin, 1999.

[18] S. Oda, M. Yamamoto, M. Seiwa, G. Kano, T. Inoue, M. Mori, H. Shimakura, M. Oyake, Semicond. Sci. Technol. 7 (1992) A215.

[19] D.T.J. Hurle, J. Phys. Chem. Solids 40 (1979) 613, 627, 639, 647.

[20] H. Petroff, R.L. Hartman, J. Appl. Phys. 45 (1974) 3899.

[21] J.E. Northrup, S.B. Zhang, Phys. Rev. B 50 (1994) 4962.

[22] J. Nishizawa, J. Crystal Growth 99 (1990) 1.

[23] P. Rudolph, Progr. Crystal Growth Charact. Mater. 29 (1994) 275.

[24] P. Rudolph, N. Schäfer, T. Fukuda, Mater. Sci. Eng. R 15 (1995) 85.

[25] G.K. Teal, J.B. Little, Bull. Am. Phys. Soc. 78 (1950) 547.

[26] G.K. Teal, E. Buehler, Phys. Rev. 87 (1952) 190.

[27] J. Czochralski, Z. Phys. Chem. 92 (1918) 219.

[28] A.H. Cottrell, Dislocations and Plastic Flow in Crystals, Clarendon Press, Oxford, 1951.

[29] W.K. Burton, N. Cabrera, F.C. Frank, Nature 163 (1949) 398.

[30] F.L. Vogel, W.G. Pfann, H.E. Corey, Phys. Rev. 90 (1953) 489.

[31] W.C. Dash, J. Appl. Phys. 27 (1956) 1193.

[32] A.R. Lang, J. Appl. Phys. 30 (1959) 1748.

[33] U. Bonse, Z. Phys. 168 (1958) 278.

[34] W.C. Dash, J. Appl. Phys 30 (1959) 459.

[35] Y. Shiraishi, K. Takano, J. Matsubara, T. Iida, N. Takase, N. Machida, M. Kuramoto, H. Yamagishi, J. Crystal Growth 229 (2001) 17.

[36] A. Steinemann, U. Zimmerli, in: H.S. Peiser (Ed.), Crystal Growth, Pegamon Press, Oxford, 1967, p. 81.

[37] E. Schmid, W. Boas, Plasticity of Crystals, Chapman & Hall, London, 1968.

[38] A.S. Jordan, A.R. von Neida, R. Caruso, J. Crystal Growth 70 (1984) 555.

[39] E. Billig, Proc. R. Soc. London 235 (1956) 37.

[40] V.L. Indenbom, Krist. Tech. 14 (1979) 493.

[41] H. Alexander, P. Haasen, in: F. Seitz, D. Turnbull, H. Ehrenreich (Eds.), Solid State Physics, Academic Press, New York, 1968, p. 27.

[42] J. Völkl, in: D.T.J. Hurle (Ed.), Handbook of Crystal Growth, vol. 2b, North-Holland, Amsterdam, 1994, p. 821.

[43] M. Naumann, P. Rudolph, M. Neubert, J. Donecker, J. Crystal Growth 231 (2001) 22.

[44] J. Amelinckx, W. Dekeyser, Solid State Phys. 8 (1959) 441.

[45] M. Zaiser, Mater. Sci. Eng. A 309–310 (2001) 304.

[46] W.G. Pfann, Trans. AIME 194 (1952) 747.

[47] H.E. Bridgers, J.H. Scaff, J.N. Shive (Eds.), Transistor Technology, vol. 1, Van Nostrand Reinhold, New York, 1958.

[48] J.A. Burton, R.C. Prim, W.P. Slichter, J. Chem. Phys. 21 (1953) 1987.

[49] V.G. Levich, Physicochemical Hydrodynamics, Prentice-Hall, Englewood Cliffs, 1961.

[50] A. Ostrogorsky, G. Müller, J. Crystal Growth 121 (1992) 587.

[51] J. Barthel, M. Jurisch, Krist. Tech. 8 (1973) 199.

[52] A. Müller, M. Wilhelm, Z. Naturforsch. 19a (1964) 254.

[53] A.F. Witt, H.C. Gates, J. Electrochem. Soc. 113 (1966) 808.

[54] D.T.J. Hurle, Philos. Mag. 13 (1966) 305.

[55] H.A. Chedzey, D.T.J. Hurle, Nature 210 (1966) 933.

[56] H.P. Utech, M.C. Flemings, J. Appl. Phys. 37 (1966) 2021.

[57] D.T.J. Hurle, E. Jakeman, C.P. Johnson, J. Fluid Mech. 64 (1974) 565.

[58] A. Eyer, H. Leister, R. Nitsche, J. Crystal Growth 71 (1985) 173.

[59] R. Monti, J.J. Favier, D. Langbein, in: H.U. Walter (Ed.), Fluid Science and Material Sciences in Space, A European Perspective, Springer, Berlin, 1987, p. 637.

[60] R.W. Smith, Microgravity Sci. Technol. 11 (1998) 78.

[61] J. Friedrich, Y.S. Lee, B. Fischer, C. Kupfer, D. Vizman, G. Müller, Phys. Fluids 11 (1999) 853.

[62] K.F. Hulme, J.B. Mullin, Philos. Mag. 41 (1959) 1286.

[63] H.J. Scheel, E.G. Schulz-DuBois, J. Crystal Growth 8 (1971) 304.

[64] Y.C. Lu, J.J. Shiang, R.S. Feigelson, J. Crystal Growth 102 (1990) 807.

[65] D.T.J. Hurle, E. Jakeman, J. Crystal Growth 5 (1969) 227.

[66] E. Scheil, Z. Metallkde. 34 (1942) 70.

[67] L. Esaki, Phys. Rev. 109 (1958) 602.

[68] J.W. Rutter, B. Chalmers, Canad. J. Phys. 31 (1953) 15.

[69] D.T.J. Hurle, O. Jones, J.B. Mullin, Solid State Electron. 3 (1961) 317.

[70] W. Bardsley, J.M. Callan, H.A. Chedzey, D.T.J. Hurle, Solid State Electron. 3 (1962) 142.

[71] W. Bardsley, D.T.J. Hurle, M. Hart, A.R. Lang, J. Crystal Growth 49 (1980) 612.

[72] W.A. Tiller, K.A. Jackson, J.W. Rutter, B. Chalmers, Acta Metall. 1 (1953) 428.

[73] W.W. Mullins, R.F. Sekerka, J. Appl. Phys. 35 (1964) 444.

[74] S.R. Coriell, G.B. McFadden, in: D.T.J. Hurle (Ed.), Handbook of Crystal Growth, vol. 1b, North-Holland, Amsterdam, 1993, p. 785.

[75] K.F. Hulme, J.B. Mullin, J. Electron. Control 3 (1957) 160.

[76] K.A. Jackson, American Society of Metals review of Metal Literature (1958) 174 (see also K.A. Jackson, D.R. Uhlmann, J.D. Hunt, J. Crystal Growth 1 (1967) 1).

[77] P. Reiche, J. Bohm, B. Hermoneit, P. Rudolph, D. Schultze, J. Crystal Res. Technol. 23 (1998) 467.

[78] J.B. Mullin, K.F. Hulme, J. Phys. Chem. Solids 17 (1960) 1.

[79] W.K. Burton, N. Cabrera, F.C. Frank, Philos. Trans. R. Soc. A 243 (1951) 299.

[80] E. Bauser, in: D.T.J. Hurle (Ed.), Handbook of Crystal Growth, vol. 3b, North-Holland, Amsterdam, 1994, p. 879.

[81] E. Billig, J. Inst. Met. 83 (1954/55) 53.

[82] D.T.J. Hurle, J. Crystal Growth 147 (1995) 239.

[83] V.V. Voronkov, Sov. Phys. Cryst. 19 (1975) 573.

[84] J. Amon, F. Dumke, G. Müller, J. Crystal Growth 187 (1998) 1.

[85] H. Chung, M. Dudley, D.J. Larson, D.T.J. Hurle, D.F. Bliss, V. Prassad, J. Crystal Growth 187 (1998) 9.

[86] M. Dudley, B. Raghothamachar, Y. Guo, X.R. Huang, H. Chung, D.T.J. Hurle, D.F. Bliss, J. Crystal Growth 192 (1998) 1.

[87] R. Pool, Compound Semiconductor 8 (2002) 4 (see also advertisement in: III–Vs Review—The Advanced Semiconductor Magazine 15 (2002) 9).

[88] T. Kawase, S. Fujiwara, M. Matsushima, Y. Hosokawa, R. Nakai, in: Proceedings of the 2001 International Conference on InP and Related Materials (13th IPRM), Nara, 14–18 May, 2001, p. 13.

[89] E. Monberg, in: D.T.J. Hurle (Ed.), Handbook of Crystal Growth, vol. 2a, North-Holland, Amsterdam, 1994, p. 52.

[90] X. Liu, III–Vs Rev. 8 (1995) 14.

[91] Th. Bünger, D. Behr, St. Eichler, T. Flade, W. Fliegel, M. Jurisch, A. Kleinwechter, U. Kretzer, Th. Steinegger, B. Weinert, Mater. Sci. Eng. B 80 (2001) 5.

[92] M. Tatsumi, K. Fujita, in: M. Isshiki (Ed.), Recent Developments of Bulk Crystal Growth, Research Signpost, Trivandrum, 1998, p. 47.

[93] M. Neubert, P. Rudolph, Prog. Crystal Growth Charact. Mater. 43 (2001) 119.

Research on macro- and microsegregation in semiconductor crystals grown from the melt under the direction of August F. Witt at the Massachusetts Institute of Technology

C.A. Wang [a,*], D. Carlson [b], S. Motakef [c], M. Wiegel [a], M.J. Wargo [d]

[a] *Lincoln Laboratory, Massachusetts Institute of Technology, Lexington, MA 02420, USA*
[b] *MA/COM, Inc., Lowell, MA 01851, USA*
[c] *Cape Simulations, Inc. Natick, MA 01760, USA*
[d] *NASA Headquarters, Physical Sciences Research Division, Washington, DC 20546, USA*

Written in memory of and dedicated to Prof. August F. Witt (1931–2002)

Abstract

The birth of the modern age of semiconductor electronics in the 1950s required the production of single crystals from the melt. In the early years of this technology, crystals exhibited non-uniform distributions of chemical and structural defects, which directly affected devices produced from this material. Uniform distribution and control of dopants, unintentional impurities, and native defects were identified as critical requirements for continued advances in device technology. However, since at that time fundamental understanding of cause and effect relationships between crystal growth parameters and the ultimate properties of the materials produced was absent, such desirable properties were unattainable. Nevertheless, it was recognized that segregation plays a key role in the creation of non-uniformities. Macrosegregation, a consequence of the directional solidification process, is generally controlled by diffusion and convective melt flows. Microsegregation is governed by local perturbations at the crystal–melt interface. Time-dependent thermal and melt velocity fields at the crystal–melt interface impact the microscopic rate of growth and the solute diffusion boundary layer. Identification of the fundamental parameters that govern axial and radial macro- and microsegregation during bulk semiconductor crystal growth was the research focus during the early years. Key contributions were made by Prof. August F. Witt and his research group. One of the major contributions was the development of quantitative analytical tools for characterization of the crystal growth process. These tools provided the foundation for understanding the origin and nature of segregation phenomena during crystal growth from the melt. This paper is a brief review of his work.
© 2004 Elsevier B.V. All rights reserved.

PACS: 81.05.Cy; 81.10.Fg; 81.10.Mx

Keywords: A1. Heat pipes; A1. Impurity striations; A1. Magnetic fields; A1. Segregation; A2. Bulk crystal growth

1. Introduction

Rapid progress in the early stages of the microelectronics and optoelectronics industries was largely enabled by improvements made in melt-grown bulk

* Corresponding author.
E-mail address: wang@ll.mit.edu (C.A. Wang).

Originally published in
J. Crys. Growth 264 (4) (2004) 565–577.

semiconductor single crystals with controlled materials properties. The electrical and optical properties of semiconductor wafers, which are sectioned from the bulk crystal, are directly related to the distribution of intentional dopants, unintentional impurities, and defects. Thus, the preparation of semiconductor crystals with homogeneous properties on both the macro- and microscale is crucial for obtaining controllable and predictable device performance. The critical importance of controlling semiconductor materials properties led Bell Laboratories, IBM, RSRE, Texas Instruments, Wacker, and numerous other organizations to create major programs in the 1960s to grow very uniform single crystals from the melt.

The difficulty in obtaining uniform segregation or dopant profiles in semiconductor crystals is a consequence of destabilizing radial or/and axial temperature gradients in the melt. Such gradients give rise to buoyancy-driven convective flows that may result in time varying temperature and velocity fields. Convection in the melt, steady or unsteady, influences the distribution of the dopant at the crystal–melt interface, whereas an unsteady temperature field in the melt results in a time varying growth rate and growth interface morphology. The combined effects of these convective phenomena coupled with inherent thermal asymmetries that characterize most crystal growth systems, results in an inhomogeneous dopant distribution with complex features. Consequently, research interests were focused on identification of the origin and nature of segregation so that engineering solutions could then be implemented to control and optimize composition profiles of crystals grown from the melt.

This paper highlights the research activities on bulk semiconductor crystals grown from the melt, which were performed under the direction of Prof. August F. Witt at the Massachusetts Institute of Technology from the late 1960s to his passing in 2002. Prof. Witt was a consummate and elegant experimentalist. His efforts to elucidate mechanisms of dopant segregation using fundamental approaches have broadly influenced both the research philosophy of crystal growers whom he trained and many whom he did not, as well as the practices and methods of bulk crystal growth. As a result of his work, new engineering methods to affect and control segregation have evolved. This paper is dedicated to Prof. Witt, and we briefly review his contributions to the development of analytical tools

for determination of segregation on the microscale and its correlation with melt convection; the design and implementation of crystal growth systems that would allow unambiguous identification of the influence of a myriad of factors that influence segregation; and the use of magnetic fields for control of melt convection.

2. Historical background and early observations of segregation

Semiconductor crystals grown from the melt are directionally solidified, and the resulting impurity or dopant composition profile is governed, in part, by alloy thermodynamics. Redistribution of the solute in the solid depends on the interface distribution coefficient, k^*, assumed equal to the equilibrium distribution coefficient k_0:

$$k_0 = C_S/C_L,$$

where C_S and C_L are solute concentrations in the solid and liquid melt, respectively, corresponding to the solidus and liquidus concentrations associated with the equilibrium phase diagram. When $k_0 < 1$, which is typically the case for dopants commonly used in Si, Ge, or GaAs crystal growth, solute is rejected into the melt ahead of the crystal–melt interface, resulting in a solute boundary layer. Analytical solutions to describe segregation profiles were initially derived for two extreme cases: (1) no convection in the melt and (2) complete convection in the melt.

In the absence of convective mixing and at constant growth rate, Tiller et al. [1] showed that C_S reaches a nearly uniform value equal to that of the starting liquid C_L, after an initial transient period according to the following equation:

$$C_S = C_L\big[1 - (1 - k_0)\exp(-k_0 Rx/D_L)\big],$$

where x is the distance from the initial crystal–melt interface, R is the solidification or growth rate, and D_L is the diffusion coefficient of the solute in the liquid. D_L/R is the characteristic distance that determines the initial and final transients, D_L/Rk_0 and D_L/R, respectively.

The opposite extreme, complete mixing in the melt, was considered by Pfann [2]. In this case, a boundary layer is assumed to be vanishingly small and the solute profile can be represented by

$$C_S = k_0 C_L (1 - f_s)(k_0 - 1).$$

This condition is the basis for zone refining [2]. However, neither of these approaches could adequately describe the experimentally observed solute concentration profiles in melt-grown semiconductor crystals. Thus, the detailed role of melt convection in the solidification process became a major focus of study.

To account for varying levels of forced melt convection (through crystal rotation during Czochralski growth), Burton, Prim and Slichter (BPS) developed a model to describe segregation in systems with steady convective flows [3]. This model is based on a boundary layer in the melt of thickness δ, inside of which solute transport is dominated by diffusion, while outside of it, the solute concentration is uniform and controlled by natural thermal convective mixing. An effective distribution coefficient k_{eff} is obtained by solving the diffusion equation and can be described by the following equation:

$$k_{\text{eff}} = C_S / C_L = k^* / [k^* + (1 - k^*)\exp(-R\delta/D_L)],$$

where k^* is the interface distribution coefficient. In most cases, one assumes that k^* is equivalent to the equilibrium distribution coefficient k_0.

For growth conditions in which R and δ are constant, C_S will vary continuously along the macroscopic length scale of the crystal, i.e. axial macrosegregation occurs. However, any instantaneous change in R or δ will lead to microsegregation where the composition varies over microscopic dimensions. Furthermore, if the crystal–melt interface is not ideally planar or the thickness of the momentum boundary layer varies due to radially non-uniform flows, δ will vary across the growth front and radial macro- and microsegregation will result. Thus, both R and δ are critical factors that determine segregation profiles.

Numerous studies of Czochralski-grown crystals show that the solute distribution in single crystals exhibits periodic profiles, in the form of striations, in axially cross-sectioned samples. Characterization of macrosegregation profiles in doped semiconductor crystals are routinely determined by Hall measurements, autoradiography, and resistivity measurements. Chemical wet-etching is relatively quick and provides good spatial resolution of microsegregation but is only

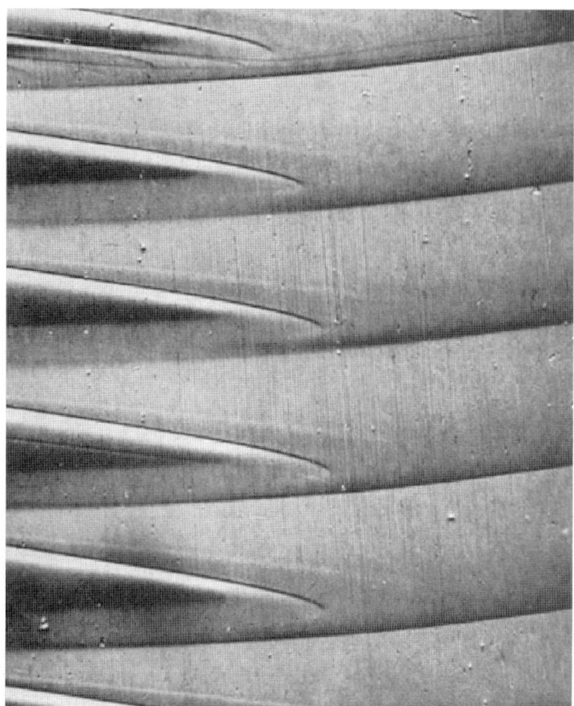

Fig. 1. Rotational striations in Czochralski-grown InSb crystal. (From Ref. [5], reproduced with the permission of The Electrochemical Society, Inc.)

qualitative in nature. The etch rate is sensitive to the solute concentration, and the resulting change in surface topography, which can be observed with Nomarski interference contrast microscopy, reflects compositional changes in the crystal [4].

Witt and co-workers reported impurity striations, such as those experimentally observed in Fig. 1, and attributed those to thermal asymmetry in the melt as the crystal was rotated during growth [5]. Hurle et al. [6] modeled the effect of sinusoidal temperature variations in the melt and showed that the growth rate could vary as a result of temperature fluctuations at the crystal–melt interface. These temperature variations at the crystal–melt interface were thought to originate from either rotation in a thermally asymmetric system or from thermal convective melt flows. Such variations were believed to induce variations in the microscopic growth rate, which in turn affected dopant incorporation, and was subsequently reflected in compositional striations. Therefore, the dopant striations in axial cross-section samples suggested deviations either in the boundary layer thickness, as a result of unsteady convection, or deviations in microscopic growth rate resulting from time-dependent and/or spatially non-uniform heat flow in the melt [5–7]. However, since microscopic growth rates could not be determined at that time, it was often assumed that the growth rate was nearly constant and equal to the pulling rate or extraction rate in Czochralski-grown (with consideration of the reduction in the melt height) or Bridgman-grown crystals, respectively. Thus, the limited resolution of composition striations in these early studies made it difficult to unambiguously determine the exact origin and nature of such striations.

3. Quantitative analyses of microsegregation

Realizing the importance of transient phenomena associated with segregation profiles, Witt and co-workers established experimental methods to quantitatively measure microscopic growth rates and solute profiles. This knowledge eventually led to significant breakthroughs in understanding the origin and nature of microsegregation. Mechanical vibrations of known frequency were intentionally introduced into the melt during crystal growth. These vibrations locally perturbed solute incorporation, and were revealed as dopant striations in the crystal. Precise measurement of the spacing between striations was then used to determine microscopic growth rates [8]. Alternatively, striations, such as those shown in Fig. 2, were introduced by the application of short duration current pulses across the crystal–melt interface. This induced an instantaneous change in the growth rate due to Peltier heating (or cooling) [9]. With current pulsing, the recovery time after the growth rate perturbation was negligible compared to mechanical vibrations. Consequently, current pulsing was more extensively used as it proved to be more versatile, and was applied to InSb, Ge, Si, and GaAs crystal growth in Witt's lab as well as others who studied segregation phenomena.

'Interface demarcation', as it came to be known, provided a detailed snapshot of the crystal–melt interface morphology and provided a history of the crystal growth process. Changes in the instantaneous growth rate within one rotation cycle of Czochralski-grown crystals were precisely determined [8,9]. Those exper-

Fig. 2. Cross-section of Te-doped InSb grown with current pulsing, as shown in the inset. Peltier cooling associated with current pulsing induces an abrupt change in growth rate and dopant incorporation. (From Ref. [9], reproduced with the permission of The Electrochemical Society, Inc.)

iments showed that the microscopic growth rate exhibited a periodic variation, which was clearly attributed to thermal asymmetry in the system. In addition, local remelting of specific regions of the crystal–melt interface during the cycle was observed. This work clearly shows that one could no longer assume that the microscopic growth rate was equal to the macroscopic growth rate. In fact, the two rates differed by as much as a factor of 10 [8].

The origin of growth rate fluctuations was clarified in a study aimed at intentionally affecting melt convection. This work shows the paramount importance of thermo-hydrodynamics in crystal growth [10]. InSb crystals were grown in a non-rotating vertical gradient-freeze geometry with destabilizing axial temperature gradients. Thermocouples were used to monitor the melt temperature during growth, providing a record of the thermal history of the melt during the growth process. It was found that the intensity of melt convection varied as solidification proceeded, corresponding to a decrease with melt height. Three convective regimes were identified: turbulent convection, oscillatory convection, and finally stable (laminar) flow. Each regime could be characterized by unique variations in the microscopic rate of growth and related dopant incorporation. The observed behavior was characterized by the Rayleigh number

$$Ra = \frac{g\beta \Delta T l^3}{\nu \alpha},$$

where g is acceleration due to gravity, β is the volumetric coefficient of thermal expansion, ΔT is the axial temperature difference, l is the characteristic length or melt height, ν is the melt kinematic viscosity, and α is the melt thermal diffusivity. In the turbulent regime, $4 \times 10^3 < Ra < 3 \times 10^5$; while in the oscillatory regime, $2 \times 10^3 < Ra < 3 \times 10^3$; and in the region of thermal stability, $0 < Ra < 10^3$. Thus, the effect of melt flow regimes on the growth behavior and ultimate properties of the crystal was demonstrated. It should be noted that the axial thermal gradient must exceed a critical value for the growth of doped semiconductors, in order to prevent constitutional supercooling and subsequent morphological instability of the solidification front [1,11].

A major breakthrough in identifying the origin of microsegregation was made when Witt and co-workers established the correlation between microscopic dopant distribution profiles, as measured by spreading resistance, and microscopic growth rates [12]. Measurements of composition and growth rate are shown in Fig. 3 for a Czochralski-grown Ga-doped Ge crystal. The growth rate varies periodically due to thermal asymmetry. Similarly, periodic fluctuations in the doping concentration are also observed. For the first time, an exact correlation between mi-

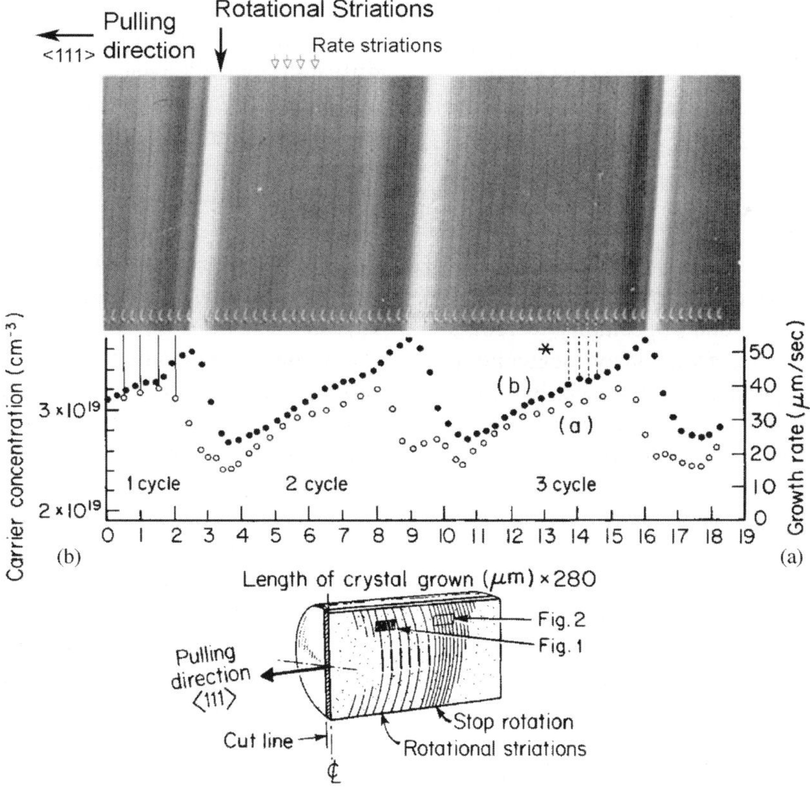

Fig. 3. Axial cross-section of a Czochralski grown Ga-doped Ge crystal with interface demarcation. The crystal was differentially chemically etched to reveal dopant striations associated with current pulsing. Markings along the lower portion of the micrograph correspond to indentations from spreading resistance measurements (filled circles), which are shown along with the microscopic growth rate (open circles). The microscopic growth rate varies periodically within a rotation cycle along with the dopant concentration. (From Ref. [12], reproduced with the permission of The Electrochemical Society, Inc.)

croscopic growth rate variations and solute segregation profiles were obtained. This study unambiguously demonstrated the segregation dependence on growth rate on the microscale.

The quantitative characterization of dopant distribution on the microscale by spreading resistance coupled with the use of interface demarcation was applied to both Czochralski- and Bridgman-grown Ge and Si crystals with destabilizing axial temperature gradients [13]. These results clarified that rotational and non-rotational striations were associated with fluctuations in the instantaneous growth rate associated with thermal and fluid dynamic perturbations. It was observed that systems with steeper thermal gradients, present for growth of higher melting point semiconductors, were subject to higher levels of thermal convection with flows that could be turbulent, oscillatory, or steady. This convection was reflected in the resulting segregation profiles. Uncontrolled and random segregation was associated with turbulent convection. Consequently, approaches to minimize convective interference with growth and segregation became a subject of great interest.

4. Space experiments

The environment of continuous freefall in low Earth orbit provides a unique opportunity to reduce the interference of melt convection on segregation, since the effects of gravity are reduced by up to six orders of magnitude. Thus *Ra*, the driving force for convection, is commensurately reduced. In the 1970s, NASA conducted a program to determine the potential of reduced gravity conditions for basic research and applications in materials processing. Witt was one of several researchers to grasp the significance of performing crystal growth in space to further our understanding of fundamental processes, and led several efforts aboard Skylab and Apollo-Soyuz [14,15].

Te-doped InSb and Ga-doped Ge crystals were directionally solidified in a gradient freeze furnace aboard Skylab [14] and Apollo-Soyuz [15], respectively. Subsequent electrical characterization of Te-doped InSb, by using Hall measurements on segments taken from the length of the crystal, indicated a dopant profile that initially increased and then reached

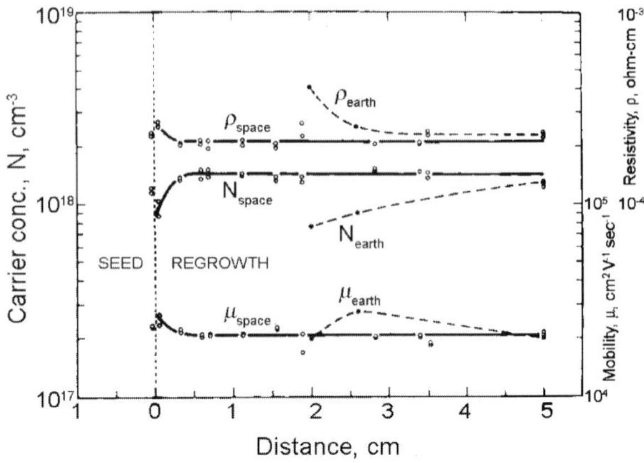

Fig. 4. Axial segregation profiles of Te-doped InSb grown by gradient freeze in space and on earth. The concentration profile for the space-grown crystal is controlled by diffusion, while that on earth is controlled by melt convection. (From Ref. [14], reproduced with the permission of The Electrochemical Society, Inc.)

a constant value. For the first time, ideal steady-state diffusion-controlled segregation was achieved in semiconductors. A ground-based control experiment performed in a similar geometrical configuration, showed a continuously increasing dopant concentration, which is associated with complete convective mixing in the melt. This comparison, shown in Fig. 4, clearly demonstrated that buoyancy-driven convection is the dominant driving force for segregation in Earth-grown crystals.

Similarly, segregation results of Ga-doped Ge crystals grown in space were indicative of diffusion-controlled growth. In this experiment, interface demarcation was performed and these results revealed that the crystal–melt interface was concave into the solid, asymmetric, and varied throughout the growth because of severe thermal asymmetries in the growth system. Under diffusion-controlled conditions, the axial and radial segregation showed no microsegregation behavior, but were controlled by the crystal–melt interface morphology. Variations in radial macrosegregation were as large as 300% for space-grown crystals, while it was just 20% for the ground-based crystals. These results suggest that melt convection could homogenize the melt, and also that under diffusion-controlled segregation, the morphology of the crystal–melt interface determines radial segregation profiles, which subsequently was modeled analytically [16]. Furthermore, variation in the crystal–melt interface shape was largely due to inadequate heat transfer control.

Along with the achievement of diffusion-controlled dopant profiles, other phenomena that had not been theoretically predicted were observed. For some dopants, the melt did not wet the quartz ampoule due to surface energy considerations, and even though solidification occurred with a free surface, Witt and co-workers showed that convection due to Marangoni forces did not affect the distribution of dopants in the bulk of the crystal. Furthermore, the observation of peripheral facets was identified to be characteristic of unconfined growth.

5. Segregation control

It had become apparent that segregation effects were strongly influenced by thermal convection, thermal asymmetry, and uncontrolled thermal gradients. To continue the progress in the understanding and ultimate control of these effects, experimental conditions that established reproducible thermal geometries and controlled thermal symmetry were required. Once these conditions were established, further control could be achieved by attempting to adjust parameters that affect thermal gradients and convection in the melt.

5.1. Heat pipes for improved thermal control

The application of heat pipes in crystal growth to provide improved heat transfer control was first proposed by Steininger and Reed [17]. Witt and co-workers implemented a heat pipe in the hot zone of a Czochralski system to provide axisymmetric and repeatable azimuthal, radial, and axial thermal gradients [18]. Ga-doped Ge was used as a model system to quantitatively analyze the effect of this improved thermal geometry on segregation, and the results were compared to those of crystals grown in a conventional Czochralski hot zone geometry. It was found that application of the heat pipe resulted in virtual elimination of rotational striations. Periodic variations in the microscopic rate of growth and composition profile were negligible, as shown in Fig. 5. This dramatic increase in axial uniformity

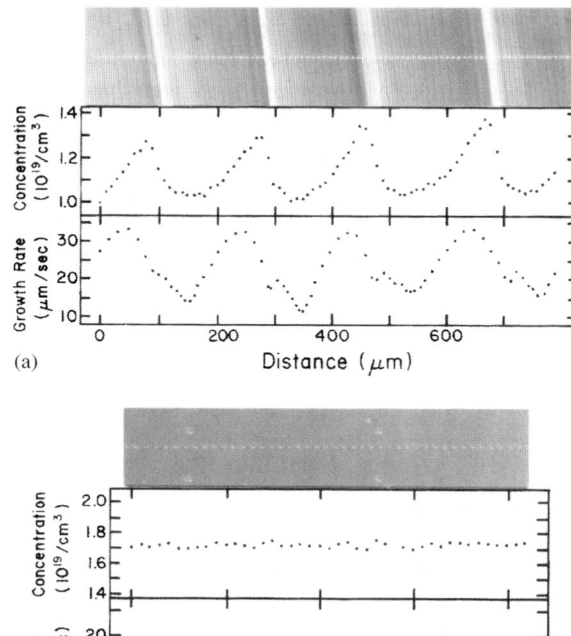

Fig. 5. Microscopic longitudinal growth rate and axial concentration profiles in Czochralski-grown Ga-doped Ge crystals. Crystal grown in (a) a conventional hot zone exhibits rotational striations and periodic growth rate while that growth in (b) an isothermal zone using a heat pipe has significantly reduced variations. (From Ref. [18], reproduced with the permission of The Electrochemical Society, Inc.)

translated directly into increased azimuthal and radial uniformity.

The elimination of microscopic growth rate variations associated with crystal rotation in an asymmetric thermal field resulted in significantly reduced radial segregation. Thus, it was possible to utilize growth rate and axial segregation data to directly compare the solute boundary layer thickness as determined from EPS theory, to that found using Cochran's analysis [19]. Under these experimental conditions the two analyses were found to be in excellent agreement. By rigorously establishing reproducible and known thermal boundary conditions, direct comparison between experimental results and theoretical models could be performed.

Witt and co-workers also demonstrated advantages of heat pipes for providing controllable thermal conditions in a vertical Bridgman configuration. This geometry was of particular interest to Witt because of its simplicity, axisymmetric geometry, constant crystal diameter, and ultimately quantifiable boundary conditions. Furthermore, this geometry lends itself to stabilizing axial temperature gradients, which minimize melt convection.

In a conventional vertical Bridgman configuration, the charge is lowered out of the furnace through a temperature gradient. This gradient zone is a critical region since the associated axial and radial temperature gradients dictate not only the microscopic growth rate, but also the intensity of melt convection and the morphology of the crystal–melt interface. In a conventional Bridgman system, it was shown that the growth rate was not only significantly different from the charge lowering rate but also that it never reached a steady-state value [20]. As the charge was extracted from the furnace, the heat flux through the crystal changed continuously. This transient growth rate was reflected in a continuous change in the location of the crystal–melt interface within the gradient zone. Consequently, the morphology of the crystal–melt interface was also transient. Furthermore, it was shown that growth rate transients were extremely sensitive to thermal end effects associated with both the charge and the furnace. Compared to the extraction rate, growth rates were either larger or smaller, depending on the boundary conditions. Thermal end effects typically will dominate growth in the conventional Bridgman geometry.

To provide better thermal control in the vertical Bridgman configuration, a three-zone furnace with isothermal heater and cooler sections that are separated by a thermal gradient zone was investigated. The isothermal zones could, in principle, be achieved by the use of heat pipes. This configuration, with its cylindrical geometry and constant, well-defined thermal boundary conditions, is extremely amenable to thermal modeling without the use of restrictive or oversimplifying assumptions. Therefore, several research groups studied various aspects of the heat transfer in this system [21]. The results indicate that temperature gradients can be adjusted by controlling thermal coupling between the furnace and charge and by variation of heater and cooler temperatures and the length

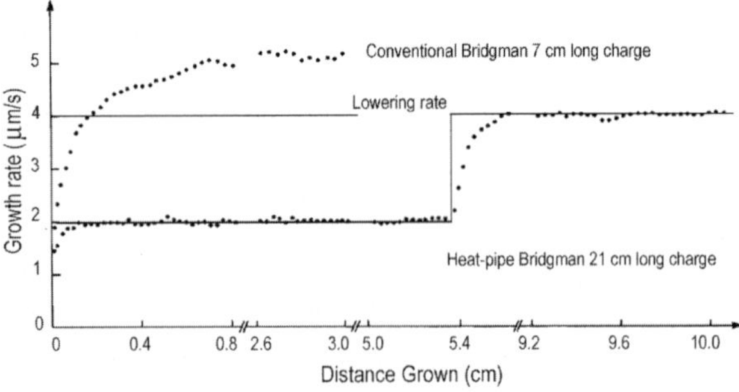

Fig. 6. Microscopic growth rates in crystals grown in conventional and heat-pipe Bridgman configurations. Ideal steady-state growth rates (growth rate equal to charge lowering rate) are obtained in heat-pipe Bridgman system with near ideal boundary conditions. (From Ref. [22].)

of the gradient zone. Under Witt's direction, a three-zone furnace with nearly ideal thermal boundary conditions was constructed and extensively characterized [22]. Experimentally determined thermal profiles were shown to be in excellent agreement with thermal models and axial and radial gradients could be controlled to a much greater extent than previously observed in conventional systems. Ga-doped Ge crystals grown in this furnace exhibited for the first time, growth without thermal end effects with steady-state growth rates equal to the charge lowering rate, as shown in Fig. 6. Furthermore, the location of the crystal–melt interface, and thus its morphology, could be controlled by adjusting the temperature of the heater and cooler and the length of the gradient zone.

With these additional growth parameters available for independent variation in the heat-pipe Bridgman system, the crystal–melt interface morphology and its influence on radial segregation were investigated. Studies using the model Ga-doped Ge system showed that the sign of the interface curvature was dominated by the thermal conductivity of the crucible material, and the magnitude of the curvature was dependent on heater and cooler temperatures. The interface was concave into the solid when a crucible with high thermal conductivity such as boron nitride was used. Conversely, it was convex into the solid when a crucible with low thermal conductivity, such as quartz was used, as shown in Fig. 7a [23]. Heat transfer studies indicated that since the melt thermal conductivity is greater than that of the solid, which is typically the case for semiconductors, and the thermal conductivity

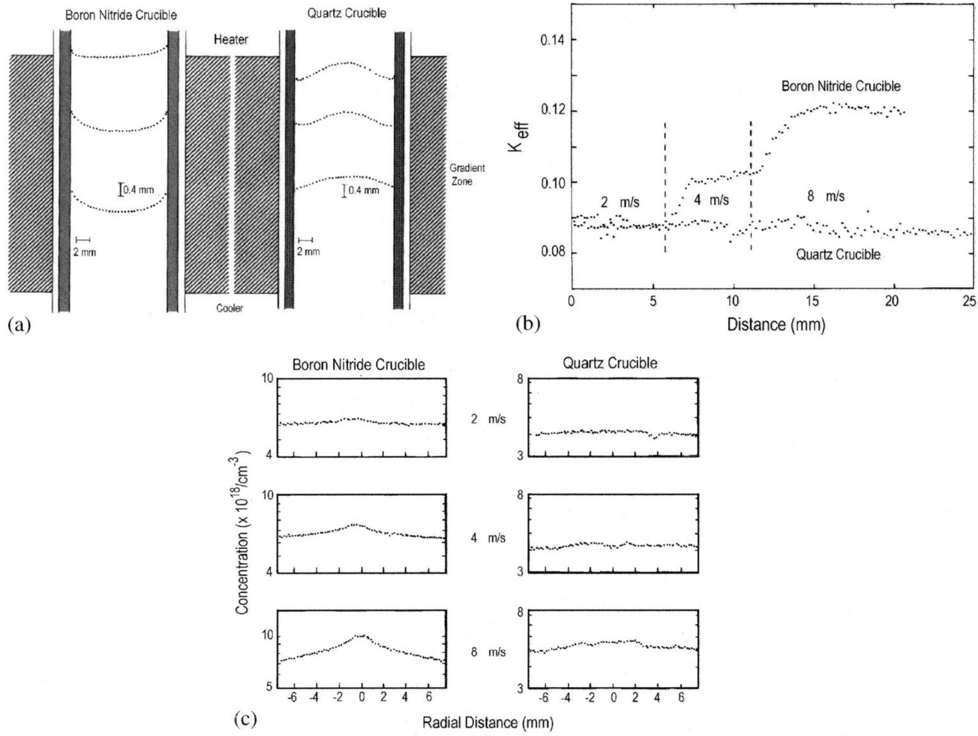

Fig. 7. (a) Crystal–melt interface morphology, (b) k_{eff} for various growth rates, and (c) radial segregation profiles of Ga-doped Ge crystals grown in the heat-pipe Bridgman system. The curvature of the growth interface depends on the crucible thermal conductivity and the charge location within the gradient zone. Complete mixing in the melt is suggested by the independence of k_{eff} on growth rate. Radial segregation profiles are dominated by melt convection rather than interface curvature. (From Ref. [23].)

of the crucible is continuous at crystal–melt interface, the heat flux from the melt at the interface must also be conducted laterally by radial thermal gradients [24]. The interface is predicted to be concave into the solid when the crucible has a similar or higher thermal conductivity, as is the case for boron nitride. Corresponding axial segregation profiles indicated that there was pronounced laminar convection in spite of stabilizing axial thermal gradients, and therefore radial temperature gradients controlled the intensity of melt convection. Values of k_{eff} were calculated, and the results shown in Fig. 7b indicate that axial segregation was dependent on the growth rate for the crystal grown in the boron nitride crucible. On the other hand, segregation was independent of growth rate for that in the quartz crucible, which suggested complete mixing in the melt during growth. The radial segregation profiles, shown in Fig. 7c, showed a higher concentration in the center of the crystal for all conditions. It was sensitive to the growth rate for the boron nitride

crucible, but not for the quartz crucible. These results indicate that radial segregation is dominated by melt convection, and not by interface curvature.

5.2. Importance of numerical modeling

Although the impact of various types of convection on segregation was long recognized, the traditional tools for quantifying convective effects in crystal growth were unable to provide the quantitative information needed for analyzing space and ground based experiments on segregation. The thermal conditions in crystal growth furnaces as well as the convection pattern in the molten charge were too complex to be handled by closed-form analytical solutions to the governing heat, momentum, and mass conservation equations. Although some of the problems with thermal boundary conditions were alleviated by using heat pipes, detailed and accurate characterization of transport in the crystal growth systems remained beyond reach.

In the early 1980s detailed numerical simulation of thermo-fluid transport was emerging as a new analysis tool, primarily because of rapid advances in the power of computers and numerical algorithms. Witt embraced and strongly supported the application of numerical simulation to crystal growth processes at MIT. Among the first works in this area was the development of a detailed model for the MIT heat-pipe based vertical Bridgman growth system [25]. Chang and Brown provided, for the first time, a detailed view of the flow structure in the melt and the interference of convection with solute rejection at the growth front. The numerical results matched the experimental observations on shape of the growth interface and dopant distribution [26]. Subsequent modeling studies were extended to include effects of external forces such as magnetic fields on segregation [27]. Oreper and Szekely [28] predicted the magnetic field strengths needed to achieve diffusion controlled conditions during Bridgman growth of Ga-doped Ge later conducted by Witt and co-workers at MIT [29]. Motakef [30,31] extended these analyses to predict the effectiveness of the microgravity environment of space and application of magnetic fields in achieving diffusion controlled growth for a large class of semiconductors.

These studies and the many that followed, both inside and outside of MIT, helped establish a quantitative framework for analyzing the relationship between, on the one hand, the geometric features of the furnace and growth parameters such as translation rate, and on the other, factors which influence segregation and crystal quality such as the shape of solid–liquid interface and intensity of melt convection. The early successes of numerical modeling of Bridgman growth provided the impetus for using this technique in nearly all crystal growth methods. For an early review see Ref. [32].

5.3. Application of magnetic fields

Efforts to reduce convection by the application of magnetic fields to electrically conducting melts were first reported by Utech and Flemings [33]. A vertical magnetic field was applied to horizontal Bridgman growth of Te-doped InSb. Temperature fluctuations in the melt were suppressed. Furthermore, the grown crystal did not exhibit the compositional striations,

which are generally present in conventionally grown crystals and are associated with back-melting of the crystal by the turbulent melt. In 1966, Chendzey and Hurle grew Te-doped InSb in a horizontal magnetic field and observed that as the field strength was increased, temperature fluctuations in the melt changed from chaotic to periodic variations, and then to stable and steady values [34]. In 1970, Witt and co-workers [35] applied a horizontal (transverse) magnetic field to Czochralski growth of Te-doped InSb. In addition to near elimination of the so-called non-rotational striations that are associated with turbulent melt convection, they observed an amplification of periodic striations caused by rotation of the crystal in an asymmetric thermal field. The importance of establishing axial symmetry during the growth process was clear. All of these magnetic field growth experiments, and the many that followed, demonstrated that application of magnetic fields could be used to modify and suppress convection during crystal growth.

Moving from the simple model systems of Te-doped InSb and Ga-doped Ge required the further development of characterization techniques [36,37]. It was found that application of a 1200 G axial magnetic field during the Czochralski growth of Si resulted in pronounced rotational striations, but a complete absence of non-rotational microsegregation effects [36]. By coupling spreading resistance analysis with high resolution Fourier Transform Infrared absorption scans, it was found that oxygen, carbon and antimony microsegregation profiles were spatially aligned [37].

Application of magnetic fields during the Czochralski growth of both GaAs and InP also received considerable attention (see for example Ref. [38]). Results based on macroscopic measurements of the crystal properties were generally interpreted within the bounds of the BPS theory, suggesting that k_{eff} approached the value of one with the application of a magnetic field during growth. Through application of microscopic composition analysis it was found that the micro- and macrosegregation results were quantifiable on the basis of BPS theory. However, the applicability of Cochran's analysis to this experimental configuration was questioned [39]. In the absence of quantitative growth rate data, it was impossible to unambiguously identify variations in the microscopic growth rate or variations in the solute boundary layer as the factor primarily responsible for the complex composition profiles created with application of a magnetic field. During magnetically stabilized growth, segregation appears to be controlled by bulk melt flows and the related characteristics of the solute boundary layer.

In 1988, Witt and co-workers [29] used magnetic fields to sufficiently suppress convective mixing in the melt to achieve diffusion-controlled segregation. To obtain such levels of quiescence, very large magnetic field strengths are generally required. Thus, they applied a 30 kG vertical magnetic field to growth of Ga-doped Ge in vertical Bridgman configuration. Diffusion-controlled growth was achieved for the first time on Earth, and pioneered application of strong magnetic fields to crystal growth of a large number of materials systems [40–42].

6. Summary

Witt's insights and contributions to the understanding of convective mechanisms as they affect the macro- and microsegregation of semiconductor crystals have significantly influenced the science and technology of crystal growth. His painstaking efforts to deconvolve the various driving forces that affect convection and segregation allowed for direct comparison of crystal growth theory with experimental results in an unambiguous fashion. His rigorous pursuit of experimental results that would confirm or refute theoretical models in many ways pushed the field of crystal growth to aggressively pursue numerical modeling to gain a more complete understanding of the growth process.

Interface demarcation provided accurate determination of the crystal–melt interface morphology with excellent time resolution and accurate determination of the microscopic rate of growth. Coupled with quantitative dopant concentration profiles on the microscale, his work led to a clear understanding of the roles of thermal asymmetry, thermal profiles, and natural thermal convection on macro- and microsegregation observed in semiconductor crystal growth. This knowledge was essential to gain control of the growth process.

Witt embraced the interdisciplinary nature of crystal growth in order to identify and understand the relationships between critical crystal growth parameters and the structure and properties of the grown crystals.

He was one of first scientists to realize the potential that was offered by conducting experiments in a microgravity environment. With his space experiments, he was able to test the veracity of theoretical predictions that were not rigorously testable on Earth, as well as to elucidate additional phenomena not present in terrestrial experiments.

He also embraced the power of numerical simulation for crystal growth analysis. The interaction between Witt's research group and those involved with numerical simulation helped transform numerical modeling from a tool describing experiments already conducted to a predictive tool that is now routinely used in design of crystal growth equipment and processes.

August F. Witt was passionate about understanding the fundamental forces at play during the growth of crystals from the melt. He dedicated his career to searching for the causes underlying the experimentally observed effects.

Acknowledgements

The authors are especially indebted to Robert Feigelson of Stanford University for his patience and encouragement in the preparation of this manuscript.

References

[1] W.A. Tiller, K.A. Jackson, J.W. Rutter, B. Chalmers, Acta Metall. 1 (1953) 428.

[2] W.G. Pfann, Acta Metall. 1 (1953) 763;
W.G. Pfann, Zone Melting, Wiley, New York, 1966.

[3] J.A. Burton, R.C. Prim, W.P. Slichter, J. Chem. Phys. 21 (1953) 1987.

[4] A.F. Witt, H.C. Gatos, J. Electrochem. Soc. 113 (1966) 808.

[5] K. Morizane, A.F. Witt, H.C. Gatos, J. Electrochem. Soc. 114 (1967) 738.

[6] D.T.J. Hurle, E. Jakeman, E.R. Pike, J. Crystal Growth 34 (1968) 633;
D.T.J. Hurle, in: H.S. Peiser (Ed.), Crystal Growth, Pergamon, Oxford, 1967, p. 659.

[7] J.R. Carruthers, in: W.R. Wilcox, R.A. Lefever (Eds.), Preparation and Properties of Solid State Material, vol. 3, Marcel Dekker, New York, 1977, p. 1.

[8] A.F. Witt, H.C. Gatos, J. Electrochem. Soc. 114 (1967) 413;
A.F. Witt, H.C. Gatos, J. Electrochem. Soc. 115 (1968) 70.

[9] R. Singh, A.F. Witt, H.C. Gatos, J. Electrochem. Soc. 115 (1968) 112;

M. Lichtensteiger, A.F. Witt, H.C. Gatos, J. Electrochem. Soc. 118 (1971) 1013.

[10] K.M. Kim, A.F. Witt, H.C. Gatos, J. Electrochem. Soc. 119 (1972) 1218.

[11] J.W. Rutter, B. Chalmers, Can. J. Phys. 31 (1953) 15.

[12] A.F. Witt, M. Lichtensteiger, H.C. Gatos, J. Electrochem. Soc. 120 (1973) 1119.

[13] A.F. Witt, M. Lichtensteiger, H.C. Gatos, J. Electrochem. Soc. 121 (1974) 787;
A. Murgai, H.C. Gatos, A.F. Witt, J. Electrochem. Soc. 123 (1976) 224;
K.M. Kim, A.F. Witt, M. Lichtensteiger, H.C. Gatos, J. Electrochem. Soc. 125 (1978) 475.

[14] A.F. Witt, H.C. Gatos, M. Lichtensteiger, M.C. Lavine, C.J. Herman, J. Electrochem. Soc. 122 (1975) 276.

[15] A.F. Witt, H.C. Gatos, M. Lichtensteiger, M.C. Lavine, C.J. Herman, J. Electrochem. Soc. 125 (1978) 1832.

[16] S.R. Coriell, R.F. Sekerka, J. Crystal Growth 46 (1979) 479;
S.R. Coriell, R.F. Boisvert, R.G. Rehm, R.F. Sekerka, J. Crystal Growth 54 (1981) 167.

[17] J. Steininger, T.B. Reed, J. Crystal Growth 13/14 (1972) 106.

[18] E.P. Martin, A.F. Witt, J.R. Carruthers, J. Electrochem. Soc. 126 (1979) 284.

[19] W.G. Cochran, Proc. Cambridge Philos. Soc. 30 (1934) 365.

[20] C.A. Wang, A.F. Witt, J.R. Carruthers, J. Crystal Growth 66 (1984) 299.

[21] T.F. Fu, W.R. Wilcox, J. Crystal Growth 48 (1980) 416;
R.J. Naumann, J. Crystal Growth 58 (1982) 569;
T. Jasinski, W.M. Rosenhow, A.F. Witt, J. Crystal Growth 61 (1983) 339.

[22] C.A. Wang, Crystal Growth and Segregation in Vertical Bridgman Configuration, Ph.D. thesis, Massachusetts Institute of Technology, 1984.

[23] C.A. Wang, J.R. Carruthers, A.F. Witt, in: Seventh American Conference on Crystal Growth, Monterey, CA, July 1987.

[24] T. Jasinski, A.F. Witt, J. Crystal Growth 71 (1985) 295;
R.J. Naumann, S.L. Lehoczky, J. Crystal Growth 61 (1983) 707.

[25] C.J. Chang, R.A. Brown, J. Crystal Growth 63 (1983) 343.

[26] P.M. Adornato, R.A. Brown, J. Crystal Growth 80 (1987) 155.

[27] D.H. Kim, P.M. Adornato, R.A. Brown, J. Crystal Growth 89 (1988) 339.

[28] J.M. Oreper, J. Szekely, J. Crystal Growth 67 (1984) 405.

[29] D.H. Matthiesen, M.J. Wargo, S. Motakef, D.J. Carlson, J.S. Nakos, A.F. Witt, J. Crystal Growth 85 (1988) 53.

[30] S. Motakef, J. Crystal Growth 102 (1990) 197.

[31] S. Motakef, J. Crystal Growth 104 (1990) 833.

[32] R.A. Brown, AIChE J. 34 (1988) 881.

[33] H.P. Utech, M.C. Flemings, J. Appl. Phys. 37 (1966) 2021.

[34] H.A. Chedzey, D.T.J. Hurle, Nature 210 (1966) 933.

[35] A.F. Witt, C.J. Herman, H.C. Gatos, J. Mater. Sci. 5 (1970) 822.

[36] K.H. Yao, A.F. Witt, J. Crystal Growth 80 (1987) 453.

[37] D.J. Carlson, A.F. Witt, J. Crystal Growth 91 (1988) 239.

[38] J. Osaka, H. Kohda, T. Kobayashi, K. Hoishikawa, Jpn. J. Appl. Phys. 23 (1984) L195;
H. Kohda, K. Yamada, H. Nakanishi, T. Kobayshi, J. Osaka, K. Hoshikowa, J. Crystal Growth 71 (1985) 813.

[39] D.J. Carlson, A.F. Witt, J. Crystal. Growth 116 (1991) 461.

[40] J.C. Han, P. Becla, S. Motakef, J. Crystal Growth 121 (1992) 394.

[41] D.A. Watring, S.L. Lehoczky, J. Crystal Growth 167 (1996) 478.

[42] Y.J. Park, S.-K. Min, S.-H. Hahn, J.-K. Yoon, J. Crystal Growth 154 (1995) 10.

The origin of Czochralski growth through B_2O_3 glass: a step in the evolution of LEC growth

R. Mazelsky

Westinghouse R&D, Pittsburgh, PA 15235

It has been over 25 years since we first reported on the growth of single crystals using B_2O_3 as an encapsulant. The process was developed to address the specific problems involved with the growth of IV–V compounds and has, since then, evolved into an industrial process and the major method for growing GaAs and other III–V Czochralski crystals. As the technology has matured, the new equipment, techniques, and knowledge have led to enormous growth. Liquid encapsulation has become a commonly used technique and I was pleased, therefore, when Bob Feigelson asked me for a retrospective on the genesis of the liquid-encapsulated Czochralski technique at Westinghouse.

Westinghouse was interested in thermoelectric materials and among these were lead telluride and lead selenide. We were trying to grow crystals of controlled and reproducible composition in order to study their electronic transport properties. As you know, nature tends to be uncooperative and, when melted, lead and the chalcogenides have finite and different partial pressures. It occurred to us that if we could cover the surface with a liquid and maintain a pressure on the outer surface of the liquid in excess of the partial pressure of the volatile species, evaporation would be suppressed and we could grow homogeneous crystals. We tested this idea using boron oxide as a cover liquid and successfully grew crystals. Our paper was submitted to the *Journal of Applied Physics* in November 1961 and

Previously published in AACG Newsletter Vol. 17(2) July 1987.

published the following year [1]. This work was performed before acronyms were as common as they are presently and throughout the paper we referred to the technique as "the technique." J.B. Mullin and coworkers [2,3] were pioneers in applying liquid boron oxide to GaAs in the mid-sixties and provided the first demonstration that the liquid encapsulation technique (LEC) could be used as a commercial process. The evolution of LEC CZ growth to today's technology provides an interesting contrast to our first experiments.

Figure 1 shows the starting materials, a melt charge and a GaAs crystal (courtesy of Noel Thomas at

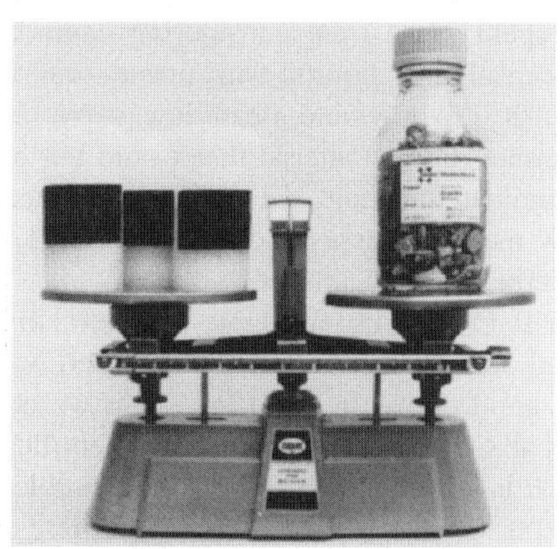

Fig. 1. Starting materials.

Fig. 2. Pyrolytic boron nitride CPBN crucible with premelted charge.

Fig. 3. Removal of GaAs boule from furnace after growth.

the Westinghouse R&D Center). A 6″ (or larger) pyrolytic boron nitride (PBN) crucible is placed in the chamber of a Cambridge Instruments Melbourne HP-LEC puller. The operator oversees the operation remotely using video cameras. In contrast, our initial work was performed in a 2″ diameter graphite crucible heated by 450 kHz generator with a power controller. The crucible was contained in a quartz chamber 4″ in diameter which was sealed and had provision for both crucible and seed rotation. A schematic of our system is shown in Figure 2. It accommodated approximately 150 g of PbTe and 10 g of B_2O_3, sufficient to provide an encapsulating layer of approximately 3 mm. The crucible was loaded from the bottom and required a reasonable degree of flexibility to set up the run.

We spent some time considering candidates for the encapsulating liquid. Our prime criteria were low density so it would float, low vapor pressure, transparency, and a material that would not be likely to contaminate the melt. We looked at various halides and low melting glasses and decided that B_2O_3 met our criteria best.

Pure B_2O_3 was not available commercially so we prepared our glass from boric acid, which was available in good purity. In our 1962 paper we described the preparation process with typical scientific detachment: "It is advantageous to heat the crucible rapidly to $\sim 1100\,^\circ$C and hold the temperature there until the frothing ceases." Our early experiments in the dehydration of boric acid were, in fact, less straightfor-

ward. We learned very quickly that moderate heating rates resulted in a foam-like solid and a very large volume increase. Continuing to heat to melting resulted in glass spilling over the crucible and covering the crucible pedestal and its support posts. Although these experiments provided useful information on the tendency of B_2O_3 to adhere to other materials, it did not put the B_2O_3 where we wanted it. Fortunately, B_2O_3 is soluble in hot water and the parts were salvageable. We finally developed a procedure involving low-temperature preheating in which we shut off the power when the foam reached crucible height; then repeated the process until a portion of the water from hydration was removed. We were then able to rapidly heat the glass and melt it before it spilled over the crucible. In this way we were able to produce the B_2O_3 discs which are now purchased so readily.

Preparing the charge for crystal growth was done similarly to current practice. The elements were weighed and reacted in situ with B_2O_3. Care had to be used in cleaning the crucible and starting materials to maintain clarity of the B_2O_3 encapsulant. We melted lead telluride covered by molten B_2O_3 glass at a pressure slightly over 1 atmosphere in the system. The melt was quiescent and clearly visible through the glass encapsulant; there was no visible evaporation and condensation of lead or tellurium. We had believed that our idea would work. Now we had a convincing demonstration that the concept was sound.

Our initial attempts at seeding and crystal growth were frustrating. The refractive index difference be-

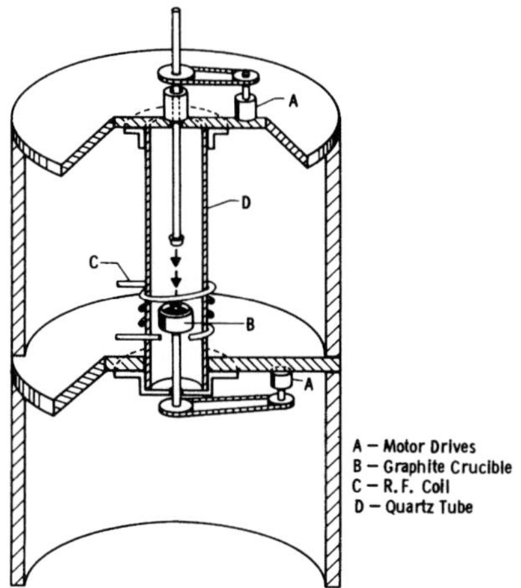

Fig. 4. Schematic of Czochralski crystal growth furnace.

tween B_2O_3 and air made contacting and meltback of the seed difficult to observe. We observed shear effects at the interface when the relative rotation of the seed and melt was varied by a few rpm. And it was difficult to see the interface during widening and growth. The fact that we were finally able to grow crystals consistently was due in large measure to Ernie Metz's patience, perseverance, and ability. When Bob Miller developed the idea for setting up a circuit between the seed and the crucible and used an ohmmeter for detecting contact between the seed and the melt, the procedure became more routine. We then had a sensitive technique for determining the touching of the seed to the melt by observing when the ohmmeter showed a change from open circuit to low resistivity. Considerable art was still required to grow crystals and, in fact, some of our earlier attempts yielded ingots that were more pleasing aesthetically than scientifically. Nevertheless, the procedure was developed for the growth of traditional cylindrical ingots. These grew with a film of B_2O_3 which encapsulated the solid as well as the liquid; the glass was later removed in hot water. The boules we grew were good-quality single crystals and,

within the limits of our measurements at that time, uncontaminated by B_2O_3.

In some respects we were fortunate to have achieved the degree of success that we did. The melting point of PbTe is in the range where the viscosity of glass is high but acceptable; the vapor pressure of the system is low and suppressed by bucking pressures of a little over an atmosphere allowing us to use a relatively simple growth system; and in spite of thermal complexity introduced by the low thermal conductivity, optically transparent glass, conditions satisfactory for crystal growth were achieved. Mullin and coworkers' dramatic advance of the technique to the higher temperature, high-pressure III–V systems has been a noteworthy achievement and has been the significant factor in the advancement of gallium arsenide technology. It is also now recognized that the thermal effects associated with the liquid encapsulant are significant and affect crystal perfection. Rapid progress continues to be made. I am gratified that we were able to make an early contribution to this exciting field.

Technical articles focus on data. This article provides me with the opportunity to recognize the success of a less tangible element. From this point of view I am grateful for the synergism and teamwork of Ernie Metz and Bob Miller, which was a learning experience for me, and to have worked in an environment where management and staff were enthusiastic and supportive of new ideas, all of which made this project a particularly rewarding one.

References

[1] E.P.A. Metz, R.C. Miller, R. Mazelsky, JAP 33 (6) (June 1962) 2016–2017.
[2] J.B. Mullin, B.W. Straughan, W.S. Brickel, J. Phys. Chem. Solids 26 (1965) 82.
[3] J.B. Mullin, R.J. Heritage, C.H. Holliday, B.W. Straughan, J. Crystal Growth 3 (4) (1968) 286.

Progress in the melt growth of III–V compounds

J.B. Mullin [a,b,*]

[a] Electronic Materials Consultancy, 22 Branksome Towers, Pools BH13 6JT, UK
[b] University of Durham, UK

Abstract

Semiconductor behaviour was predicted and discovered in the III–V semiconductor compounds almost 50 years ago at the beginning of the 50s. The demand for high-purity single crystals was axiomatic in view of the prior pioneering research on germanium as a semiconductor device material. However, while the development of semiconductor grade Ge took about a decade, the comparable development of the readily dissociable III–V compounds have followed a very much slower evolution, which is still in progress. This review discusses some of their intrinsic material problems and some of the solutions that have required so much research and development effort to produce single crystal compounds suitable for high-performance devices. The primary focus of the paper is the compounds of major commercial importance, GaAs and InP. While most of the problems that arise in the growth of high-quality single crystals of these materials have been endured over the years, new and more refined technical and scientific solutions to these have been developed. The growth techniques that will be considered include Liquid Encapsulation, Vertical Gradient Freeze, Vapour Pressure Controlled Czochralski and Hot Wall Pulling techniques including Pressure Balancing. The practical constraints to implementing these techniques and their advantages and limitations are considered. Problems of twinning, cellular structure, dislocation formation, lineage, constitutional supercooling and defects are reviewed. The significance of technique in relation to recent developments in commercial exploitation, especially with regard to size, is noted. In addition, related areas of scientific interest which have either not been researched significantly or could be of future significance such as the potential of electromagnetic stirring are highlighted.
© 2004 Elsevier B.V. All rights reserved.

PACS: 81.05.Ea; 81.10.-h; 81.10.Fq

Keywords: A2. Bridgman technique; A2. Czochralski method; A2. Growth from the melt; A2. Single crystal growth; B2. Semiconducting III–V materials

1. Introduction

This review is a revised version of the Seattle talk. It is not a definitive review. Nevertheless, it aims to identify and discuss some of the significant developments in the melt growth of III–V compounds. It represents the author's perspective on the topic and will concentrate on the more important developments involving Liquid Encapsulation Czochralski (LEC), particularly as applied to GaAs, InP and GaP. In keeping with the spirit of the 50 years symposium, it will highlight in a less formal way some of the problems encountered in the melt growth of III–Vs and trace their solutions in more recent work. More

---

* Corresponding author. Electronic Materials Consultancy, 22 Branksome Towers, Poole BH13 6JT, UK.

E-mail address: brian.mullin@btinternet.com (J.B. Mullin).

0022-0248/$ – see front matter © 2004 Elsevier B.V. All rights reserved.
doi:10.1016/j.jcrysgro.2003.12.036

Originally published in
J. Crys. Growth 264 (4) (2004) 578–592.

formal reviews can be found in the literature [1–6], where many earlier reviews are referenced.

The discovery of transistor action in germanium by Brattain and Bardeen [7] in 1948, a result stimulated by the predictions of Shockley [8], brought about the evolution of the semiconductor era. The major international effort following the discovery of transistor action was focussed on germanium. The benign chemistry of germanium enabled it to be used as a splendid vehicle for materials research. Much of our crystal growth knowledge stems from the research on this material. Indeed, the framework of the melt growth of semiconductors was created by the pioneering work of Pfann [9] on zone melting, which encompassed zone refining, and by Teal and Little [10] on the vertical pulling of single crystals of germanium.

Two overriding requirements for semiconductors emerged: the need for high purity and the need for single crystals. The first requirement resulted in the creation of new standards of purification, the significant evolution being that of the concept of semiconductor purity, a specification demanding unprecedented low levels of impurities, typically less than 10 parts per billion atomic (ppba) of electrically active impurities. The second requirement resulted in the development of new technologies for producing completely single crystals free from defects including dislocations.

The intense interest in semiconductors sparked a quest for new materials that exhibited semiconductor properties. This quest resulted in the invention of the III–Vs. Welker and his colleagues [11,12] in Germany were pioneers in the early 50s in this activity. Their participation, however, was short lived. Siemens made what proved to be the injudicious choice of AlSb, a material with cheap components, as a program target. It was doomed to failure by the ease of oxidation of he polycrystalline material. It fell to pieces by oxidation at the grain boundaries.

Theoretical work in the early 50s at the Radar Research Establishment (RRE subsequently renamed RSRE, then DERA, and then split to form QinetiQ) Malvern, England had predicted potential IR absorption capability in the III–Vs. There had been studies during the war on the infrared, but a strategic decision was taken to drop this work in order to concentrate on their pioneering work on radar. A program of work on infrared detectors, which included the lead salts, was initiated following the end of the 1939–1945 war.

Table 1
Property constraints to crystal growth

III–V compound	Melting point ($^{\circ}$C)	Vapour pressure at m.pt. (atm)	CRSS at m.pt. (MPa)
InSb	525	4×10^{-8}	
GaSb	712	1×10^{-6}	
InAs	943	0.33	
GaAs	1238	~ 2.2	0.7
InP	1062	27.5	0.36
GaP	1465	32	

The realisation that the energy gap of InSb covered the 3–5 μm window in the atmosphere resulted in a major program of work on InSb. This was the source of much of the author's early research activity. Like germanium, InSb proved to be an excellent vehicle for basic material studies on purification, segregation and crystal growth [13–15], resulting in the rapid accumulation of a very valuable fund of basic knowledge.

InSb followed the pattern of development of Ge and completely dislocation-free single crystals of InSb containing less than 10^{13} carriers cm^{-3} (1 ppba is equivalent to 2.9×10^{13} atoms cm^{-3}) were routinely grown both by the horizontal and vertical pulling techniques within a decade. This rapid development was not matched by the melt growth of the other III–Vs. All of them, with the exception of the antimonides, possess significant vapour pressures of the group V component at the melting point, as can be seen in Table 1. Vapour pressure is probably the most crucial parameter that determines the technology needed to grow them from the melt. As a result, advances in the growth technologies for the difficult III–V compound semiconductors, the arsenides, phosphides and nitrides, have taken over three decades and they are still in a development phase.

2. Evolution and limitations of early growth techniques for GaAs, InP and GaP

The III–Vs of major commercial interest are GaAs, InP and GaP. I have excluded the nitrides from this review. The bulk material can be grown from solution under very high pressure. The driving force for the development of GaAs, InP and GaP is their key role in semiconductor devices and modern electronics. Up to the dot com debacle, the compound semiconductor market had doubled about every 6 years and was increasing to some 25–33% per year. That is an enormous increase. Very roughly, the market is divided into two similarly sized fields of application, essentially the electronic field driven by high-frequency application, and optoelectronics which is driven by optical applications. They require differently specified material, which has affected the evolution and relative importance of different growth technologies. In connection with the development of the III–Vs, indeed in semiconductors in general, one must distinguish between the development of a commercial technique and the development of the most scientifically advanced one. The former is driven by the need to fulfil a commercial specification at the minimum cost. Cost is an increasingly important factor in material and device technologies. Cost is directly related to yield. But here, industrial information is not normally freely available. So, predicting where a technology will go in the future is not straightforward. But it is interesting to speculate.

2.1. Horizontal Bridgman

The Horizontal Bridgman (HB) technique was the first growth technology to be exploited commercially for GaAs. The crystal was grown in a sealed silica tube with or without a simple boat. The technology was used in the production of doped n-type GaAs, and gained a significant foothold in the LED market. The interesting feature of this type of technology is that it still is a major source of material for the huge run of the mill LED market. The small individual devices do not demand a crystal of precisely controlled shape. Hence, boat-grown material is acceptable.

Horizontal growth has important advantages. The growth geometry readily lends itself to the establishment of low-temperature gradients at the solid–liquid interface, without creating a control problem. By contrast, relatively high-temperature gradients are needed to maintain control of the shape of the crystal in the pulling process. Low-temperature gradients are extremely important in minimising stress-induced slip during crystallisation, and hence in minimising dislocation formation. In the case of HB growth, GaAs has a low dislocation density, typically 10^2–10^3 dislocations cm^{-2}, which is a factor of at least a hundred less than is found in routinely grown LEC crystals. High dislocation densities can be a problem for laser diodes based on GaAs where even a single dislocation can readily bring about device failure.

There are also disadvantages of HB. These are both fundamental and preparation related, such as growth orientation, contamination, strain and shape. One of the fundamental problems that is not widely recognised is the problem of constitutional supercooling caused by a non-stoichiometric melt. This may occur from a lack of accurate starting composition or the independent vapour pressure control of the group V component. It is important to realise that the equilibrium vapour pressure of As over a stoichiometric melt of GaAs was believed to be 1.0 atm [16], whereas more recently it has been shown to be close to 2 atm [17]. The topic is discussed in a related review on defects by Hurle [18]. Constitutional supercooling can be especially troublesome when growth occurs with low-temperature gradients. The most troublesome problems, however, generally concern the contact of the crystal with the boat. Misnucleation from the walls of the tube or boat can give rise to twinning and more often polycrystallinity. Also, crystallisation in the confining shape of the boat with materials like III–Vs, if combined with localised sticking, will inevitably lead to stress, slip and dislocation formation. However, provided non-wetting surfaces are used for the containing boats and a non-confining boat shape is used, this problem can be minimised.

Attempts to grow semi-insulating (SI) material for the electronics market were bedevilled by the introduction of silicon (an n-type dopant) into the grown crystal by Ga attack on the SiO$_2$ boats [19]. This made production less reliable for SI material required for electronic applications. The development of more reliable technologies came with the introduction of boron nitride (BN) crucibles [20], and subsequently the introduction of the LEC process. An important aspect here is the fact that B$_2$O$_3$ can remove low concentrations of Si in GaAs melts. This stimulated the use of LEC for SI GaAs.

2.2. Vertical Bridgman

The vertical translation of ingots can provide practical advantages over horizontal growth. The Stockbarger technique, often referred to as the vertical Bridgman method, involves the lowering or withdrawal of the ingot vertically from a furnace. This technique was not initially considered to be suitable for semiconductor crystal growth. Semiconductors expand on freezing. The resulting constraint could be expected to cause polycrystallinity, especially for ingots adhering to their containers. More recent developments [21] highlight the need for appropriate crucible treatment to improve singularity and yield. The vertical Bridgman technique does, however, offer scope for reduced cost, and Sumitomo [22] have reported the successful development of 6 in (150 mm) crystals suitable for devices in the fast developing communications field.

InP and GaP were also the subject of intense scientific and technological study using closed-tube technologies. A number of these technologies, for example, sealed systems with magnetic coupling, are discussed in earlier reviews [23]. The very high pressures developed in these systems posed very real hazards for these intrepid pioneering crystal growers.

3. Liquid-encapsulation technologies

The invention of the Liquid Encapsulation technique [24] subsequently called the Liquid Encapsulation Czochralski (LEC) process, made a decisive impact on the growth technologies for all the III–Vs. The process in its low-pressure mode is illustrated in Fig. 1. It was used for the growth of both InAs and GaAs.

The low-pressure system used a silica growth chamber that was some 150 mm in diameter. The inert gas pressure required to suppress evaporation of the group V component was thus limited to a few atmospheres on grounds of safety.

The growth of single crystals of both InP and GaP [25], having dissociation pressures of 28 and 32 atm, respectively, at the melting point, clearly required a pressure vessel. A diagram of the initial design is shown in Fig. 2. An experimental model fitted with a weighing machine and a TV-viewing system is shown in Fig. 3. The development of the pressure vessel design was extremely interesting and raised many unexpected problems from the limitations of castings technology to the use of silica windows in compression and the strain relief of indium seals at room temperature. The human problems, however, were the more stressful. There was an imperative need to demonstrate that appropriately designed pressure

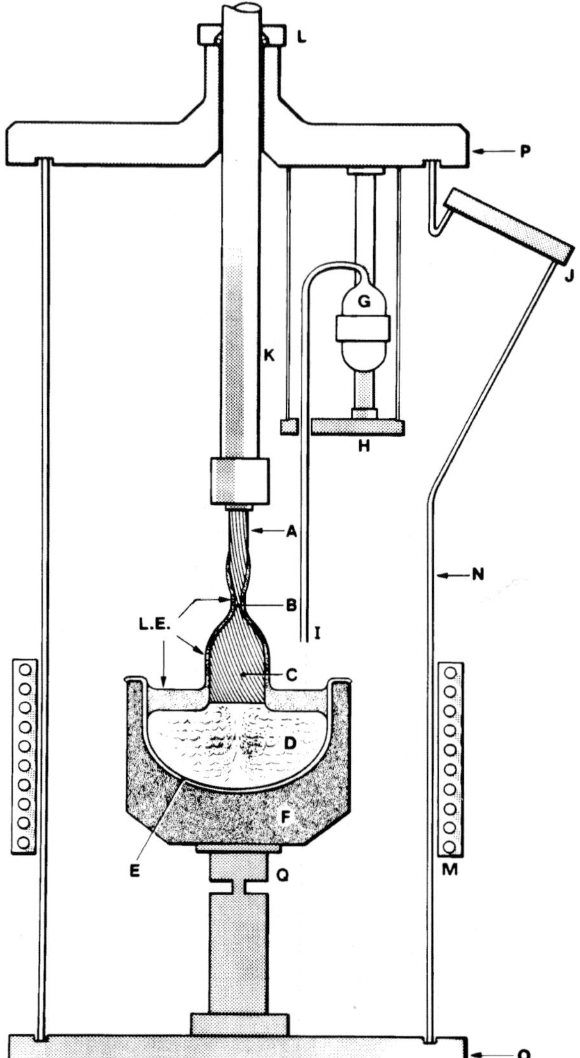

Fig. 1. Vertical pulling apparatus for Low Pressure Liquid Encapsulation. The silica outer vessel N with viewing port J is held between end plates O and P. The induction heating coils couple into the graphite surround F mounted on Q. The seed A is fixed in the chuck on the pull rod K that rotates and moves through the bearing and seal L. The crystal C grows from the seed through a necking process at B, and on withdrawal pulls out a layer of B_2O_3 over its surface. Loss of the volatile group V component from the seed, crystal and melt is prevented if the inert gas pressure in the chamber is greater than the group V dissociation pressure.

vessels could be safe to operate for the growth of these III–Vs.

The solution to this problem was relatively straightforward. A growth chamber was taken to an exper-

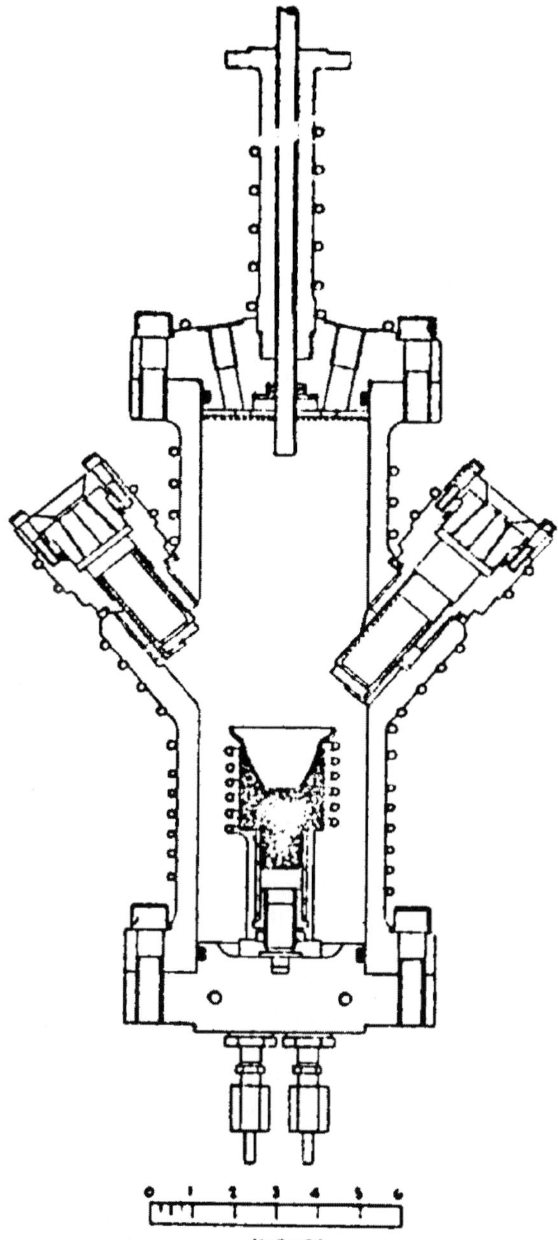

Fig. 2. Two hundred atmosphere steel high-pressure chamber used for pressure testing and crystal growing. The inner silica window viewing assembly is retracted on the left port, but lowered on the right.

imental explosives establishment, completely filled with cordite (effectively gun powder), and when suitably located in a safe bunker the charge was ignited. The pressure in the chamber rose to some 10,000 atm

Fig. 3. High-pressure LEC crystal puller developed at RSRE, showing water-cooled steel pressure vessel and two optical ports for viewing, one fitted with a video camera. Below the steel pressure vessel is a large chamber containing the weighing cell for diameter control.

in a few milliseconds. The chamber bulged, but held the pressure. The experiment clearly demonstrated that the vessels offered a considerable margin of safety. There were many other technical problems which I will not go into. The 200 atm was intended to grow II–VIs, but the development did not materialise because of the incompatibility with II–VIs with B_2O_3. Various growth problems were encountered in the development of the LEC technology, and these and some of their solutions are presented in the following section.

4. Growth constraints in LEG and related technologies

4.1. Evaporative group V loss from the crystal emerging from the B_2O_3

Melts of the III–Vs rapidly lose their group V component, unless there is a pressure above the melt at least equal to the equilibrium vapour pressure of the group V component over the melt. This applies to all the arsenides and phosphides. As the crystal emerges above the surface of the B_2O_3, it can lose its inert protective layer. The inert gas pressure then cannot prevent group V evaporation from the exposed surface of the crystal. In the case of GaAs, a problem arises with the Ga which is liberated. It forms droplets that migrate under the applied temperature gradient into the crystal by Temperature Gradient Zone Melting (TGZM). Ultimately, since the crystal grows faster than the droplet motion, the latter freezes in the solid. The resulting differential contraction between the frozen liquid droplet and the surrounding GaAs causes dislocation clusters.

The problem with InP is exacerbated by the relatively high inert gas pressure (in excess of 28 atm). Pressures in excess of 50 atm were often used in the mistaken belief that the higher the pressure, the greater the reduction in evaporative loss. Unfortunately, the higher pressures result in turbulent gaseous convection. The high gas velocities are the result of Rayleigh convection, which is driven by the high pressure, the large temperature differences and relatively large dimensions of the Bénard cells. The magnitude of the convection in pressure-pulling systems correlates with the value of the Rayleigh number Ra, as related in

$$R_a = (T_h - T_c)gd^3 P^2/[T_m K_0 v_0], \qquad (1)$$

where $(T_h - T_c)$ is the vertical temperature difference across the volume of the convecting gas (the temperature difference between surfaces driving the Bénard cell), and T_m is an average gas temperature, d is the depth of volume of the convecting gas, K_0 is the thermal diffusivity, v_0 is the kinematic viscosity and P is the gas pressure. The critical significance of R_a is that its magnitude depends on the square of the gas pressure, the cube of d and the temperature difference between surfaces driving the convective Bénard cell. It is important, therefore, in the pulling systems, to avoid large free volumes with large temperature differences between hot and cold surfaces.

Increasing gaseous convection enhances evaporative loss from the surface of the crystal, as it emerges from the surface of the B_2O_3. This can be a big problem for InP.

Two types of technology have been developed to deal with it. In addition to Liquid Encapsulation, the so-called hot wall technology has been used. In hot wall technology, the walls of the containing vessel surrounding the charge of the semiconductor are kept sufficiently hot to prevent condensation of the group V element on the walls of the container. In the case of arsenic or phosphorus, this requires a temperature of $\sim 600\,^{\circ}C$ or $\sim 700\,^{\circ}C$ for the respective elements. Clearly, this constraint is much easier to apply in the case of a horizontal crystal-growing apparatus than in the case of a vertical pulling apparatus. In the latter case, all the pull rod seals need be kept at a temperature determined by the need to maintain an adequate pressure of group V vapour.

4.2. Vapour pressure-controlled Czochralski (VCz)

One solution to evaporative loss in the case of InP was reported in a patent by Azuma [26] in 1983. Here, it was proposed to arrange for phosphorous to be deposited on the heated walls of the upper part of the chamber. The temperature of this P_4 source could be adjusted so as to maintain a balancing vapour pressure of P_4 that prevented evaporative loss from the crystal as it emerged through the surface of the B_2O_3. In addition, the pull rod was pulled through a cup-shaped reservoir of B_2O_3 that acted as a liquid seal. However, 10 years prior to this, the author and his colleagues [27] reported a pressure-balancing technology that predated the Japanese patents. The concept is illustrated in Fig. 4 where the crystal is grown in a vessel using the hot wall concept. The versatility of the system is discussed in the paper [27], where the ability to form the III–V in situ and by implication the ability to control the group V pressure independently from a separate source are discussed. A novel feature of the system is the use of a liquid seal of B_2O_3, which allows one to equilibrate the pressure in the system. In addition, a thread on the BN bearing acts as an Archimedian screw and prevents the B_2O_3 from running down the pull rod as it rotates. This worked extremely well. The growth of crystals of InAs and GaAs [27] demonstrated the system's efficacy.

Tada and Tatsumi [28] reported a system without the B_2O_2 upper seal in a patent in 1984. A version of this VCz technology has been developed and reported by Rudolph et al. [3,29]. A diagram of their apparatus is shown in Fig. 5. Rudolph et al. [4] developed their technique and used it for the growth of single

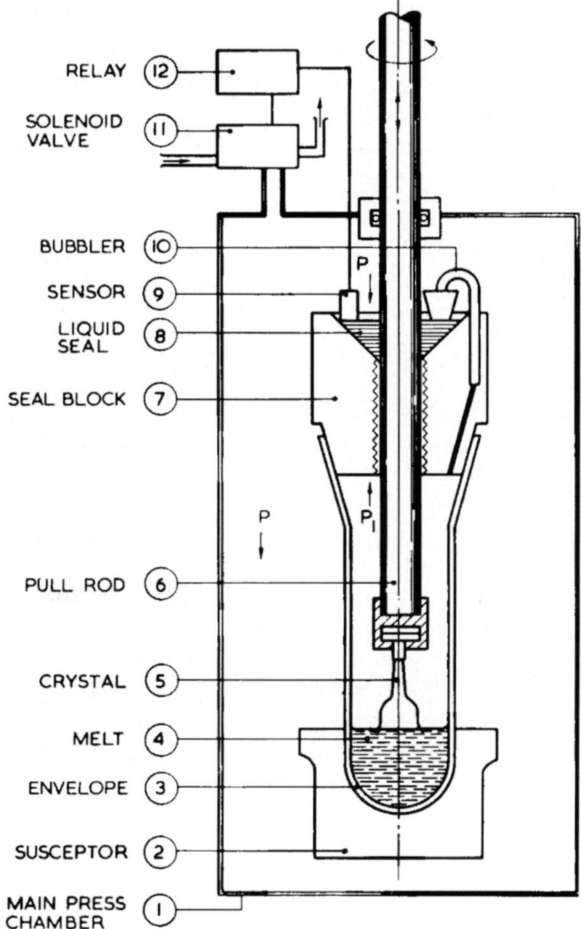

Fig. 4. A schematic diagram showing the principle of pressure balancing. The walls of the inner quartz envelope are maintained at temperature to prevent group V condensation. The inert gas pressure P in the outer chamber is maintained at a pressure to balance the dissociation pressure of the group V component, which is sensed by the surface position of the B_2O_3 liquid seal.

crystals of GaAs. The lower temperature gradients that are inherent when using this technology make the control of crystal growth diameter more problematic compared with conventional technology. Very careful thermal design is needed.

The more recent developments in the case of LEC InP relate to minimising convection and its deleterious effects. Low dislocation density is very important for device applications, and there is an imperative need to reduce their densities well below the norm of 10^4–10^5 cm^{-2} found in undoped and

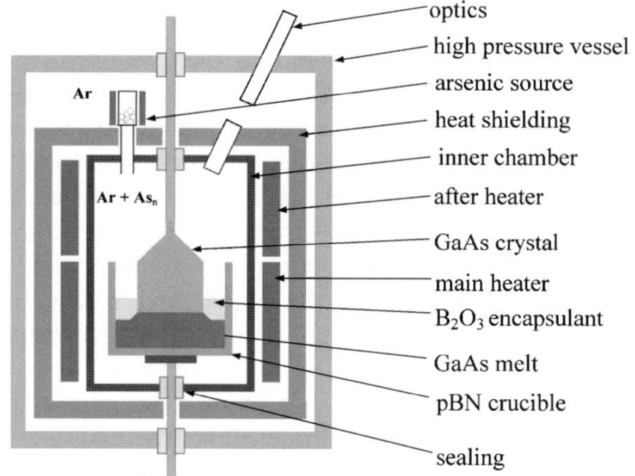

Fig. 5. A schematic diagram of the Rudolph et al. [3] VCz system which is designed to provide a controlled but changeable pressure of arsenic vapour.

lightly doped material to between 10^3 and 10^2 cm^{-2} for many device applications. Attempts to reduce the temperature gradients and reduce the dislocation densities have been reported by Hirano et al. [30]. They used a system of double heat shields or baffles. This was done in a way that minimised P_4 loss, presumably by minimising gaseous convection.

A measure of the commercial exploitation of GaAs and InP is the relentless drive to grow larger diameter crystals. For GaAs, there is currently (in 2003) a general movement to the growth of 150 mm diameter crystals [3,31,32], and Seidl [33] has most recently reported the growth of 200 mm SI LEC single crystals of a quality comparable with those of 150 mm diameter. There is also intense development work to increase the diameter of InP LEC crystals. This is more difficult, because of the evaporative loss problem arising from enhanced convection in larger diameter systems. Improvements in overcoming this problem are leading to the growth of device-quality 100 mm-diameter InP [34].

4.3. Twinning

Twinning is a serious problem in the growth of III–Vs, and can strongly affect yield in any growth process. Early work on InSb demonstrated that the material was an excellent III–V for studying twin formation and its effect on uniformity. The results of early detailed studies using radioactive tracers are reviewed in a somewhat inaccessible book chapter [13]. The studies clearly reveal not only the effect of twinning, but also the Facet Effect [35] and the astonishing anisotropic segregation effects of many dopants in InSb. In the case of Te, the effective distribution coefficient on a facet can be ~ 4, whereas in a non-facetted region close to the $\langle 1\,1\,1 \rangle$ direction it can be ~ 0.5. $\{1\,0\,0\}$ facets also show effects for some dopants. Facet formation is generally a potential source of non-uniformity for all the III–V compounds for many dopants.

The development of facets at the edge of single crystals (see Fig. 6) is a principal source of growth twins. Facet formation appears to be a necessary but not sufficient requirement for twinning. The avoidance of twinning appears to correlate with the absence of facets. However, any convex growth surface which is tangential to a $\{1\,1\,1\}$ plane or to a lesser extent a $\{1\,0\,0\}$ plane is prone to develop a facet. Under equivalent growth conditions, the lower the temperature gradient the bigger the facet, since a specific supercooling is required for nucleation on a growing facet. Since low-temperature gradients are a basic requirement to minimise dislocation formation, facet formation and consequently twinning is often difficult to avoid.

Growth twins occur on $\{1\,1\,1\}$ planes, which are the twin composition planes. The twin can be described as a rotation of $60°$ about the $\langle 1\,1\,1 \rangle$ direction. First nearest neighbour atoms are not affected by the rotation, only second nearest neighbours. The interaction energy associated with the marked increase in distance of the second nearest neighbours is thus quite small, a factor which enhances the twinning probability. Thus, the edge facet on a $\langle 1\,1\,1 \rangle$ crystal will cause the twinned region to propagate in a $\langle 5\,1\,1 \rangle$ direction and frequently lead to more twins. $\{1\,1\,1\}$ facet planes can "terminate" in group III or group V atoms. In InSb, $\{1\,1\,1\}$ Sb facets develop more readily than $\{1\,1\,1\}$ In facets and twin more readily. They have different stabilities and twinning probabilities. Twinning is less of a serious problem now with GaAs, and can be overcome with controlled growth conditions such as the avoidance of facet formation and the elimination of large temperature fluctuations.

In the case of InP, a more common phenomenon is the presence of pairs of lamellar twins which may or may not propagate across the whole crystal with growth. A single twin develops over part of the growth

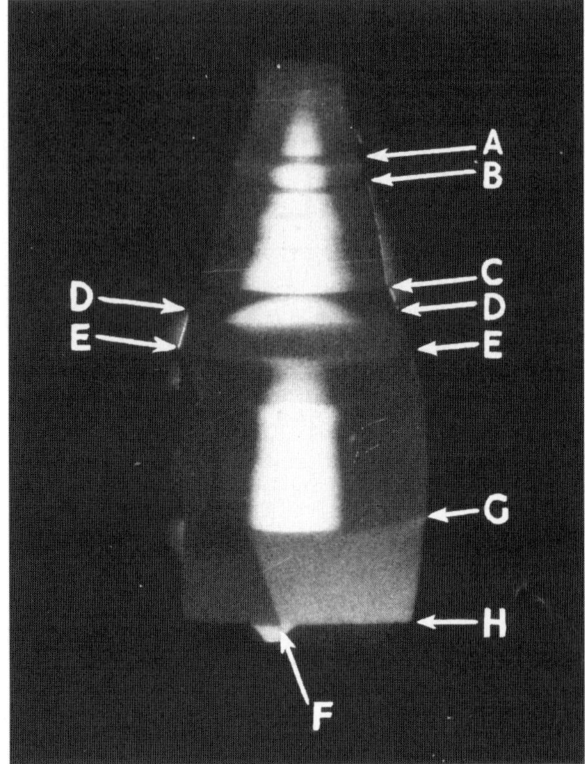

Fig. 6. Facet effect, twinning and anisotropic impurity segregation. An autoradigraph of a $(\bar{1}\,1\,0)$ section of a $[1\,1\,1]$ ^{127}Te-doped InSb crystal showing the non-uniformity of enhanced (white) ^{127}Te incorporation caused by $(1\,1\,1)$ Sb and $(1\,1\,1)$ In-type facet formation. Note the effect of the principal (centre) $(1\,1\,1)$ Sb facet and the $(1\,1\,1)$ In-type facets. Additional electromagnetic stirring of the melt was used. It imparts a convex growth surface. (A, B) Perturbation in stirring conditions. Crystal regrown from (C) after melt-back. Rotational twin on the central facet causes disappearance of right edge facet and the appearance of the left edge facet. Growth twin at EF shows the effect of ^{127}Te incorporation during $[5\,1\,1]$ growth. Electromagnetic stirring turned off at G; growth surface went concave, revealing anisotropic ^{127}Te incorporation—compare $[5\,1\,1]$ growth with growth off the $[1\,1\,1]$ direction.

front and then reverts back to its original orientation. Fig. 7 shows an example. Here again facets are implicated.

The exact mechanism of twinning is not understood as a cause-and-effect phenomenon. Anything that could allow an atom to go down on a $(1\,1\,1)$ surface misoriented in rotation by $60°$ could be implicated. Impurity atoms, temperature fluctuations and stoichiometry have all been invoked but unequivocal

Fig. 7. Etched longitudinal section of an InP crystal showing the presence of pairs of lamellar twins. As a result, the crystal growth direction is maintained.

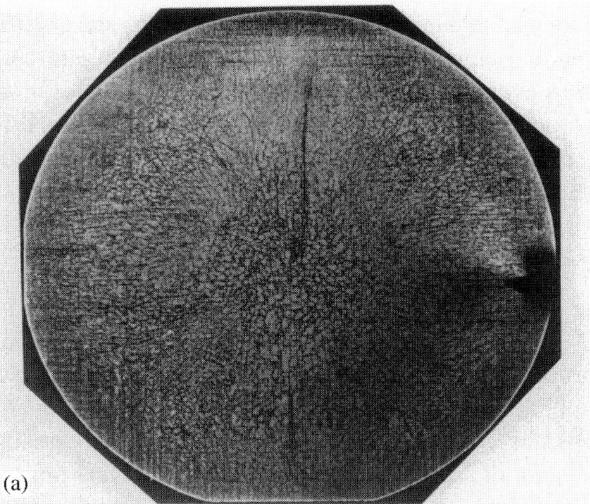

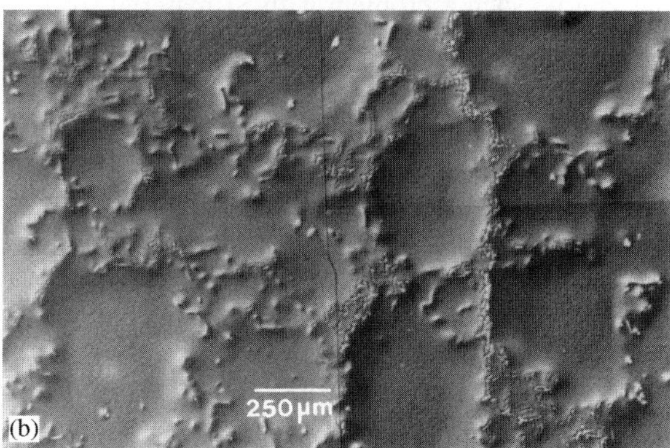

Fig. 8. Cellular dislocation structure. Etched (1 1 1) cross-sections of a GaAs crystal (a) showing the honeycomb structure of cells, the larger ones in the central region and (b) a higher magnification image (acknowledgment to D.J. Stirland's unpublished work).

proof of a specific causal relationship as opposed to strong evidence is difficult to establish.

More recently, the concept of surface free energy has been advanced by Hurle [36] to account for the stability condition, which would allow a twinned nucleus on a facet to be stable and propagate. This does not take into consideration the potentially significant effect of impurity adsorption, which is a notable feature of facetted growth. Further consideration on this and related aspects of twinning have been advanced in more recent literature [37–40]. On the practical side, twinning continues to be one of the more frustrating and annoying yield-limiting phenomena in crystal growth.

4.4. Cellular dislocation structure

One of the troublesome defect structures found in the larger GaAs crystals is the presence of the so-called cellular structure. This honeycomb structure is revealed by etches which show up the dislocation structure. An example of an etched cross-sectional specimen is shown in Fig. 8. The cell-like formation is the result of the polygonisation of the dislocations (Fig. 8). They propagate as the crystal grows. Their presence is a potential factor that can give rise to electrical non-uniformity, which is highly undesirable from a device perspective. Rudolph et al. [3] have argued that they are not the result of constitutional supercooling. They do not show the characteristic cellular patterns [41] characteristic of the presence of constitutional supercooling. More significantly, annealing can significantly change their dimensions. The cell dimensions revealed in the centre of wafers are larger than those at the edge of crystals. The lower temperature gradients found in VCz crystals, which reduce dislocation densities below 10^3 cm^{-3}, can virtually eliminate their formation. However, they are found in VGF crystals, which also have very low dislocation contents, but here it should be noted that the times for crystal growth and consequently polygonisation are longer. But, perhaps even more significant is that cellular dislocation arrays do not appear to occur in InP crystals. This difference could be directly related to the difference in the point defect character of the two materials.

4.5. Dislocation formation

Dislocation formation can be one of the major problems in the LEC growth of III–Vs. Dislocation densities normally tend to be in the region of 10^5 cm^{-3}, whereas in VGF growth they are typically a factor of a 100–1000 less. This poses a severe limitation for device materials used in junction devices, but may not be such a problem with SI material for ion-implanted structures. A principal cause of the formation of the relatively high dislocation densities found in LEC crystals is the hoop stresses which are generated by the large temperature gradient as the crystal emerges from the B_2O_3 encapsulant. These hoop stresses can cause the critical resolved shear stress to be exceeded, and this causes slip on the $\{1\,1\,1\}$ planes and consequent dislocation formation. A detailed modelling of the mechanism of formation of the dislocation distribution as found in LEC GaAs crystals was originally reported by Jordan et al. [42]. These authors identified the origin of the so-called w-dislocation distribution shown by tracing the dislocation density distribution across etched wafers. More refined modelling attempts have subsequently been made.

Reduction of the dislocation density in LEC crystals has been achieved by (a) increasing the depth of the B_2O_3, (b) by reducing the convective heat losses from the gaseous environment through the use of lower pressures, (c) better baffling and the use of thermally less conducting gases [43]. Jacob [44] has also demonstrated the use of total encapsulation, growing the crystal completely immersed in B_2O_3. VCz technology can reduce the temperature gradients at the solid–liquid interface. While this can also reduce the resulting dislocation densities, the difficulty of controlling crystal diameter limits the potential advantages of the lower temperature gradients. The really significant reduction in dislocation density has occurred with the advent of VGF technology.

4.6. Other defect-forming mechanisms

4.6.1. Lineage

Lineage has been less studied than other defects, but it is an important defect that can significantly limit the yield. It originates at the edge of crystals. Its exact cause has not been established, but it appears to be associated with the nucleation and growth of dislocations on edge facets. They develop as a bundled dislocation structure that propagates and spreads into the bulk of the crystal, ultimately limiting the effective usable length of a grown crystal.

4.6.2. Constitutional supercooling

Constitutional supercooling in crystal growth was one of the earliest phenomena studied in the growth of Ge [41,45] and InSb [46]. It only manifests itself as a problem in III–V crystals when the growth involves non-stoichiometric melts. Even in the growth of SI GaAs from a slightly As-rich melt, constitutional supercooling is not a problem, provided the growth rates are not significantly high.

4.6.3. As precipitation and defect formation

These defects can be considered as post-growth phenomena, although growth conditions need to be taken into account in understanding their formation. It is also very important to appreciate that GaAs has a concentration of some 10^{19} As interstitials [18] at the melting point. The condensation of these interstitials is a major factor in the post-growth environment, which affects the resulting defect interactions, such as the EL2 concentration, the movement of dislocations and the quality of the crystals generally.

5. Vertical gradient freeze and related techniques

Crystallisation of ingots in a vertical container by the Stockbarger or Vertical Bridgman techniques was and still is associated with the growth of high-quality single crystal optical materials like CaF_2. But, these technologies were not considered suitable for the growth of good-quality III–V compounds in view of their expansion on freezing and their marked tendency to stick to the walls of a silica container.

But, in 1986, a breakthrough in the use of vertical container technology came with the introduction of the Vertical Gradient Freeze technique by Gault et al. [47].

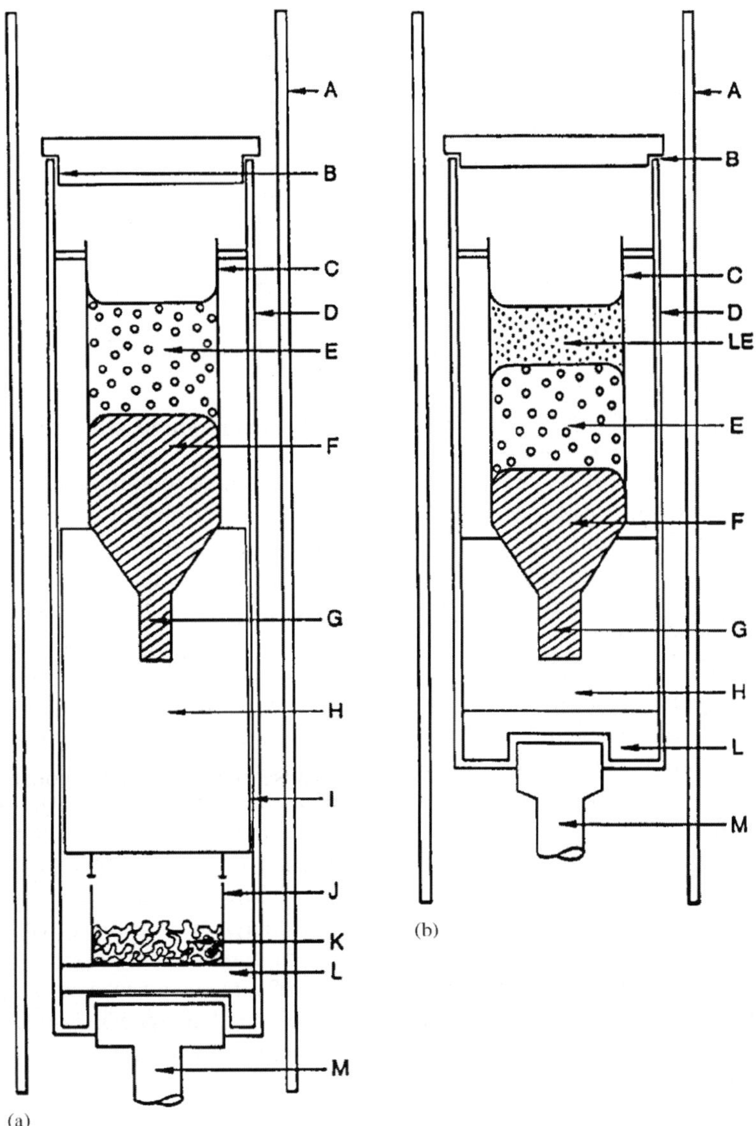

Fig. 9. VGF: Schematic diagrams of crucibles used in the Vertical Gradient Freeze technique: (a) "conventional" VGF showing compound F, melt E and separate holder J containing group V component K at a controlled temperature in order to maintain sufficient pressure of V to avoid the dissociation of the compound. Plug B allows pressure equilibration between the crucible and the outer chamber. Loss of group V into the outer chamber is inevitable even when the pressure of the group V vapour is greater than the inert gas pressure. This is one of the drawbacks of the technique: A, furnace; C, BN crucible; D, main containing vessel; G, seed; H, crucible support; I, gap for group V transport; J, crucible for holding V; K, source of group V; L, base support; M, holder, (b) Liquid Encapsulation VGF with the inert gas pressure over the B_2O_3 being greater than the pressure of the group V vapour. The symbols have same meaning as above. B_2O_3 encapsulant LE covers the melt and prevents the loss of the volatile component.

The development of the technique has been reviewed by Bourret [48]. Gault et al. [47] reported the growth of large-diameter GaP, InP and GaAs which did not involve the use of a B_2O_3 encapsulant.

The technique is diagrammatically illustrated in Fig. 9. The growth is simple in concept, but is critically sensitive to crucible preparation and system stability. The VGF technique involves the controlled freezing from the bottom up of a molten charge of material held in a tube-shaped vertical container. The freezing is best brought about, not by the movement of the furnace relative to the tube, but by the use of a furnace comprising separate independently controlled heating elements. Adjustment of the heating elements controls the position of the isotherms so that the movement of the L/S interface can be raised smoothly to bring about the crystallisation of an ingot.

An important factor affecting the yield is the capricious formation of growth twins. More recent developments [49] indicate that, for the reproducible growth of GaAs, it is advantageous but not essential to use a B_2O_3 encapsulant in a BN crucible. Such an arrangement is illustrated in Fig. 9. It appears that B_2O_3 is now also generally used for InP. It enables the use of a simpler safer system. Additionally, the encapsulant can wet the container wall and this reduces the twinning probability. This is also true with GaAs although its use is not essential.

As with LEC, a vast VGF effort is involved in the development of larger diameter crystals. In the case of GaAs, VGF is fully commercialised; indeed, some manufacturers rely on it entirely. One hundred and fifty millimetre diameter crystals are the norm and 200 mm single crystals have recently been reported by Stenzenberger et al. [50]. 100 mm diameter VGF is clearly becoming a feature of the commercialisation of InP [51,52].

The development of VGF for GaAs occurred in response to the need to find a cost-effective solution to the production of uniform GaAs wafers compatible with integrated circuit technology. Here there is a requirement for circular wafers having precise dimensions and very good electrical uniformity. "Conventional" wisdom would consider that crystallisation in a vertical rigid container would give rise to unacceptable stress due to the expansion of the liquid GaAs on freezing. In the event, this apparently has not been a problem.

The technique provides two favorable growth conditions. It naturally lends itself to low-temperature gradients, which in turn favour low dislocation densities [52,53] and, secondly, it provides an ingot of ideal shape and of the required diameter. Provided the interface is planar or at least the growth surface is slightly convex, the expansion on freezing does not appear to be serious and any stress can be annealed out. The

main problems appear to be those of twinning, furnace design difficulties and the choice of boat material. The ingots are usually encapsulated with B_2O_3. The problem here is that one cannot tell whether a crystal has twinned until the end of the growth run. The advantage of pulling is that the operator can generally identify the formation of a twin and melt back for a regrowth. This improves the yield for LEC growth. Another disadvantage is that the growth rate may have to be reduced by upto a factor of ten compared with that of LEC growth in order to obtain good crystals.

6. Electromagnetic stirring

A dominant requirement in most device specifications is uniformity of doping and/or some physical property like resistivity. While this uniformity is featured in a high proportion of studies on the melt growth of III–Vs, there is less published evidence correlating directly the relationship between the role of the stirring and uniformity. In the case of early studies on melt-grown InSb, it was apparent that electrical uniformity was very poor with carrier concentrations across slices varying by factors of ten or more. This, and the thought that a 5 mm seed would have only a weak stirring effect on melts some 50 mm in diameter, prompted alternative strategies. The author and his late colleague Ken Hulme embarked on a study of the potential of electromagnetic stirring of melts of InSb both in zone refining and in vertical crystal pulling. This early work is briefly reviewed in a Historical Introduction [54] to the excellent special issue in Progress in Crystal Growth and Characterisation by Professor Benz on the "The Role of Magnetic Fields in Crystal Growth" [55], where the topic is visited by a range of authors.

In the case of zone refining, it was shown that the optimum purification in the zone-refining process for the studied dopants could be achieved at much higher zoning rates than was the case without electromagnetic stirring. However, it was possible to achieve the optimum zone-refining effect by slow growth; hence, the work was not developed further. In the case of the Czochralski growth of InSb, a pulling system was developed for routine use and over 400 single crystals of high purity and doped InSb were grown generally in the $\langle 3\,1\,1\rangle$ or $\langle 1\,0\,0\rangle$ directions. The orientation significance is related to $\{1\,1\,1\}$ Sb and $\{1\,1\,1\}$ In facet development on the growth surface of crystals. Studies were made on the effect of electromagnetic stirring using melts doped with radioactive ^{127}Te. They were instrumental in identifying facet formation and the discovery of The Facet Effect. These facets were shown to be the principal cause of the non-uniformity which was mentioned earlier.

Electromagnetic stirring technology for semiconductors was not followed up in the literature for a number of years. Clearly, there was a lack of perceived need, larger diameter seeds and crystals, as well as crucible rotation appeared to be adequate. In addition, there was the hurdle of developing a suitable system. Nevertheless, as Benz's review shows, there is a resurgence of interest in the topic. An aspect surprisingly ignored until relatively recently [56] is the use of electromagnetic stirring in VGF growth. The latter technology is generally reliant on natural convection or the avoidance of convection altogether, with the use of expensive systems providing high magnetic fields.

The potential virtue of electromagnetic stirring is that it can, in principle, be tailored to provide controlled directional liquid transport at a predetermined depth. But, sophisticated developments will probably only emerge with the advent of appropriately designed commercial equipment.

References

[1] J.B. Mullin, Compound Semiconductor Processing, R.W. Cahn, P. Haasen, E.J. Kramer (Eds.), Materials Science and Technology, vol. 16, Processing of Semiconductors, K.A. Jackson (Ed.), VCH, Weinheim, 1996, p. 63.

[2] J.B. Mullin, Melt growth of III–V compounds by the liquid encapsulation and horizontal growth techniques, in: R.J. Malik (Ed.), III–V Semiconducting Materials and Devices, Elsevier, Amsterdam, 1989, p. 1 (Chapter 1).

[3] M. Neubert, P. Rudolph, Prog. Crystal Growth Character. Mater. 43 (2–3) (2001) 119.

[4] P. Rudolph, M. Jurisch, J. Crystal Growth 198–199 (1999) 325.

[5] M. Tatsumi, K. Fujita, in: Isshiki (Ed.), Recent developments in bulk crystal growth, vol. 47, Research Signpost, Trivandrum, 1998.

[6] R.N. Thomas, H.M. Hobgood, P.S. Ravishankar, T.T. Braggins, Prog. Crystal Growth Character. Mater. 26 (1993) 219.

[7] J. Bardeen, W.H. Brattain, Phys. Rev. 74 (1948) 203.

[8] W. Shockley, Bell. Syst. Tech. J. 28 (1949) 435.

[9] W. Pfann, Zone Melting, second ed., Wiley, New York, 1966.

[10] G.K. Teal, Preparation of germanium single crystals by the pulling method, in: H.E. Bridges, J.H. Scaff, J.N. Shive (Eds.), Introduction in Transistor Technology, Van Nostrand, New York, 1958 (Chapter 4).

[11] H. Welker, Z. Naturforsch. 7a (1952) 744.

[12] H. Welker, Z. Naturforsch. 8a (1953) 248.

[13] J.B. Mullin, Segregation in InSb, in: R.K. Willardson, H.L. Goering (Eds.), Compound Semiconductors, vol. 1, Preparation of III–V Compounds, Reinhold, New York, 1962, p. 365.

[14] K.F. Hulme, J.B. Mullin, Solid State Electron. 5 (1962) 211.

[15] K.F. Hulme, J. Electron. Control 6 (1959) 397.

[16] J.R. Arthur, J. Phys. Chem. Solids 28 (1967) 2257.

[17] D.H. Matthiesen, J. Crystal Growth 137 (1994) 255.

[18] D.T.J. Hurle, J. Appl. Phys. 85 (1999) 6957.

[19] H.G.B. Hicks, P.D. Greene, in: Proceedings of the Third International Symposium on GaAs and Related Compounds, Aachen, 1970, Institute of Physics Conference Series No. 9, Institute of Physics, Bristol, 1971, p. 92.

[20] E.M. Swiggard, S.H. Lee, F.W. von Batchelder, in: Proceedings of the Seventh International Symposium on Gallium Arsenide and Related Compounds, St Louis, 1978, Institute of Physics Conference Series No. 45b, Institute of Physics, Bristol, 1979, p. 125.

[21] K. Sonnenberg, E. Kuessel, Developments in vertical Bridgman growth of large diameter GaAs, III–Vs, Review, vol. 10(5), 1997, pp. 30, 32–34.

[22] S. Sawada, T. Kawase, Y. Hagi, H. Miyajima, T. Sakurada, M. Yamashita, H. Yoshida, M. Kiyama, R. Nakai, Development of a 6-inch diameter VB-GaAs wafer, SEI Technical Review, 2000, pp. 111–115.

[23] R. Gremmelmaier, Czochralski technique, in: R.K. Willardson, H.L. Goering (Eds.), Compound Semiconductors, vol. 1, Preparation of III—V Compounds, Reinhold, New York, 1962, p. 254.

[24] J.B. Mullin, B.W. Straughan, W.S. Brickell, J. Crystal Growth 65 (1965) 782.

[25] J.B. Mullin, R.J. Heritage, C.H. Holliday, B.W. Straughan, J. Crystal Growth 3–4 (1968) 281.

[26] K. Azuma, Japanese Patent 60-11299, Furukawa Industries, 1983.

[27] J.B. Mullin, W.R. Macewan, C.H. Holliday, A.E.V. Webb, J. Crystal Growth 13–14 (1972) 629.

[28] K. Tada, M. Tatsumi, Japanese Patent 60-264390, Sumitomo, 1984.

[29] P. Rudolph, M. Neubert, S. Arulkumaran, M. Seifert, Crystal Res. Technol. 32 (1) (1997) 35–50.

[30] R. Hirano, T. Kanazawa, M. Nakamura, in: Fourth International Conference on InP and Related Materials, Newport, IEEE Publications, New York, 1992, p. 546.

[31] R.M. Ware, W. Higgins, K. O'Hearn, M. Tiernan, in: Technical Digest-GaAs 1C Symposium, Gallium Arsenide Integrated Circuit, 1996, pp. 54–57.

[32] T. Flade, M. Jurisch, A. Kleinwechter, A. Köhler, U. Kretzer, J. Prause, Th. Reinhold, B. Weinert, J. Crystal Growth 198–199 (1999) 336.

[33] A. Seidl, S. Eichler, T. Flade, M. Jurisch, A. Köhler, U. Kretzer, B. Weinert, J. Crystal Growth 225 (2001) 561.

[34] A. Noda, K. Suzuki, A. Arakawa, H. Kurita, R. Hirano, in: Conference Proceedings—International Conference on Indium Phosphide and Related Materials, 2002, pp. 397–400.

[35] K.F. Hulme, J.B. Mullin, Phil. Mag. 4 (1959) 1286.

[36] D.T.J. Hurle, J. Crystal Growth 147 (1995) 239.

[37] Y. Han, L. Lin, Solid State Commun. 110 (1999) 403.

[38] V.A. Antonov, V.G. Elsakov, T.I. Olkhovikova, V.V. Selin, J. Crystal Growth 235 (2002) 35.

[39] H. Chung, M. Dudley, D.J. Larson Jr., D.T.J. Hurle, D.F. Bliss, V. Prasad, J. Crystal Growth 187 (1998) 9.

[40] M. Dudley, B. Raghothamachar, Y. Guo, X.R. Huang, H. Chung, D.T.J. Hurle, D.F. Bliss, J. Crystal Growth 192 (1998) 1.

[41] W. Bardsley, J.S. Boulton, D.T.J. Hurle, Solid State Electron. 5 (1962) 395.

[42] A.S. Jordan, J. Crystal Growth 49 (1980) 631.

[43] H. Emori, T. Matsumura, T. Kikuta, T. Fukada, Jpn. J. Appl. Phys. 22 (1983) 1652.

[44] G. Jacob, J. Crystal Growth 58 (1982) 455.

[45] D.T.J. Hurle, Solid State Electron. 3 (1961) 37.

[46] D.T.J. Hurle, O. Jones, J.B. Mullin, Solid State Electron. 3 (1961) 317.

[47] W.A. Gault, E.M. Monberg, J.E. Clemans, J. Crystal Growth 74 (1986) 491.

[48] E.D. Bourret, Am. Assoc. Crystal Growth Newslett. 20 (3) (1990) 8.

[49] E.D. Bourret, B.C. Merk, J. Crystal Growth 110 (1991) 395.

[50] J. Stenzenberger, T. Bünger, F. Börner, S. Eichler, T. Flade, R. Hammer, M. Jurisch, U. Kretzer, S. Teichert, B. Weinert, J. Crystal Growth 250 (2003) 57.

[51] T. Asahi, K. Kainosho, T. Kamiya, T. Nozaki, Y. Matsuda, O. Oda, Jpn. J. Appl. Phys. 38 (1999) 977.

[52] Y. Hosokawa, Y. Yabuhara, R. Nakai, K. Fujita, in: SEI Technical Review, 1999, pp. 53–55.

[53] J. Amon, J. Härtwig, W. Ludwig, G. Müller, J. Crystal Growth 198–199 (1999) 367.

[54] J.B. Mullin, An historical perspective, K.W. Benz (Ed.), The Role of Magnetic Fields in Crystal Growth, in: Prog. Crystal Growth Character. Mater., vol. 38, 1999.

[55] K.W. Benz, Prog. Crystal Growth Character. Mater. 38 (1999) 1.

[56] O. Pätzold, I. Grants, U. Wunderwald, K. Jenkner, A. Cröll, G. Gerbeth, J. Crystal Growth 245 (2002) 237.

Early history of lithium niobate: personal reminiscences

Kurt Nassau

AT&T Bell Laboratories, Murray Hill, New Jersey 07974

Early studies of LiNbO$_3$ indicated potential uses based on ferroelectric, piezoelectric, electrooptic, and related properties, all dependent on the availability of single domain material. The actual steps which led to the preparation of such crystals and the subsequent understanding of poling in this unique material are described.

1. Introduction

We are accustomed to writing up experiments in inverted historical sequence. First we describe what was observed as if this was our original anticipation and then describe the experimental results as confirmation of our inspired insight. All too often, however, the actual progress to insight is a result of serendipity, the begrudging acceptance of fortuitous results frequently unrelated to or even the opposite of our initial expectations. We should not apologize too much for such an indirect path to our successes, since the data on which we based our experiments are frequently wrong or so incomplete as to be hopelessly misleading. Such, indeed, was the case with LiNbO$_3$.

The major steps in the development of insight into the control of the peculiar ferroelectric characteristics of LiNbO$_3$ occurred at the Bell Laboratories, beginning with the growth and first "report" of ferroelectricity by Bernard T. Matthias and Joe P. Remeika in 1949 [1]. At the suggestion of Robert A. Laudise, the growth of single crystals was studied by Al A. Ballman

in 1965 [2]; this was followed by the preparation of single domain material by the author and co-workers [6–8], and the beginning of a series of studies [9,10] in the special properties of this unique material. An excellent review by Räuber [11], which contains 209 references, should be consulted for general details and additional references.

Crystal growth by Czochralski pulling was reported in 1965 by Ballman [2] and by Fedulov et al. [12]. The Ballman work was a continuation of an extended study of niobates and tantalates as part of the search for new nonlinear and laser host materials; it made large LiNbO$_3$ crystals readily available for investigation for the first time.

2. The problem

Early in 1963 Gary D. Boyd was searching for materials with a large birefrigence and a low dispersion, aimed at devices using phase-matched second harmonic generation and other parametric phenomena. At that time KDP, potassium dihydrogen phosphate was the only suitable material known and improved properties were sought. Crystals of LiTaO$_3$ and LiNbO$_3$ were shown by Ballman to be strongly pyroelectric and thus worthy of further investigation. Crystals moved from Ballman to Boyd via George E. Peterson, who also showed an early interest in such material, in April 1963 and optical properties were examined by Walter L. Bond. Only a small birefringence was found

Previously published in AACG Newsletter Vol. 14(3) November 1984.

for $LiTaO_3$ but $LiNbO_3$ had the desired large birefringence coupled with a small dispersion.

Using Ballman's crystals, Boyd and Robert C. Miller performed calculations and preliminary experiments which showed the expected properties, albeit with some problems, attributed by them to the presence of a domain structure, which would of course interfere with parametric effects. Since Ballman was becoming interested in other studies at about this time, I acquired the problem of studying $LiNbO_3$ in February of 1964. At this time some electrooptic measurements by Peterson [9] showed anomalous "patches"; these had a curious behavior which could not be explained; with the benefit of hindsight, we can now recognize this as a combination of ferroelectric domains with regions of optical damage.

3. The search for single domain crystals

My first $LiNbO_3$ notebook entry of February 14, 1964 listed several possible approaches: Bridgman growth as an alternative to Czochralski growth; the variation of stoichiometry; oxygen annealing; and one long-shot approach. Since ferroelectricity in the titanates is strongly affected by impurities, I wondered if the presence of a few present of additives might have a beneficial effect. By analogy with the coupled substitution work I had recently completed on $CaWO_4$ [13], I viewed $Li^+Nb^{5+}O_3$ as a "1:5" compound oxide, in which isostructural substitution of roughly equal size ions of "1:5" compounds such as $NaNbO_3$, $NaVO_3$, or $NaTaO_3$, "2:4" compounds such as $MgTiO_3$ or $MgZrO_3$, or "0:6" compounds such as WO_3, MoO_3 or CrO_3 should be readily accommodated.

At first this work occupied only one day or so a week, since several other studies were under way; after about three months all else was neglected in trying to solve this stubborn problem. Several weeks were spent in performing Bridgman growth in a platinum crucible (the domains were still present), annealing crystals in air, preparing a stereographic projection of $LiNbO_3$, learning to orient and etch crystals so as to be able to determine the precise relationship between domain structure and the growth direction, and annealing the crystals in oxygen and even under pressure to try to affect the domain pattern and also to remove a frequently obtained brown coloration.

Numerous attempts were made to observe hysteresis loops and to pole specimens, applying high voltages and raising the temperature in the traditional way; this was found to be a good way to destroy crystals by electrocution, but achieved little else.

Material preparation and crystal growth were performed by me assisted by Gabriel M. Loiacono and etching and electrical experiments were performed by Hy J. Levinstein who had a strong background in dislocation studies in a variety of materials, assisted by Caesar D. Capio.

At this point a close reading of the Matthias and Remeika paper [1] revealed a source of confusion. Crystals of $LiNbO_3$ and $LiTaO_3$ had first been grown from the flux by Remeika and tested for ferroelectricity by Remeika and Matthias. Even a careful reading of their paper may leave the impression that hysteresis loops were seen in both materials, although this is not actually stated explicitly. In fact this phenomenon had been seen only in $LiTaO_3$, as was confirmed by conversations in 1965 with Remeika and the late Matthias. In view of this inability to obtain hysteresis loops and poling, the question was even raised at this point if $LiNbO_3$ was indeed ferroelectric!

Crystal growth with additives were begun in June: first two runs with WO_3 and then on June 9, 1964 two crystals with 0.5 and 1.0 wt% of added MoO_3. Much to our surprise neither of these MoO_3– added crystals showed significant domain structure; this was confirmed by examination of harmonic generation by Miller and electrooptic behavior by Peterson.

With the most urgent material needs met by these two and two more single domain crystals grown the following week, many other additives and additive combinations were tried over the next few weeks, but none produced single domain material. Moreover, a recheck of the MoO_3 addition growth late in July also failed to produce single domain crystals! Having succeeded in our second serious attempt, we now experienced a total inability to duplicate the process. As a notebook entry on August 12, 1964 put it: "There seems to be lack of control over one or more growth factors in this system" and on September 4, 1964: "Nothing seems to work."

By October a "pseudo-symmetry" was recognized. Crystals which spontaneously changed their growth orientation during growth, say from perpendicular to

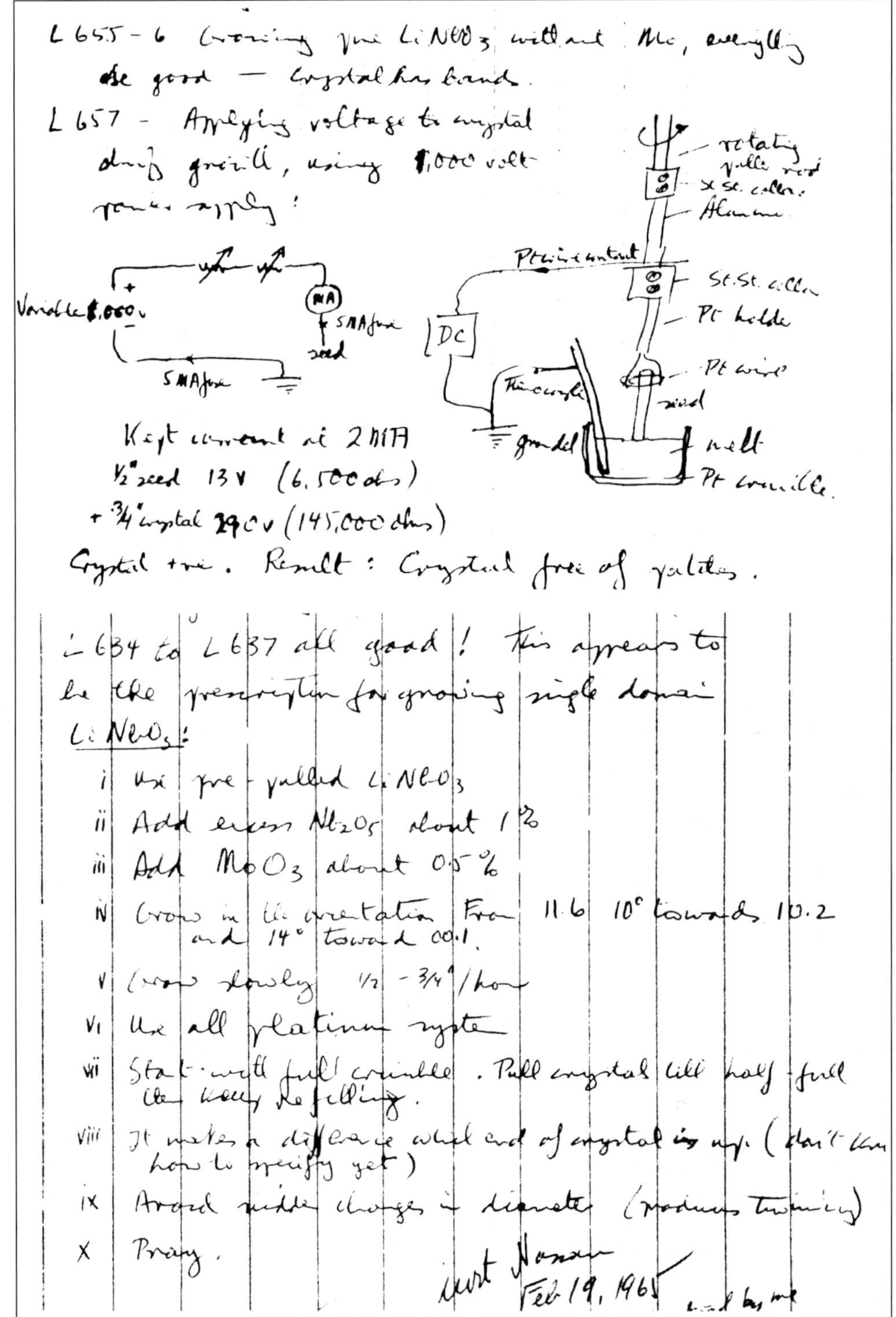

Fig. 1. Notebook listing of the main factors important to the consistent production of single domain crystals.

(14.0) to (13.8) and then to (11.12) would give essentially superimposable Laue X-ray diffraction patterns, with almost identical intensity distributions and differences between zones varying only by $1°$ in $20°$. These orientations did not, however, produce single domain material. Finally in November an occasional single domain crystal could be obtained and by February the careful control of no less than ten factors, nine technological and one other, reproduced in Figure 1, was found to result in a consistent yield of single domain crystals.

4. The electrical work

Early in this work we had considered, and a number of others had also suggested, the possibility of applying an electric field during crystal growth. The following reasoning process, faulty in retrospect but clearly sensible on the basis of general knowledge at that time, prevented us from performing this experiment at that time: Consider applying, say, 1,000 volt across a growing crystal. The cooler parts of the crystal distant from the melt have a very high resistivity, tens to thousands of ohms. The potential drop across this region would accordingly be about 999 volts, compared to the highly conductive region near the interface, which might have a drop of about 1 volt. And, of course, everyone immediately conceded that it would be useless to apply one volt to a ferroelectric well below its Curie temperature. At about this time there was additional confusion created by erroneous values of the melting point by Fedulov et al. [12] who reported $1170\,°C$ and Smolenskii et al. [15] who reported $1140\,°C$. This data together with a report by Shapiro et al. [16], which gave the Curie temperatures of the $LiTaO_3$– $LiNbO_3$ system up to 80% $LiNbO_3$ and yielded an extrapolated Curie temperature for $LiNbO_3$ of about $1220\,°C$, resulted in the conclusion that the Curie temperature of $LiNbO_3$ was above the melting point. Subsequently it has been shown that the melting point of both stoichiometric and congruent-melting $LiNbO_3$ is about $1240\,°C$, close to the $1253\,°C$ reported in earlier phase diagram work [17], and therefore above the Curie temperature. More detailed phase diagrams came later [18–20].

About this time Boyd and Miller reported occasional small scattering particles in the $LiNbO_3$ crystals, which we identified as being platinum particles derived from the crucible. A paper had recently been published by Brian Cockayne [21] in which he reported the control of similar metal particles during the growth of $CaWO_4$ crystals by making the crystal positive with respect to the melt, using a current of 0.1 mA obtained by applying 240 V.

Accordingly I built a system based on a 1000 V power supply which provided a constant 2 mA current (Figs. 2 and 3). The Mo additive was omitted, and the first crystal, grown in March 1965, was found to be single domain. Varying the current produced no change, but reversing the polarity introduced a domain wall and turning off the current resulted in a patchy domain structure. There was, incidentally, no particular improvement in the number of metal inclusions.

Levinstein and Capio immediately tried poling with just a few volts applied at very high temperatures. Lack of initial success was found to originate from parallel conducting paths in the fused silica or ceramic supporting the crystal. When this was avoided [5], single domain crystals were produced by poling at temperatures ranging from $1090\,°C$ to a little over $1210\,°C$, the temperature accepted at that time as the Curie temperature.

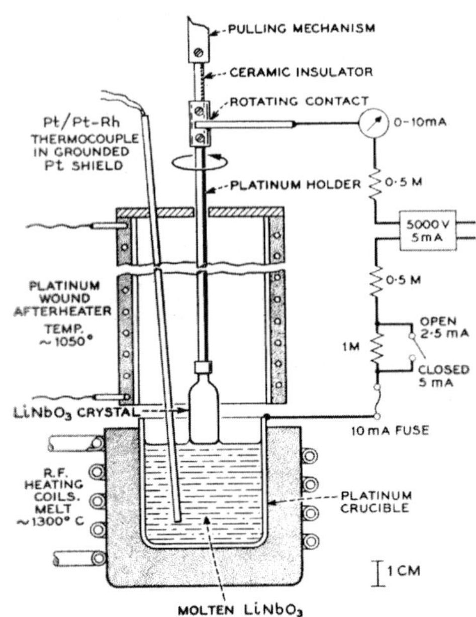

Fig. 2. Lithium niobate system.

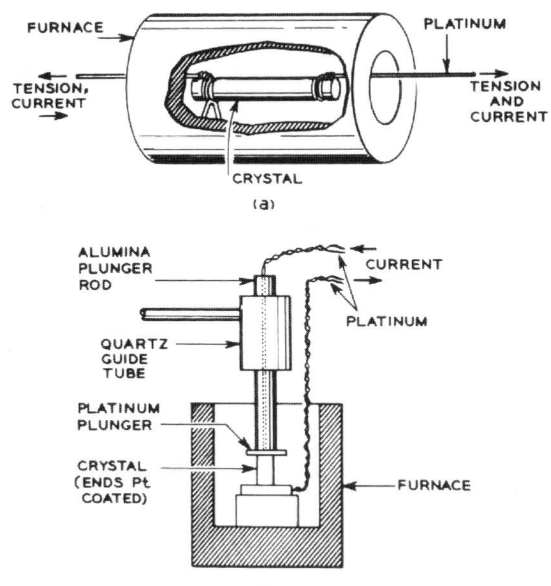

Fig. 3. Arrangements for poling $LiNbO_3$ crystals; long crystals as in (a), short crystals as in (b).

The materials and structural work on $LiNbO_3$ was summarized in a five paper series in the Journal of Physics and Chemistry of Solids, published in 1966 [4–8]. Many problems still had to be solved, including control of the Curie temperature via an understanding of stoichiometry and the phase diagram [19], the understanding of optical damage [22], the diffusion techniques, as well as the ever-present cracking problem. Nevertheless this was clearly the end of the beginning for $LiNbO_3$.

Adapted from a talk presented at the SPIE Symposium, Los Angeles, CA, Jan. 23, 1984 (to appear in the SPIE Proceedings).

References

[1] B.T. Matthias, J.P. Remeika, Phys. Rev. 76 (1949) 1886.

[2] A.A. Ballman, J. Am. Ceram. Soc. 48 (1965) 112.

[3] K. Nassau, H.J. Levinstein, G.M. Loiacono, Appl. Phys. Lett. 6 (1965) 228.

[4] K. Nassau, H.J. Levinstein, G.M. Loiacono, J. Phys. Chem. Solids 27 (1966) 983.

[5] K. Nassau, H.J. Levinstein, G.M. Loiacono, J. Phys. Chem. Solids 27 (1966) 989.

[6] S.C. Abrahams, J.M. Reddy, J.L. Bernstein, J. Phys. Chem. Solids 27 (1966) 997.

[7] S.C. Abrahams, W.C. Hamilton, J.M. Reddy, J. Phys. Chem. Solids 27 (1966) 1013.

[8] S.C. Abrahams, H.J. Levinstein, J.M. Reddy, J. Phys. Chem. Solids 27 (1966) 1019.

[9] G.E. Peterson, A.A. Ballman, P.V. Lenzo, P.M. Bridenbaugh, Appl. Phys. Lett. 5 (1964) 62.

[10] G.D. Boyd, R.C. Miller, K. Nassau, W.L Bond, A. Savage, Appl. Phys. Lett. 5 (1964) 234.

[11] A. Räuber, Chemistry and Physics of Lithium Niobate, in: E. Kaldis (Ed.), Current Topics in Materials Science, vol. 1, North-Holland Publ. Co., Amsterdam, 1978, pp. 481–601.

[12] S.A. Fedulov, Z.I. Shapiro, P.B. Ladyzhenskii, Kristallografiya 10 (1965) 268.

[13] K. Nassau, G.M. Loiacono, Phys. Chem. Solids 24 (1963) 1503.

[14] K. Nassau, Phys. Chem. Solids 24 (1963) 1510.

[15] G.A. Smolenskii, N.N. Krainik, N.P. Khuchua, V.V. Zhdanova, I.E. Myl'nikova, Phys. Stat. Solidi 13 (1966) 309.

[16] Z.I. Shapiro, S.A. Fedulov, Yu.N. Venevtsev, L.G. Rigerman, Bull. Acad. Sci. USSR, Phys. Series 29 (1965) 1047.

[17] A. Reisman, F. Holtzberg, J. Am. Chem. Soc. 80 (1958) 6503.

[18] P. Lerner, C. Legras, J.P. Dumas, J. Cryst. Growth 3 (1968) 231.

[19] J.R. Carruthers, G.E. Peterson, M. Grasso, P. Bridenbaugh, J. Appl. Phys. 42 (1971) 1846.

[20] L.O. Svaasand, M. Eriksrud, G. Nakken, A.P. Grande, J. Cryst. Growth 22 (1974) 230.

[21] B. Cockayne, Solid State Comm. 2 (1964) 381.

[22] A. Ashkin, G.D. Boyd, J.M. Dziedzic, R.G. Smith, A.A. Ballman, H.J. Levinstein, K. Nassau, Appl. Phys. Lett. 9 (1966) 72.

Czochralski growth of oxides

C.D. Brandle

CrysTex, Basking Ridge, NJ, USA

Abstract

The first oxide grown using the Czochralski technique ($CaWO_4$) was in 1960. Since that time, the Czochralski technique has become the method of choice for the growth and production of many bulk oxide materials used in the electronics and optical industries, e.g. lasers, substrates, scintillators, nonlinear and passive optical devices. This paper will trace the development of the initial process from its beginnings in the early 1960s to its present state and highlight some of the significant advances over the years that have allowed the oxide crystal growth industry to develop. It will conclude with an assessment of the current status and outline directions for further research and development.
© 2004 Published by Elsevier B.V.

Keywords: Al. Phase diagrams; A2. Czochralski method; Bl. Oxides

1. Introduction

The use of the Czochralski [1] process to grow semiconductor crystals (Si and Ge) was well established by the mid-1950s [2–4]. The first reported oxide material grown using the Czochralski technique ($CaWO_4$) was in 1960 [5]. This initial paper was followed quickly by numerous other papers dealing with the growth of a variety of oxide materials such as $LiNbO_3$ [6], $LiTaO_3$ [7], BGO [8], YAG ($Y_3Al_5O_{12}$) [9], Nd:YAG [10] and Al_2O_3 (Sapphire) [11,12]. By the mid-1960s, the Czochralski process for the growth of oxide materials was becoming well established. Today, the Czochralski technique, also known as crystal pulling, has become the method of choice for the growth and production of many bulk oxide materials used as components for the electronics and optical industries, e.g. lasers, substrates, scintillators, nonlinear and passive optical devices.

___
E-mail address: brandle@optonline.net (C.D. Brandle).

To achieve this progress, numerous advances to the understanding of the role that liquid and crystal composition, interface shape/fluid dynamics, thermal geometry and puller design have on the quality of the resulting crystal had to be determined. This paper will trace the development of the initial process from its beginnings in the early 1960s to its present state and highlight some of the significant advances over the years that have allowed the oxide crystal growth industry to develop so that today the size and variety of materials has expanded to include 125 to 150 mm diameter oxide single crystal that can weigh as much as 50 kg. It will then conclude with an assessment of the current status and outline directions for further research and development.

2. Crystal composition

One of the first problems encountered during the growth of various crystals was that the properties

0022-0248/$ – see front matter © 2004 Published by Elsevier B.V.
doi:10.1016/j.jcrysgro.2003.12.044

Originally published in
J. Crys. Growth 264 (4) (2004) 593–604.

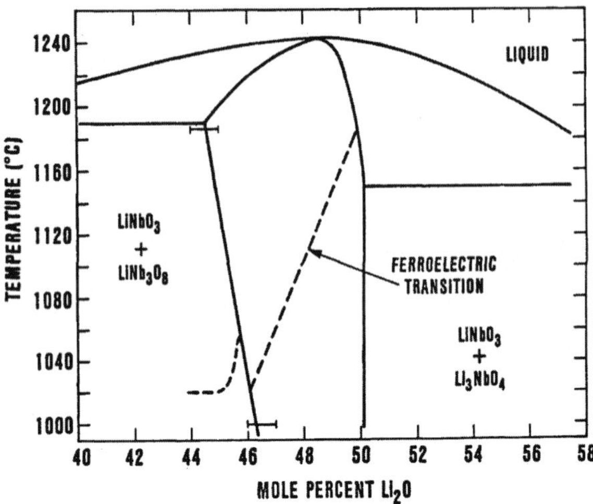

Fig. 1. LiNbO$_3$ phase diagram showing the change in composition from stoichiometric to congruent. From O'Bryan et al., Ref. [16].

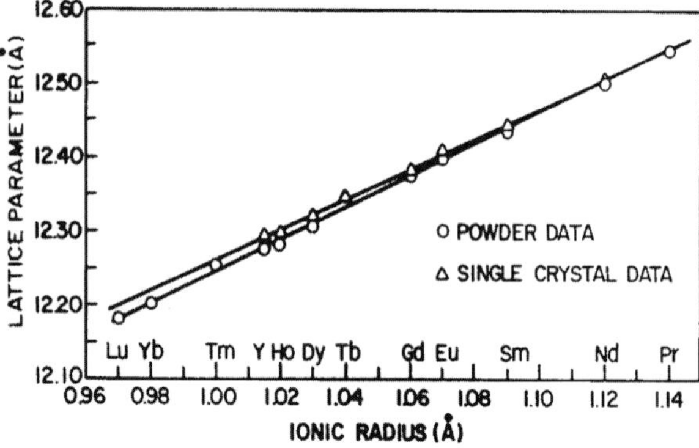

Fig. 2. Lattice parameters of single crystals grown from stoichiometric melts and literature powder data vs. rare-earth dodecahedral ionic radius. From Brandle and Barnes, Ref. [21].

were found to vary through the crystal length as well as from crystal to crystal and from one researcher to another. It was always assumed that the composition of these oxide materials was stoichiometric; hence, the initial starting melt composition was also stoichiometric. The first oxide material identified to show variations in properties due to stoichiometric variations of the liquid was LiTaO$_3$ [13]. Fay et al. [14] found similar results for LiNbO$_3$ and that the Curie temperature and birefringence would vary over a wide range of starting melt compositions. A detailed investigation of the phase diagram in the region of the stoichiometric composition for LiNbO$_3$ by Carruthers et al. [15] showed that the congruently melting composition was 48.6% Li$_2$O and crystals grown from this melt composition showed much lower variations in the optical properties. The phase diagram (Fig. 1) shows the shift in the composition in the Li$_2$O–Nb$_2$O$_5$ phase diagram from stoichiometric to congruent [16]. Using a similar approach (Curie temperature vs. composition) for LiTaO$_3$, Miyazawa and Iwasaki [17] showed that the congruently melting composition was also nonstoichiometric. Up until the early 1970s, these two materials were believed to be the only examples of an oxide material that had a nonstoichiometric, congruently melting composition.

With the discovery in the early 1970s that rare-earth iron garnets had sufficient growth-induced anisotropy to support small, circular magnetic domains (magnetic bubbles) and that thin films of these materials could be grown using a liquid phase epitaxy (LPE) technique [18], the need for a suitable substrate for the deposition of thin films of these materials became apparent. The rare-earth gallium garnets provided an excellent lattice match for the various rare-earth iron garnet films and therefore became the material of choice for substrate use [18]. The first gallium garnet that was grown for substrate use was GGG (Gd$_3$Ga$_5$O$_{12}$). When the lattice parameter of the Czochralski grown material was compared to that of flux grown material, a significant difference was found (12.376 Å for the flux material vs. 12.384 Å for the Czochralski grown material) [20]. Brandle and Barns [19] showed that as the ionic radius of the rare-earth ion decreased, the departure of the lattice parameter for flux grown material vs. Czochralski grown material increased (Fig. 2). Analysis of the Czochralski grown material showed that it contained an excess of rare-earth that was substituting on the octahedral site in the garnet structure and that the congruent melting composition was gradually shifting to the rare-earth rich region of the phase diagram as the ionic size of the rare-earth ion decreased (Fig. 3) [21]. The actual congruently melting composition is represented by {RE}$_3$[Ga$_{2-x}$RE$_x$](Ga)$_3$O$_{12}$ where { } signifies the dodecahedral site, [ ] signifies the octahedral site, and ( ) signifies the tetrahedral site in the garnet structure. Thus, this class of materials became the second known group to exhibit nonstoichiometric behavior at its melting point.

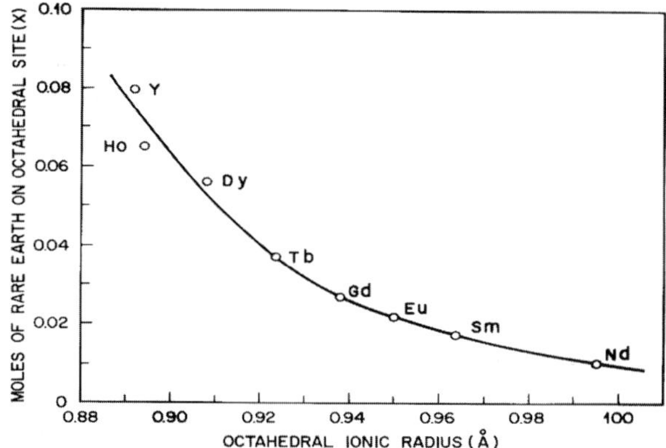

Fig. 3. Moles of rare-earth on the octahedral site in RE_3-$\{RE_xGa_{2-x}\}Ga_3O_{12}$ versus octahedral radii difference. From Brandle and Barnes, Ref. [21].

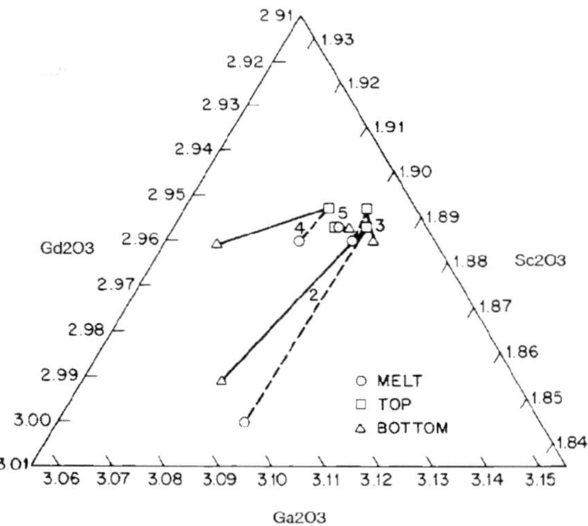

Fig. 4. Section of the phase diagram for Gd_2O_3–Sc_2O_3–Ga_2O_3 showing the shift in the GSGG crystal composition from top to bottom of the crystal for various melt compositions. From Fratello et al., Ref. [22].

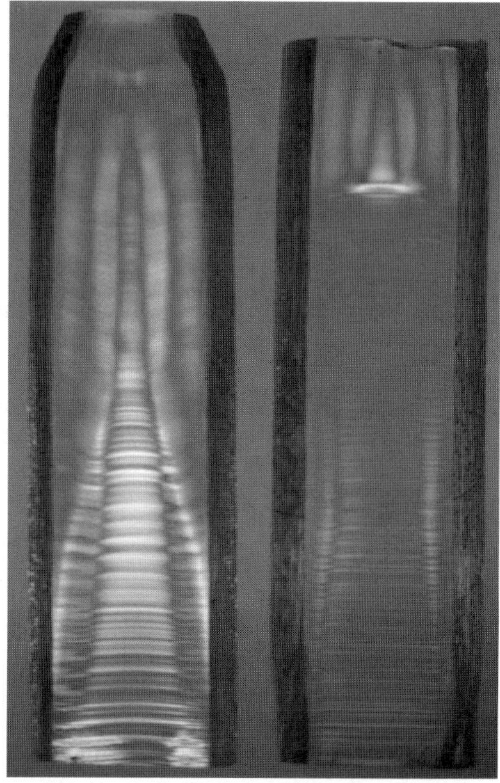

Fig. 5. Polished crystal sections showing the change in the growth striations for a crystal grown from a stoichiometric melt (a) and grown from a congruent melt (b). From Fratello et al., Ref. [22].

Another example of a further shift from the stoichiometric composition is given by GSGG ($Gd_3Sc_2Ga_3O_{12}$). In the mid-1980s, this material was of interest as a laser host for Nd and Cr; however, growth striations associated with optical variations in the material limited its use. Fratello et al. [22] showed that the congruent composition of GSGG was $\{Gd_{2.957}Sc_{0.043}\}[Sc_{1.862}Ga_{0.138}](-Ga)_3O_{12}$. As one can see, this material shows that multiple substitutions can occur on the various garnet sites. Fig. 4 shows a section of the phase diagram in the region

of the congruent melting composition [22] whereas Fig. 5 shows a comparison of the growth striations in a crystal grown from a stoichiometric melt composition and that grown using a congruent melt composition. Similar results have been shown for other Sc-based garnets [23]. Although these materials are still garnets, they clearly show that as the complexity of the cation substitution increases, the probability for multiple substitutions on various sites also increases and hence, the shift from a stoichiometric melting composition to a congruently melting nonstoichiometric composition is more likely to occur.

Other examples of nonstoichiometry continue to be reported in the literature [24–26]. Unlike what was believed in the 1960s and 1970s, a nonstoichiometric, congruently melting composition appears to be the rule rather than the exception and very few oxide materials exist as line compounds. Thus today, our current understanding of composition is:

- For most oxides, the congruently melting composition most probably is not the stoichiometric composition.
- The more complex the crystal structure, the higher the probability for nonstoichiometry to exist at the melting point.
- Materials with a crystal structure that have similar cation coordination sites are most likely to show nonstoichiometry at its melting point.
- Only simple oxides such as Al_2O_3 tend to have congruent melting composition that is stoichio metric.

3. Interface shape/fluid dynamics

With the growth of oxide materials in the early 1960s, very little attention was paid to the shape of the growth interface and its influence on the quality of the resulting crystal. The growth systems were small and the crystals usually were 1–2 cm in diameter. Given such small growth systems, most early crystals were grown with a high thermal gradient and the shape of the growth interface was conical. Furthermore, it was assumed that a conical interface was necessary to maintain stable growth and that a flat or convex interface shape would result in unstable growth. One of the first attempts to modify the shape of the growth interface was by Cockayne et al. [27]. They showed that under the proper conditions, the shape of the growth interface could be controlled by the crystal rotation rate. At sufficiently high rotation rates (150 rpm), the normally convex interface in YAG could become very shallow. Using a similar approach, Brandle and Valentino [19] were able to grow GGG ($Gd_3Ga_5O_{12}$) with a flat interface and maintain stable growth conditions.

For GGG, the main purpose of changing the interface shape was to reduce or eliminate the facets and the strain associated with these facets that normally form on the growth interface as well as minimize growth striations. Since facet strain and growth striations were replicated by the epitaxial iron garnet film [28,29], their reduction or elimination was necessary to produce more uniform magnetic garnet films.

By changing the shape of the growth interface, the position of the facets on the growth interface is changed thereby moving them from the central section of the crystal to the outside surface. In this manner, the strain associated with the facets can be removed during the fabrication step yielding a strain free central section that can be processed into substrates or optical parts. Fig. 6 is a schematic of the position of the facets on the growth interface of a garnet crystal at a slow and fast rotation rate.

For a given thermal environment, changing the shape of the growth interface through varying the rotation rate of the crystal implies one is altering the fluid flow within the crucible. Carruthers [30] and Carruthers and Nassau [31] were among the first to simulate the fluid hydrodynamics of a Czochralski growth system. By the early to mid-1970s, numerous papers had been published dealing with the influence of the crystal rotation rate on the fluid flow during the growth process and its impact on the crystal interface shape [32–36]. Fig. 7 shows schematics of typical flows observed in an oxide melt at various crystal rotation rates [37]. This type of surface structure as illustrated in Fig. 7 has been studied extensively in the $Bi_{12}SiO_{20}$ system [38]. The outer annular area consists of fluid flow driven by natural convective forces of the heated fluid whereas the main driving force for the inner area is forced convection driven by the rotating crystal. Fig. 8 shows an example of the sudden melt back of the conical growth interface of GGG that is produced by the reversal of the fluid flow as a result of the crystal rotation [39]. Also, visible at the top of the crystal in Fig. 8 is the facet strain associated with the formation of the (2 1 1) facets that form on the conical portion of the growth interface as illustrated in Fig. 6.

Another factor that has been shown to have a pronounced effect on the fluid flow is the liquid depth within the crucible. By placing baffles within the liquid, effectively changing the liquid depth and disrupting the natural convection, Whiffin and Brice [40] have pointed out that the ratio of liquid depth to crucible height has a most pronounced effect on the thermal oscillations in the liquid. Their experiments in zinc tunstate melts in the late 1960s and early 1970s showed that the melt stability is increased when a baffle in positioned within the liquid or as the liquid level within a given crucible size is reduced; however, growth under these conditions becomes more difficult.

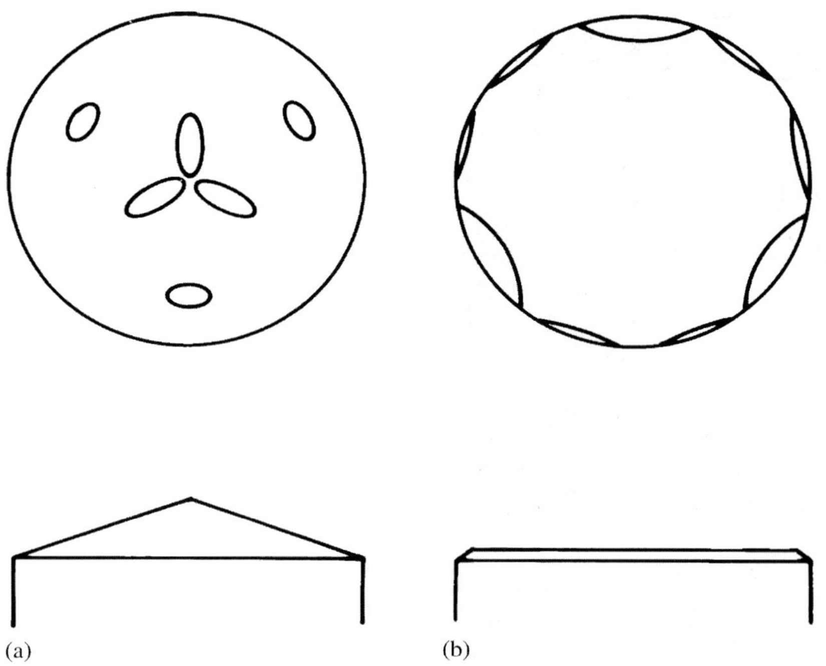

Fig. 6. Change of the position of interface facets in $\langle 1\,1\,1 \rangle$ grown garnet as a function of interface shape (a) conical, (b) flat.

Fig. 7. Surface and bulk flow observed in water/glycerin simulations (a) slow rotation, (b) moderate rotation, (c) fast rotation. From Brandle, Ref. [37].

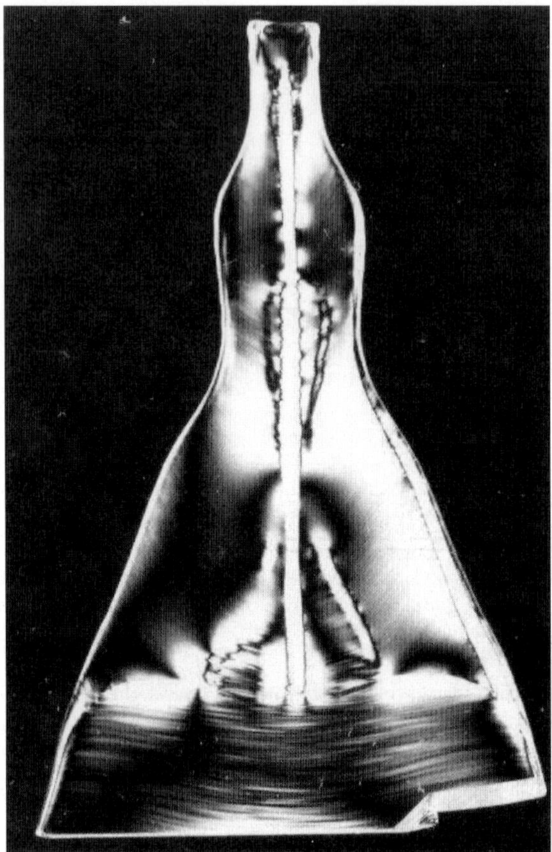

Fig. 8. The change in interface shape of a GGG crystal as indicated by growth striations that resulted from a change of the liquid flow going from natural convection to forced convection.

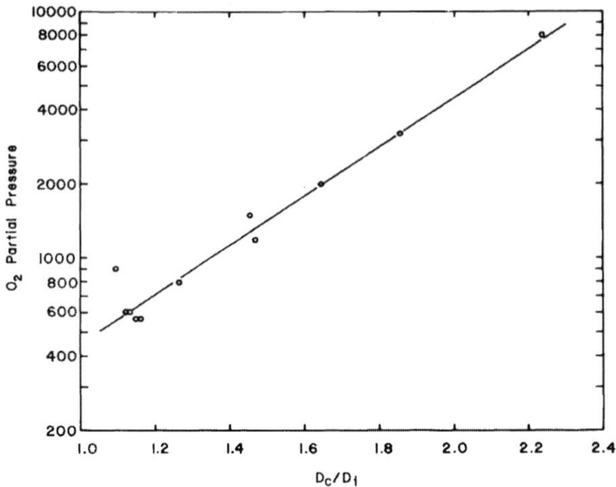

Fig. 9. Aspect ratio (crystal diameter to interface depth (D_c/D_i)) for Al_2O_3 single crystals as a function of O_2 partial pressure. From Brandle, Ref. [39].

Numerous researchers [32,33,36,41] have found that using a crucible with approximately a 1:1 aspect ratio, i.e. crucible diameter to crucible height provides the best compromise between melt stability and growth difficulty.

Given the success of the earlier fluid flow simulations, numerous authors [42–45] simulated the flow that resulted in the rapid interface melt back and transition from a conical to flat interface with the general conclusion that such interface transitions were the result of going from natural convection to forced convection under the region of the rotating crystal interface.

By the mid-1970s, numerical simulations of fluid dynamics became practical and as computational power increased so has the complexity of the mod-

els. Kobayashii [46] was among the first to examine the features of forced convection generated by a rotating crystal or crucible such as seen in a Czochralski growth configuration. A more recent example of modeling effort is that of Derby and Brown [47]. This trend has continued and been quite successful for Si-based systems and has been successful in explaining the general behavior of molten oxide-based systems.

Also, included in the many factors that influence the interface shape is the composition of the growth atmosphere. By changing the O_2 partial pressure, one can vary the concentration of various oxide species in the liquid and hence change some of the liquid properties such as viscosity and surface tension. Fig. 9 gives an example of the change in interface shape of Al_2O_3 as a function of O_2 partial pressure [39]. In this case, it is believed that by changing the O_2 partial pressure, the equilibrium concentration between Al_2O_3 and Al_2O in the liquid is shifted resulting in a change in the surface tension of the melt and perhaps its viscosity. This change in fluid properties then impacts the fluid flow within the crucible and hence the interface shape.

Based on the information gathered over the past 30 years, our current understanding is:

1. Hydrodynamics and fluid properties are crucial elements in defining the interface shape and crystal quality.

2. The interface shape can be altered by changing the thermal environment, growth atmosphere and/or the crystal rotation/growth rates.
3. The interface shape can be controlled and tailored to the end use of the crystal.

4. Diameter control

Of the many mechanical advances that were made in the growth system and furnace construction for oxide materials, the one that has made the most contribution to the growth and quality of oxides has been the development of a useful method for active control of the diameter of the growing crystal. Initial control systems were based solely on control of the power input based on the assumption that if the power input into the furnace was constant then the resulting temperature during growth would also be constant and hence the diameter would also be constant. For the early low melting oxides such as $CaWO_4$, a thermocouple was placed in the liquid and provided the signal for diameter control [48]. A schematic of a furnace using this form of control is shown in Fig. 10 whereas an actual photo of the growing crystal is shown in Fig. 11. Note that the furnace assembly is very open and thus the crystal is growing in a very high thermal gradient. Diameter control was achieved by manually changing the temperature control set point based on visual observations of the change in diameter in the crystal. This early system of diameter control had the advantage that the control circuit was simple and based on readily available instrumentation. However, its disadvantages were many:

1. The power had to be controlled manually requiring constant operator attention.
2. Diameter control was based on visual observation of the growing crystal, thus the crystal/liquid boundary had to visible. This in turn required a growth furnace in which the crystal/liquid boundary was easily visible throughout the entire growth process.
3. Power/temperature corrections for diameter control were made after a visible change in diameter. This in turn required a skilled operator to detect these changes as early as possible.

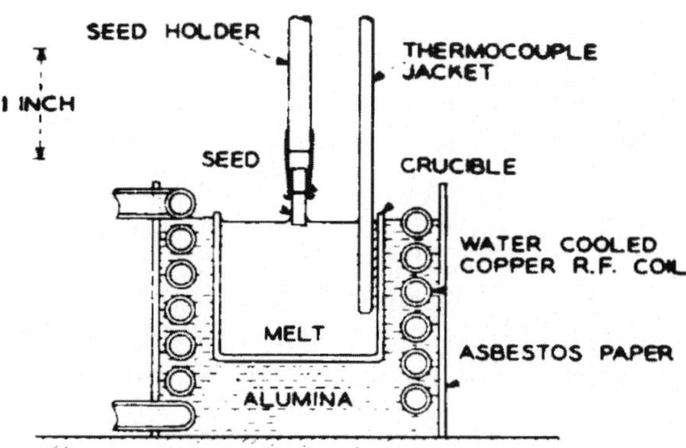

Fig. 10. Early furnace design and control for the growth of $CaWO_4$. From Nassau and Broyer, Ref. [48].

Fig. 11. Early growth of $CaWO_4$ crystal. Note the thermocouple in the melt at the right of the photo.

4. Because each type of crystal had different thermal properties, it behaved differently during growth and thus many growth tests were required to gather the necessary data to produce a crystal of reasonably uniform diameter, i.e. to predict the temperature changes necessary to compensate for the changing thermal environment during crystal growth.
5. For the higher melting oxides, a means, other then a thermocouple had to be used for power control. This required either RF power pickup loops or such devices as optical pyrometers focused on the bottom of the crucible.

Examples of early crystals grown using these types of control are shown in Fig. 12.

Fig. 12. Early garnet crystals grown using RF power pickup and diameter control based on visual observations.

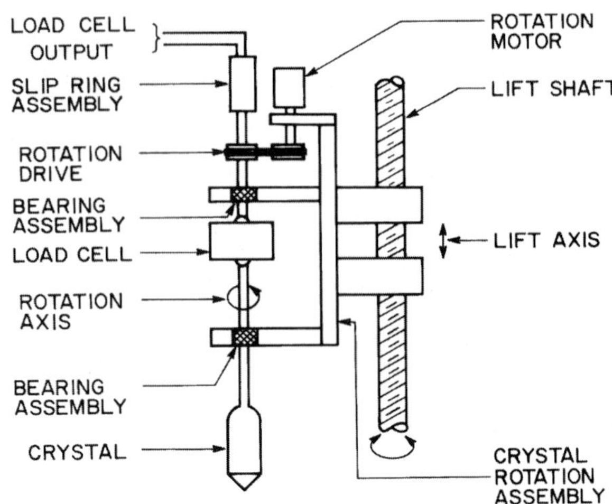

Fig. 13. Schematic showing a typical diameter control assembly based on the crystal weighing approach for diameter control. From Brandle, Ref. [39].

With the demand for larger crystals and the need for higher quality, work began in the late 1960s and early 1970s on alternate methods of diameter control. One early system used an optical sensor to detect the position of the crystal/liquid interface. This system was mechanically linked to the puller so that it would gradually change the viewing angle during growth to compensate for the liquid level drop [49–51]. This type of diameter control has the unique advantage that the control system is truly sensing the diameter changes of the crystal and not some other variable. However, this system still required a clear, unobstructed view of the crystal/liquid interface during the entire growth process.

Another approach taken for diameter control was based on a weighing method. The first type of weighing control was based on a melt weighing technique [52–57]. The weight of the crystal was not measured directly but determined through the loss of weight of the melt. The entire furnace assembly was placed on an electronic balance. During the growth process, the rate of weight loss from the balance was monitored and the balance output was then used as the diameter control signal. This signal was then processed to provide the final control signal to control the furnace power and therefore the crystal diameter.

The melt weighing technique had numerous advantages over the manual technique:

1. It provided active control feedback of the weight (diameter) of the growing crystal.
2. The puller head and crystal rotation required no modifications for active diameter control.

3. It didn't require the attention of a skilled technician to make power adjustments for diameter control.
4. Since diameter control was no longer dependent upon the visual observation of the crystal/liquid interface, there was more freedom in designing the furnace and thermal gradients.

Despite these obvious advantages, the melt weighing technique had two serious disadvantages:

1. Except for a unique crucible position, the balance output was sensitive to power changes due to the levitation of the crucible when an RF generator was used. These levitation weight changes required a second circuit to compensate for the power changes that were required for diameter control.
2. As the size of the crystal increased, there was a corresponding increase in the weight of the furnace assembly with a corresponding decrease of sensitivity until the point was reached that this approach became impractical.

As the need for larger crystals developed and hence larger furnace assemblies, the above-mentioned disadvantages became real obstacles to the growth of large diameter crystals. In the mid-1970s, the crystal weighing technique for diameter control and

Fig. 14. Current examples of large diameter perovskite crystals that were grown using the crystal weighing technique with digital control.

the mechanical assemblies required for this approach were developed. A schematic [39] of such a system is shown in Fig. 13. With the continued advances in load cell technology and refinement of the mechanical design, coupled with digital control, this type of approach to diameter control has become the method of choice (Fig. 14).

Its advantages are many:

1. The load cell can be matched to the final expected weight of the crystal without any reduction of sensitivity.
2. It can provide a completely automatic operation from seeding to cool down.
3. Complex algorithms provide excellent diameter control as well as providing the capability for a predetermined cone shape.
4. As in all weight control techniques, there is no need to see the crystal/liquid boundary for diameter control.
5. Commercial systems are now currently available.

Despite these many advantages, this does not imply that the growth of new materials and compositions has become routine. The best diameter control system will not compensate for poor furnace design, compositional shifts or thermal problems. It can only produce high-quality crystals if these other issues have been addressed.

5. Future directions

As the demand for oxide crystals and their uses increases, the need for the growth of larger crystals (> 75 mm diameter) with improved quality will also increase, i.e. the oxide crystal growth industry will follow the same path as the semiconductor industry with cost becoming the main driver. This has already happened for several materials such as Nd:YAG, Sapphire (Al_2O_3), $LiNbO_3$ and several garnets (GGG and SGGG). However, unlike the semiconductor industry, adequate models for the design of these large oxide crystal growth furnaces did not exist and their ultimate configuration was based on a trial-and-error methodology that, for these large systems, can become very costly. For example, the investment in one large Iridium crucible could be as much as $100,000. Thus, it becomes imperative that models be developed that can be used for scaling purposes for the design of large oxide crystal growth systems. Presently, such models do not exist. The reasons for this lack of understanding are many and can be broken into two distinct groups. The first group deals with the oxide material itself while the second group deals the physical design of the furnace and the method of heating.

5.1. Hydrodynamics

As pointed out in an earlier section, computational analysis of convection in molten oxide systems has shown general agreement with the characteristics of fluid flow observed within these systems as well as the simulation experiments. However, if these models are to be used to accurately design and scale a given oxide Czochralski system to produce large diameter crystals, one must have accurate thermal properties as a function of temperature for both the liquid and its solid. Such data is very difficult to obtain because the operating temperature is in many cases above 1800 °C. Because of this difficulty, only a limited amount of data has been reported for the physical properties of a few molten oxides [58,59]. Some of the differences reported in the literature might be explained by the influence of the atmosphere on the liquid properties although for some cases the reported differences are not fully understood and could be associated with the method of measurement [60,61]. In addition, the thermal properties of the various furnace components must be known since these also enter into any calculations to determine the thermal environment.

5.2. Physical design

The high operating temperature required for the growth of many oxide materials generally requires that the crucible becomes the active source for heat generation within the furnace, i.e. the crucible is heated by RF energy. The coupling of the crucible to the RF field is strongly dependent upon the diameter and length of the RF coil as well as the shape (diameter and height) of the crucible. Also, the position of the crucible within the RF coil strongly influences the thermal gradients generated within the liquid [62]. Here again, no adequate models exist that can accurately predict the impact of changing the RF coil size or the crucible size and/or shape that can be used for the design of large systems. General "rules of thumb" exist for smaller systems ($< 3''$ diameter crystal); however, they are unproven as one goes to larger systems that are capable of producing 25–40 kg crystals. One of the most difficult tasks in designing these large systems is the establishment of a suitable radial thermal gradient in the liquid that allows good seeding and controlled growth during the "shouldering" phase when the crystal is being brought out to the required diameter. In general, as the size of the system increases, the radial liquid thermal gradient decreases [63]; hence, making control of the start more difficult. At present time, adequate models do not exist that can utilize information derived from smaller systems to predict a design of a furnace for the growth of large crystals that would establish the "correct" growth conditions.

6. Summary

Over the past 40+ years of Czochralski oxide growth, significant progress has been made in understanding the various factors involved in the growth of these materials. Each decade has addressed a series of pressing problems that have advanced the Czochralski growth of oxide materials (Table 1). We have progressed from the growth of small (1 cm diameter × 3 cm length) 100 g crystals suitable for research purposes to the growth of large oxide crystals that now are 75 mm or larger in diameter that can weigh as much as 30 kg. We have a much better understanding of the impact of stoichiometry, growth atmosphere and fur-

Table 1

Progress in Czochralski oxide crystal growth over the past 40 + years

Decade	Driver	Crystal examples	Issues addressed
1960s	Lasers	$CaMoO_4$, ruby ($Cr:Al_2O_3$)	Dopants, charge compensation
	Ferroelectric and piezoelectric	Nd:YAG	Distribution coefficients
	Nonlinear optics	$LiNbO_3$	
		$LiTaO_3$	Stoichiometry
		Tungsten bronze	
1970s	Magnetic bubbles	$Gd_3Ga_5O_{12}$ and other garnets	Stoichiometry, flat interface
	SOS	Al_2O_3, $MgAlO_4$	Growth
	Saw devices	$LiNbO_3$	Diameter control
1980s	Phosphors	$RE:Y_2SiO_5$	Dopants
	scintillators	Gd_2SiO_5	Size
		$Bi_4Ge_3O_{12}$	
1990s	HTc substrates	$LaAlO_3$, $NdGaO_3$	Scale up to larger crystals
		LSAT, other perovskites	
		Substituted garnets	Lattice matching
	Optical components (Telcom)	YVO_4, $LiNbO_3$	Optical uniformity
2000s	Substrates for intergrated		
	Optics		Substrate engineering
	GaN	Al_2O_3	Size
	Larger diameter	All substrate materials	Scaling parameters
	Passive and nonlinear optics		New materials

nace design on the quality of the resulting crystal. At the same time, significant advances have been made in the design of control systems and equipment that today allows complete automatic control of the crystal growth process. Materials first conceived and grown in the late 1960s and early 1970s, e.g. Nd:YAG, $LiNbO_3$ and Al_2O_3 to name a few have become important materials in today's modern world; however, much work still needs to be done. The design of the furnace, the aspect ratio of the crucible, the size of the RF coil, the placement of the crucible within the work coil are still based on general "rules of thumb" that may or may not be true for larger growth systems. Although a large base of knowledge has been developed for these systems, accurate predictive capability that can be used for design purposes does not exist and must still be developed.

As the optical and electronic industries merge, the demand for new materials will increase coupled with the demand for increased performance, smaller component size and lower cost. Currently, new materials are being developed that utilize the knowledge gained over the past decades. These new materials are being "engineered" for improved properties to satisfy a given set of applications by using such techniques as coupled substitution, selective ion replacement or solid solutions. With each of these new materials come new challenges which continue to drive a better understanding of the growth process.

References

[1] J. Czochralski, J. Phys. Chem. 91 (1918) 219.

[2] G.K. Teal, J.B. Little, Phys. Rev. 78 (1950) 647.

[3] G.K. Teal, E. Buehler, Phys. Rev. 87 (1952) 190.

[4] K. Lehovec, J. Soled, R. Koch, A. MacDonald, C. Sterns, Rev. Sci. Instrum. 24 (8) (1953) 652.

[5] K. Nassau, L.G. Van Uitert, J. Appl. Phys. 31 (1960) 1508.

[6] A.A. Ballman, J. Am. Ceram. Soc. 48 (1965) 112.

[7] S.A. Fedulov, Z.I. Shapiro, P.B. Ladyshinskii, Kristallografia 10 (22) (1965) 268.

[8] A.A. Ballman, J. Crystal Growth 1 (1967) 109.

[9] W. Bardsley, B. Cockayne, in: Crystal Growth Supplement to J. Phys. Chem. Sol. (1967) 109.

[10] M. Kestigan, W.W. Holloway, in Crystal Growth supplement to J. Phys. Chem. Sol. (1967) 451.

[11] B. Cockayne, M. Chesswas, D.B. Gasson, J. Mater. Sci. 2 (1967) 7.

[12] F.R. Charvat, J.C. Smith, O.H. Nestor, in: Crystal Growth supplement to J. Phys. Chem. Sol., 1967, p. 45.

[13] Z.I. Shapiro, S.A. Fedulov, Yu.N. Venevtseu, Fiz. Tverd. Tela 6 (1964) 316.

[14] H. Fay, W.J. Alford, H.M. Dess, Appl. Phys. Lett. 12 (1968) 89.

[15] J.R. Carruthers, G.E. Peterson, M. Grasso, J. Appl. Phys. 42 (1971) 1846.

[16] H.M. O'Bryan, P.K. Gallagher, C.D. Brandle, J. Am. Ceram. Soc. 69 (9) (1985) 493.

[17] S. Miyazawa, H. Iwasaki, J. Crystal Growth 26 (1974) 169.

[18] S.L. Blank, J.W. Nielsen, J. Crystal Growth 17 (1972) 302.

[19] C.D. Brandle, A.J. Valentino, J. Crystal Growth 12 (1972) 3.

[20] C.D. Brandle, D.C. Miller, J.W. Nielsen, J. Crystal Growth 12 (1972) 195.

[21] C.D. Brandle, R.L. Barnes, J. Crystal Growth 26 (1974) 169.

[22] V.J. Fratello, C.D. Brandle, A.J. Valentino, J. Crystal Growth 80 (1987) 26.

[23] C.D. Brandle, R.L. Barnes, J. Crystal Growth 20 (1973) 1.

[24] S.C. Abrahams, P. Marsh, C.D. Brandle, J. Chem. Phys. 87 (7) (1987) 4221.

[25] D. Mateika, H. Landan, J. Crystal Growth 46 (1979) 85.

[26] N. Iyi, Z. Inoue, S. Takekawa, S. Kimura, J. Solid State Chem. 54 (1984) 70.

[27] B. Cockayne, M. Chesswas, D.B. Gasson, J. Mater. Sci. 3 (1968) 224.

[28] W.T. Stacy, U. Enz, IEEE Trans. MAG-8 (1972) 268.

[29] H.L. Glass, P.J. Berrer, T.N. Hamilton, R.L. Sterner, Mater. Res. Bull. 8 (1973) 309.

[30] J.R. Carruthers, J. Electrochem. Soc. 114 (1967) 959.

[31] J.R. Carruthers, K. Nassau, J. Appl. Phys. 39 (11) (1968) 5205.

[32] B. Cockayne, B. Lent, J.M. Roslington, J. Mater. Sci. 11 (1976) 259.

[33] K. Takagi, T. Fukazawa, M. Ishii, J. Crystal Growth 32 (1976) 89.

[34] J.R. Carruthers, J. Crystal Growth 36 (1976) 212.

[35] J.C. Brice, P.A.C. Whiffen, J. Crystal Growth 38 (1977) 245.

[36] Y. Miyazawa, M. Mori, S. Honma, J. Crystal Growth 43 (1977) 245.

[37] C.D. Brandle, J. Crystal Growth 42 (1977) 400.

[38] P.A.C. Whiffen, T.M. Burton, J.C. Brice, J. Crystal Growth 12 (1976) 205.

[39] C.D. Brandle, in: E. Kaldis (Ed.), Crystal Growth of Electronic Materials, 1985, p. 101.

[40] P.A.C. Whiffin, J.C. Brice, J. Crystal Growth 13 (1976) 13.

[41] S. Miyazawa, J. Crystal Growth 49 (1980) 515.

[42] K. Shiroki, J. Crystal Growth 40 (1977).

[43] D.C. Miller, A.J. Valentino, L.K. Shick, J. Crystal Growth 44 (1978) 121.

[44] D.C. Miller, T.L. Pernell, J. Crystal Growth 53 (1981) 523.

[45] C.D. Brandle, J. Crystal Growth 57 (1982) 65.

[46] N. Kobayashii, J. Crystal Growth 30 (1975) 177.

[47] J.J. Derby, R.A. Brown, J. Crystal Growth 83 (1987) 137.

[48] K. Nassau, A.M. Broyer, J. Appl. Phys. 33 (1962) 3064.

[49] D.F. O'Kane, T.W. Kwap, L. Gulitz, A.L. Berdnortz, J. Crystal Growth 13/14 (1972) 624.

[50] K.J. Gartner, K.F. Rittinghaus, A. Seeger, J. Crystal Growth 13/14 (1972) 619.

[51] D.F. O'Kane, V. Sadagapan, E.A. Giess, J. Electrochem. Soc. 120 (1973) 1272.

[52] W. Bardsley, G.W. Green, C.H. Holliday, D.T.J. Hurle, J. Crystal Growth 16 (1972) 277.

[53] A.E. Zinnes, B.E. Nevis, C.D. Brandle, J. Crystal Growth 19 (1973) 187.

[54] T.R. Kyle, G. Zydzik, Mater. Res. Bull. 8 (1972) 443.

[55] A.J. Valentino, C.D. Brandle, J. Crystal Growth 26 (1974) 1.

[56] G. Zydzik, Mater. Res. Bull. 10 (1975) 9.

[57] W. Bardsley, B. Cockayne, G.W. Green, D.T.J. Hurle, G.C. Joyce, J.M. Roslingtom, P.J. Tufton, H.C. Webber, M. Healey, J. Crystal Growth 24/25 (1974) 369.

[58] J.A.S. Ikeda, V.J. Fratello, C.D. Brandle, J. Crystal Growth 92 (1988) 271.

[59] V.J. Fratello, C.D. Brandle, J. Crystal Growth 128 (1993) 1006.

[60] X. Chen, Q. Wang, X. Wu, K. Lu, J. Crystal Growth 204 (1999) 163.

[61] P.F. Paradis, J. Yu, T. Ishilawa, T. Aoyama, S. Yoda, J.K.R. Weber, J. Crystal Growth 249 (2003) 523.

[62] C.D. Brandle, D.C. Miller, J. Crystal Growth 24/25 (1974) 432.

[63] C.D. Brandle, J. Crystal Growth 57 (1982) 65.

The accelerated crucible rotation technique (ACRT)

Hans J. Scheel

Ferrofluidics Corporation, Nashua, NH 03061

The late 1960s saw a fast development in two areas of crystal growth which had an impact on the discovery of ACRT, namely the increasing popularity of the flux growth method and the recognition of hydrodynamics as an important growth parameter. The interest in crystal growth from high-temperature solutions (flux growth) was initiated with the growth of $BaTiO_3$ butterfly twins in 1954 and ruby laser crystals in 1964 by J.P. Remeika, and was enhanced by the preparation of magnetic garnet crystals by J. Nielsen and E.F. Dearborn in 1958. Flux growth was widely used to prepare crystals for research in solid-state physics and as an exploratory technique for the preparation of new materials, whereas the development of the flux growth method as a commercial process was either kept secret (ex. Chatham emeralds) or was in its initial phase of development (ex. magnetic garnets).

The situation with hydrodynamics as a growth parameter was (and still is) contradictory. On the one hand compositional inhomogeneities (striations) were attributed to temperature oscillations in the melts. As these temperature oscillations were related to convective instabilities the logical approach to solve the striation problem was to reduce convection by using shallow melts, magnetic fields or microgravity.

The opposite approach was taken with the application of forced convection during crystal growth in aqueous solutions. Wulff (1884) introduced stirring and thereby broke with the old tradition that crystals had to be grown from quiescent solutions. The beneficial effects of stirring in aqueous solution growth are homogeneization of the solution (prevention of uncontrolled nucleation) and reduction of Nernst's diffusion boundary layer which allows increased stable growth rates. Reciprocating stirring (periodic reversion of the rotation of the stirrer or the rotating seed crystals) is often applied, and the author is not aware of any published results on inhomogeneities caused by rotation reversal.

A stirring technique for flux growth was proposed by R.A. Laudise [1] but has not found wide application due to technological and evaporation problems. Therefore stationary crucibles were generally used in crystal growth from high-temperature solutions and resulted mostly in small crystals due to multinucleation, and in inclusions, particularly in the few exceptional cases when the crystals grew larger than usual. This was the state-of-the-art and science of crystal growth in 1968 when I was invited to join the IBM Zurich Research Laboratory in Switzerland in order to set up a crystal growth laboratory for service to the physics department. The large variety of crystals required demanded initiation of several crystal growth techniques, including flux and aqueous solution growth.

One of the tough crystal growth problems at that time concerned the perovskite gadolinium aluminate $GdAlO_3$ which was intensively studied by my former colleague Heini Rohrer. For the physical property investigations, oriented crystal rods up to 15 mm in length, and spheres up to 5 mm diameter were required which were free of inclusions, defects and twins. $GdAlO_3$ has a congruent melting point of $2070\,°C$

Previously published in AACG Newsletter Vol. 15(3) November 1985.

and thus could be grown in principle in our Malvern Czochralski puller and rf-heated iridium crucibles if there were not a phase transition at high temperatures. R. Mazelsky, W.E. Kramer and R.H. Hopkins had already tried to grow this promising Cr-doped laser host by Czochralski pulling; however, the crystals were always cracked and twinned, indicating the occurrence of destructive phase transition. We looked into this and concluded that there could well be a structural phase transition in $GdAlO_3$ in the temperature range $1300\,°C$ to the melting point. A flux-grown single crystal of $GdAlO_3$ was heated carefully to about $1800\,°C$. After slow cooling to room temperature the crystal cracked and twinned; therefore we concentrated further crystal growth attempts on flux growth which allowed us to grow crystals below the assumed phase transition. Here we first confirmed the results of other groups and obtained either small crystals (up to 3 mm) without inclusions or larger crystals (up to 10 mm) with inclusions. Optimizing the $PbO–PbF_2–B_2O_3$ solvent by having excess Al_2O_3 and some V_2O_5 helped a little, but for the growth of really large crystals a breakthrough was needed. We did not want to spend a fortune on 10-liter platinum pots and huge quantities of ultrapure chemicals and obtain large crystals full of inclusions (following J.W. von Goethe and many crystal growers who said that for large crystals you need large crucibles).

As stirring seed crystals at high temperatures from volatile fluxes would be difficult, the first attempts were made to increase buoyancy-driven convection (1969). Despite using spherical and horizontal tube crucibles, the permissible temperature gradients and crucible sizes limited convection flow (the Grashof number is proportional to ΔT and h^3), so that the experiments showed marginal improvement only. So the problem was to achieve smooth stirring within a closed container and avoid shaking or strong agitation which would lead to spontaneous multinucleation. Kirgintsev and Avvakumov [2] suggested several stirring techniques of which the rotating horizontal tube would have been a possibility. However, both our commercial Superkanthal furnace and the laboratory-designed SiC furnace did not allow installing horizontal rotation, only rotation about the vertical axis. Rotation of a crucible about its vertical axis would lead to rigid-body rotation of the contained liquid, which would not

Fig. 1. Single crystal of perovskite gadolinium aluminate ($GdAlO_3$) of size $3.5 \times 3 \times 2.5$ cm, with large inclusion-free regions.

help. Instead, relative motion between liquid and solid crucible walls was required.

One day in late summer 1969 I got an idea when I awoke early and did not want to disturb my wife. Instead of continuous crucible rotation, periodic starts and stops of crucible rotation would lead to relative motion between large liquid fractions and the crucible walls due to inertia, and thus would provide mixing. I estimated the kinematic viscosity and got the feeling that the start/stop rotation would lead to excessive agitation, especially when one imagined growing crystals fixed to crucible wall or bottom. Therefore I planned from the beginning to use a smooth acceleration and deceleration together with maximum rotation rates and periods which were later theoretically shown to be within 10% of the optimum conditions.

The next steps were clear: 1) motivating the draftsman Hans Schmid (a former model shop manager) to design a simple electromechanical device with cam-driven potentiometers and switches in order to achieve the ACRT cycle. The motor and transmission were at-

Fig. 2. Inclusion-free garnet crystal of about 2.5 cm size.

tached to a vertical ceramic tube which extended into the furnace chamber and held the platinum crucible, 2) having a special crucible made with a Bridgman tip (coolest spot for localized nucleation site, nucleus selection), and 3) planning and running the first experiment. I remember that it was about the middle of January 1970 that with the help of my technician H.R. Küpper we removed the hot crucible (900 °C) from the furnace, punched two holes into the lid, and poured the residual solution out. As the cooling rate was initially 0.3 °C/hour and later 0.6 °C/hour and the temperature range 350 °C, we had to wait about five weeks for the end of the experiment, reason enough to be ex-

cited when opening the crucible by cutting off the platinum rim. The result was fantastic: a single crystal of $3.5 \times 3 \times 2.5$ cm size and 210 g weight (representing 2/3 of the starting material) with large inclusion-free regions (see Fig. 1[*]). This result showed that the maximum stable growth rate could be significantly increased by stirring and thus opened the way to grow large inclusion-free crystals from high-temperature solutions, a fact which led to the upgrading of the flux growth method.

We also succeeded in growing garnet crystals for the magnetic bubble program, first YIG and then $Y_3Fe_{5-X}Ga_XO_{12}$ solid solutions. These crystals, typically 2.5 cm size, were inclusion-free, an example of which is shown in Fig. 2[*]. The solid solution crystals showed remarkable homogeneity.

ACRT is also applied now in other crystal growth techniques with significant success; for instance, in the Czochralski growth of Si and garnet solid solutions, in Bridgman growth and in the hydrothermal growth of $AlPO_4$.

When I explained this new technique to my wife she replied: "This is not new. My washing machine is doing the same!" This was disappointing, especially when I found out then, that rotation reversal had been applied in washing machines for more than half a century! Possibly ACRT could have been discovered earlier had I more experience with using washing machines.

References

[1] R.A. Laudise, in: J.J. Gilman (Ed.), The Art and Science of Growing Crystals, Wiley, New York, 1963, pp. 267, 270–272.
[2] A.N. Kirgintsev, E.G. Avvakumov, Sov. Phys. Crystallogr. 10 (1965) 375.

---

[*] Impressive photographs of ACRT-grown garnets have been published by W. Tolksdorf and D. Mateika, for instance in the March 1985 AACG Newsletter and on the cover page of the German DGKK Newsletter of October 1984.

Shaped crystals from the melt by EFG

A.I. Mlavsky

Executive Director, Israel–U.S. Binational Industrial Research and Development Foundation, Tel Aviv, 61390, Israel

Editor's Note: The Crystal Growth Technique known as "Edge-Defined, Film Fed Growth" (EFG) was first reported by H. LaBelle in part II of a four part series of articles on "Controlled Profile Crystals" in the Mat. Res. Bull., Vol. 6, 1971, pp. 581–591. LaBelle received a U.S. Patent (No. 3,591,348) on this method on July 6, 1971.

The evolution of "edge-defined film-fed growth" (EFG) from an observation of an apparently impossible phenomenon to a mature industrial production technique is, I think, a quite fascinating example of the success that occasionally visits those who, though bent on invention in a focused manner, are nonetheless able to receive and interpret weak signals from the magical land of Serendip.

The roots of the development are to be found in a 1960's U.S. Air Force requirement for high strength, low density, continuous filaments for reinforced composites. Boron filaments, made by CVD onto tungsten wires, had opened the possibility of a new class of high performance composites. Sapphire, in principle, was posited to have significant advantages, if economically producible in continuous lengths. (Of course, we know now that carbon filaments waltzed away with the prize, but that's another story.)

Unimpressed by proverbial angelic phobia to tread new paths, I persuaded a credulous staffer (Harry Materne) at Wright Field to sponsor a tiny project on the development of not too well specified techniques— except that they would have to be new—for producing continuous sapphire filaments directly from the melt.

Tyco Laboratories' young Harry LaBelle Jr. was the lucky (!) technician chosen to spearhead this assault on conventional crystal growing wisdom.

LaBelle, belying his lack of formal qualifications, proved to be highly inventive—and much more than just intuitive—in conceiving and implementing successive approaches to growing long skinny crystals. Together we had reasoned, not too subtly, that the way to grow a small diameter crystal—the goal was $\phi \leqslant 250$ μm—was to use a small diameter crucible. But how to overcome the small capacity of the latter? LaBelle's simple but ingenious approach was to mount a short molybdenum capillary (which is wet by molten alumina) in a sizeable crucible, and to grow the crystal from the melt at the tip of the capillary. Surface tension would provide the force necessary to keep the capillary full.

The technique, christened by me as "SFT" (self-filling tube), was immediately successful, but with a subtle pecularity in the results. Specifically, LaBelle noticed that the diameter of the sapphire filaments produced appeared to be larger than the hole in the capillary. Since the latter was tapered at the tip, this observation was by no means trivial.

We discussed this at length, covering and erasing successive blackboards full of sketches and squiggles. It then occurred to me that, given the configuration per Fig. 1A, the progression shown in Figs. 1B, C and D would materialize since, thermodynamically, the liquid meniscus would progress to the edge of the capillary and then stop. Were it to run down the outer

Previously published in AACG Newsletter Vol. 14(1) March 1984.

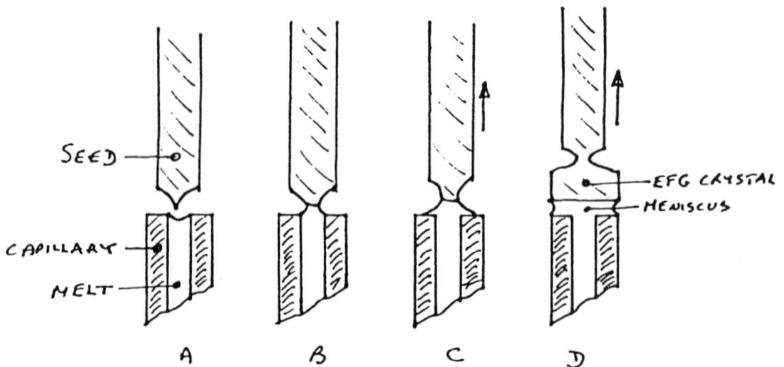

Fig. 1.

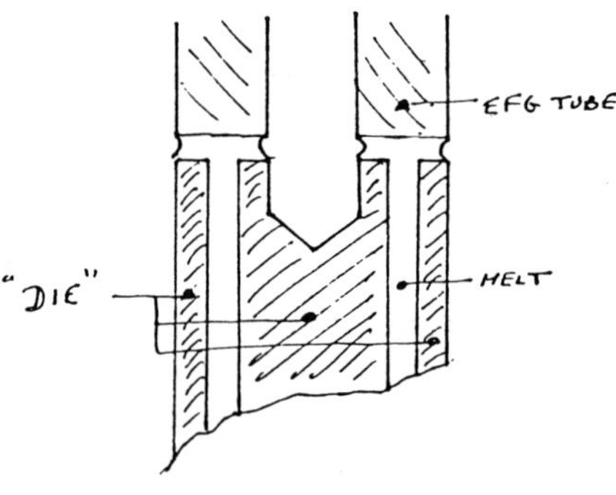

Fig. 2.

surface thereof, a perpetual motion machine would have been created—unlikely! Accordingly, the outer dimensions of the crystal would be determined by the outer diameter of the capillary.

We then immediately recognized that the meniscus would spread across the top surface of a "die" (capillary), stopping at any vertical edge which did not bound an area containing melt. One would expect, therefore, to be able to grow a tube from a die such as that shown in Fig. 2. In no time flat, LaBelle produced such a tube. This, one might say, was the first deliberate and successful demonstration of a radically new technique in shaped crystal growth. It took place around the summer of 1966 and into 1967.

This technique obviously deserved a name, and the name itself seemed obvious to me—not the ABC (or D) of crystal growth (sorry Czochralski!) but the E (edge-defined) F (film-fed) G (growth).

(Remarkably, the initial reaction of the U.S. Patent Office was that the technique could not possibly work. Fortunately, we had made a movie of the process in action. This, together with a wide selection of shaped crystals from a variety of materials, enabled me to persuade the examiner that truth, as we saw it, may indeed be stranger than his version of fiction, but true nonetheless.)

Although LaBelle and I felt we understood the basic mechanism of EFG, and proved our hypothesis by the growth of a variety of intricate shapes, we were less understanding of the thermal situation, in general, and of the observed highly stable nature of the growth, in particular.

At that point, I called on Bruce Chalmers at Harvard. In his inimicable fashion, he rapidly analyzed the thermal balance and showed that the observed stability—not intrinsically to be expected in a but modestly well controlled system operating around 2000 °C—was easily explicable in simple terms.

Although there have been many publications describing EFG and its application to, for instance, silicon ribbons and other shapes for solar cells (an activity I initiated in 1971 in anticipation of the viability of terrestrial photovoltaics), the first public presentation of EFG was, by far, the most memorable.

Harry LaBelle, a little nervous but remarkably assured, addressed a huge assembly at the International Conference on Crystal Growth in Marseilles, in July 1971 on behalf of himself and his co-authors, Bruce Chalmers and me. The impact of the description and of the home-made movie of the process in action was considerable. In fact, the editor of the proceedings of the conference, Prof. Kern was kind enough to state that the presentation was (in rough translation from the French) the hit of the show.

Experimental work leading to EFG

H.E. LaBelle

Saphikon Division of Tyco, Milford, New Hampshire

[Editor's Note: This is Part II in a series on the invention of EFG. The first was written by project leader Ed Mlavsky, and the current article by the process inventor.]

The invention of processes to grow shaped sapphire crystals and the solution to the ancillary problems along the chosen experimental path brought a great deal of personal enjoyment in the late 1960's. Each experiment quite naturally yielded a result which emphasized a problem and hence a new experiment. The edge-defined, film-fed growth was an outgrowth of this evolutionary process.

While employed at Tyco in the summer of 1965, I was anxious to work on a project of my own. The timing of my request for a project proved ideal since Dr. Ed Mlavsky told me of a contract which he had proposed that would soon be starting. In September, I began to work armed with Dr. Mlavsky's proposal which specified growth of filaments by either a dendritic approach (as was being used at Westinghouse for silicon ribbon), or electron beam melting of a pool of liquid from which a filament could be pulled (this latter approach was never tried).

We focused on dendritic growth and with the help of Dr. G.A. Wolff on questions of dendrites, twinning and reentrant corners, and Dr. Mlavsky's assistance on heat flow and high temperature materials, I began working on the dendritic growth of sapphire filament.

I first determined that sapphire could be supercooled to approximately 500 °C in molybdenum and by plunging a cold tungsten rod into a molten bath of alumina and supercooling, I was able to grow small dendrites. Dr. Wolff and Dr. T. Mariano (Ledgemont Laboratories) assisted in determining that the c-axis played a major role in these spontaneously nucleated crystals.

Experimentation then began in earnest on dendritic filament propagation of sapphire. The work proved exciting, since a number of experiments could be run each day and quickly analyzed and the next experiment designed and run. However, with all but a couple of months left to go on the nine month project, only a few short lengths (3 mm) of dendrites had been produced.

At the time, it was not clear whether the limitation in producing longer lengths was due to inadequate temperature control or the c-axis dendritic morphology. We discussed the bleak prospects for success and considered the electron beam approach, however, there was insufficient time left to set up this equipment and run experiments.

I decided to work on improving the temperature stability. Previous experiments had been performed using an archaic Transitron Czochralski puller which was the subject of substantial instability. In an attempt to find a more stable way of controlling the temperature of the melt, I attempted resistance heating in a vacuum evaporator. A small quantity of liquid alumina was melted in a tungsten evaporating boat and a .25 mm diameter tungsten wire used as a seed. The wire was suspended vertically down from a steel strip attached to the top end of a rotary push-pull vacuum seal, which

Previously published in AACG Newsletter Vol. 14(2) July 1984.

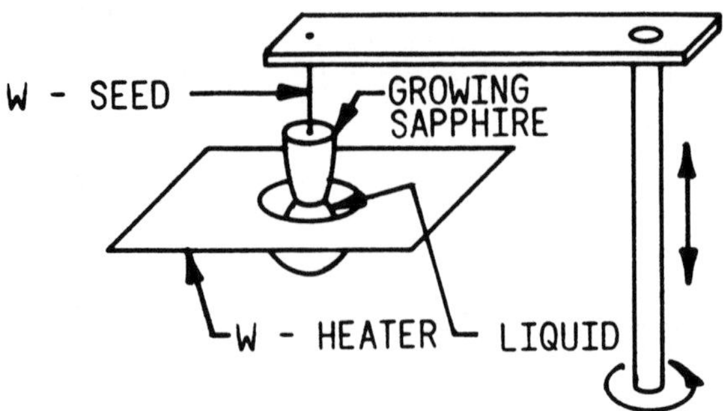

Fig. 1. First successful experiment.

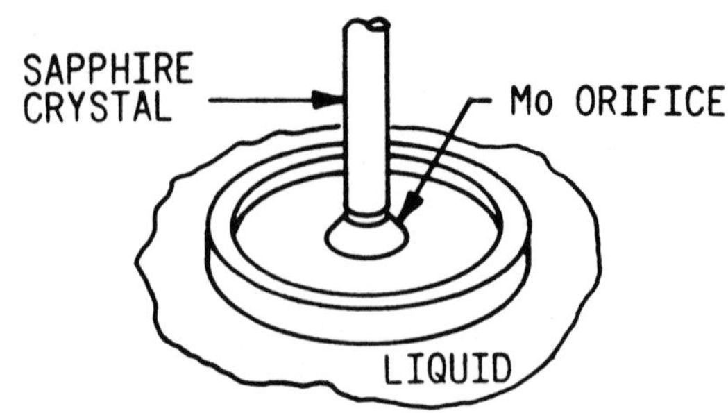

Fig. 2. First filament process.

became the pulling machine. The seed was lowered into contact with the melt by manually pulling it down and growth initiated by carefully pushing the shaft upwards. Due to the high radiation losses (no shielding) it proved easier than one might expect to actually seed and pull the molten pool empty. The system represented growth by the Czochralski technique in a miniature system. Crystals grown using this approach were of a carrot shaped geometry with the tapered tip being grown from the last of the melt to solidify (Figure 1).

In considering how to use this system to grow filaments, I made the following observation. At any given time, the diameter of the growing sapphire crystal (carrot) was related to the diameter of the liquid drop remaining. It therefore because obvious that if one wanted to grow a filament size cross section, that one could do so from a similar size droplet. While this would meet the cross sectional area requirement, the droplet clearly needed to be fed from a large reservoir of melt continuously to meet the length requirements.

The first solution meeting this requirement which I tried was a floating disk of molybdenum having a central orifice (Figure 2). Although being a bit concerned about the higher density of molybdenum making it difficult to float on liquid alumina, the experiment was attempted. It proved successful on the very first experiment and a most gratifying experience as it represented my first invention which actually worked.

The experiment was run after hours, but I recall locating a fellow colleague, Mr. Dick McNeil, to share the success with. The following day, it was with great pride that I showed Dr. Mlavsky the grown crystal and described how it was accomplished. A number

of crystals were generated in the following few weeks which were 15–30 cm long and approximately 1 mm in diameter. The cross sections were a bit irregular, however, sufficient to insure an additional 12 months funding of the project.

The project was now sufficient to support both myself and Mr. John Bailey who was added to the staff to assist technically. We began to work on improving the quality of the crystal (grown by the floating orifice technique).

With great enthusiasm experiments were run using numerous orifice configurations. After a few months it became obvious that floating a molybdenum disk on liquid alumina was tenuous, and although capable of producing crystals for research, probably would never yield a manufacturing process. We considered ways of mechanically supporting and lowering the orifice during growth, however, concluded that such a system would be difficult to use in production due to the high temperature required.

Back in the laboratory I returned to the more basic question of obtaining a system which yielded a small diameter crucible of large volume. It occurred to me that the use of a molybdenum capillary projecting out of the melt might yield the positive result that the floating orifice had, yet represent mechanical, and therefore, thermal stability as well. A molybdenum capillary tube with a conical outer top surface was made (self filling tube, SFT) and met with immediate success (Figure 3a). Filaments were produced with highly regular surfaces as compared to the earlier technique and with tensile strengths 3 to 4 times as strong. Very soon after this process was tried, I used it to produce sapphire tube and ribbon. Hence, dies were made (Figure 3b) which produced similar shaped

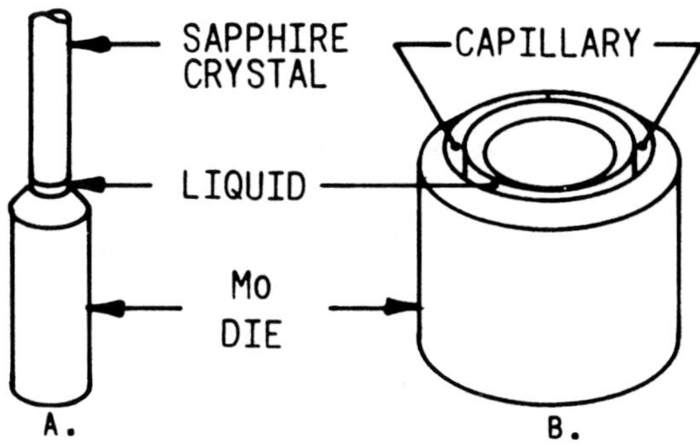

Fig. 3. First use of capillaries.

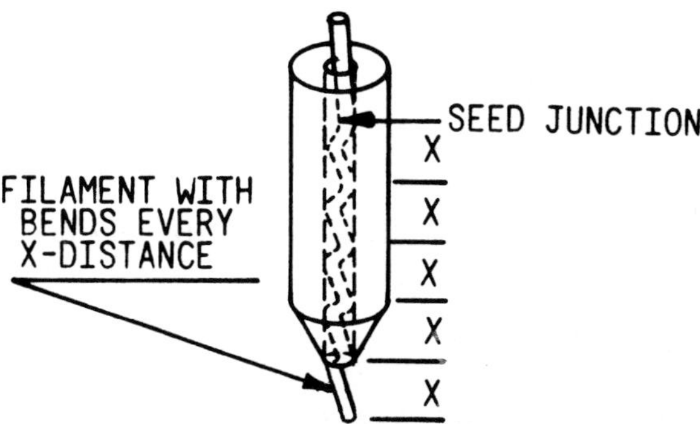

Fig. 4. Guidance problem.

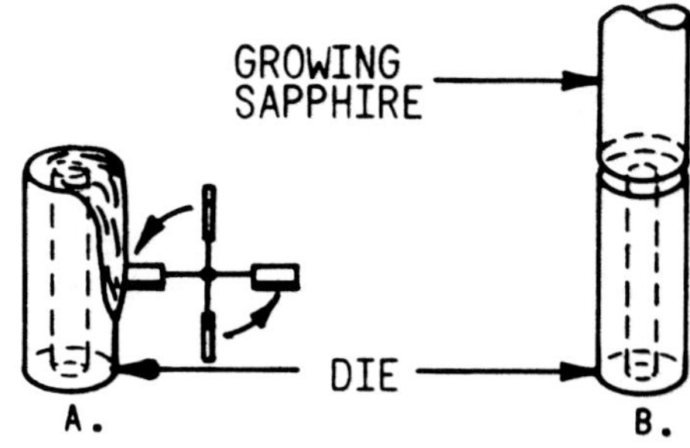

Fig. 5. First use of edge definition.

capillaries (it is of interest to note that sapphire tubes and ribbon were first grown by this process).

The next major hurdle to be encountered was the growth of continuous lengths of sapphire filament. All previous work had been done on equipment having a 30 cm pull. Also, it would be necessary to have some sort of mechanical guiding system for both the seed crystal and the growing filament, since convection during growth would otherwise cause the filament to move laterally. Initially, a small bore copper tube was used to guide the seed and subsequently the growing filament (Figure 4). The seed junction would always catch or bend as it entered the tube and thereby cause another bend at the interface. Each bend on entering the guide would cause another bend.

Considering solutions to the oversized diameter problem, I had observed that there were instances where the filament actually grew larger than the orifice. The conical orifice tip typically had either a 90° or 120° included angle. The problem of increased diameter growth was more prevalent on the 120° included angle. One obvious solution was to make the included angle very small. However, this would not be mechanically stable. An alternative solution would be to simply use a flat top surface and allow the liquid to spread to the edge where it would stop.

In trying to consider whether or not it would run over the edge, I was unable to resolve the question quickly from a surface tension standpoint, however, I was able to invoke a "Mlavskyism." After working for Dr. Mlavsky for several years, I had been well indoctrinated by him in attempting to make a perpetual motion machine out of a wide variety of systems to determine if they could work. I was, therefore, able to conclude that the liquid would not run over, otherwise I could run a paddlewheel with the liquid flowing down the outside of the capillary (Figure 5a).

John Bailey and I built a die (edge defined, film fed, growth) and found it worked as well, if not better, than was hoped (Figure 5b). I met with Dr. Mlavsky to share this success and we discussed the idea of growing other than filamentary shapes as part of the process capability. Only limited interest had been shown in the shapes made by the earlier process, which included tubes, ribbon, L-shapes, etc. The new process (EFG), however, offered an even wider selection of shapes.

Back in the laboratory I concluded that it was important to distinguish the difference in the two techniques (EFG and SFT) and designed the capillary for the first EFG tube to establish this difference. The first tube die was made (Figure 6) so that four circular capillary holes fed the solid molybdenum tube die. The uniqueness of EFG in shaping was therefore established in the first tube growth since the capillaries were of totally different cross-sectional shape than the growing crystal. A number of shapes were produced over the following year in an attempt to develop applications for the process.

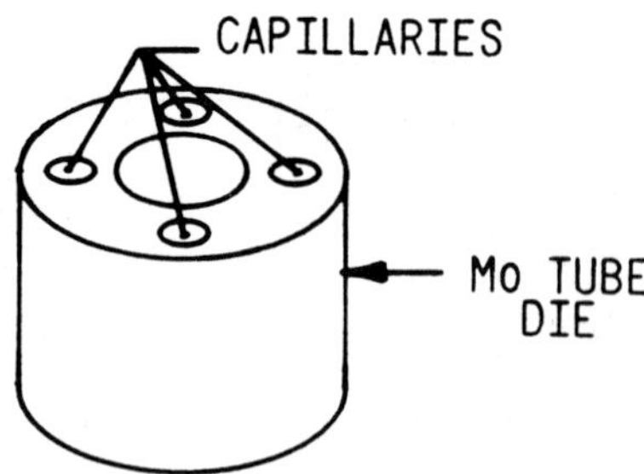

Fig. 6. First EFG tube die.

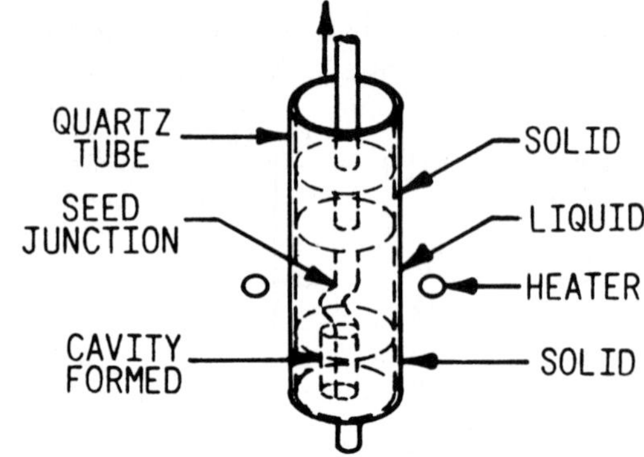

Fig. 7. Paraffin bearing.

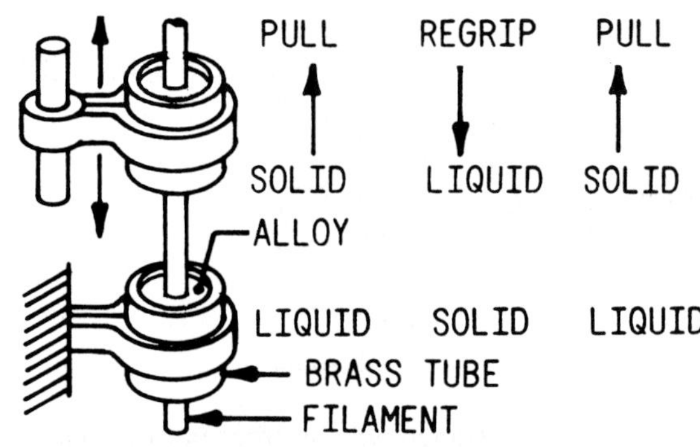

Fig. 8. First semi-continuous puller.

In returning to the problem of continuous filament growth, the question of what type pulling mechanism to use remained. This was eventually answered by Mr. Sy Mermelstein who designed a puller consisting of two continuous belts which would have the filament sandwiched between them. Although the problem of the growing filament exceeding the die diameter was solved by EFG, the problem of producing a perfectly straight seed junction remained. As a batch process, we were able to produce a maximum length of 30 cm which was straight (since this full grown length could be broken back from the seed junction). It appeared that any rigid guiding mechanism would cause a perturbation to the growing interface when the seed junction entered the guide, hence, if a rigid guide was positioned 5 cm above the die, there would be perturbations every 5 cm after pulling began. These perturbations or jogs would generally increase in amplitude and eventually jam in the guide as shown in Figure 4. (X equals the distance between the die and the tip of the guide.)

A guide was needed which kept the filament from moving by convection during growth, yet allowed the first major jog at the interface to pass without putting a transverse load on the growing filament. The only solution apparent which would not involve the use of an elaborate mechanical and sensing device was what I called a solid–liquid–solid bearing, shown in Figure 7. It is formed by loading a 1 cm diameter quartz tube with liquid paraffin and inserting the sapphire filament at its center along the axis of the tube and subsequently allowing the paraffin to solidify. A one turn resistance heating element was then raised around the center of the tube and the paraffin melted in this central region. By adjusting the temperature it was possible to soften the lower paraffin region so that when the seed junction, which was moving up vertically, reached its

softened region, it pushed the paraffin up out of its way and continued into the liquid region where a small quantity of liquid would run down in to the cavity where the soft paraffin had been removed. This bearing worked quite adequately to aid us in producing the first long lengths of filament.

As the belt pulling mechanism was still not available, there was the additional problem of how to pull a long crystal from a 30 cm pulling machine. It seemed that the perfect way to attach onto the filament without moving it would be to surround the filament with an annular ring of liquid metal which expanded upon solidification. The filament would, therefore, be held (and not moved) by solidifying the metal. The apparatus used consisted of two small brass tubes filled with an alloy (Figure 8). Using a small diameter hole and limiting the length of the alloy prevented it from dropping down due to gravity. The top tube was attached to a moving platen on the puller and the lower one rigidly fixed below it and above the paraffin guiding bearing.

During the pulling cycle, the upper metal was solidified while the lower one remained liquid so that the filament could be pulled up smoothly through it. When the top of the stroke was reached, the bottom metal was solidified so as to hold the filament in place. In a regripping cycle, the top metal was made liquid and the platen brought down to its lower position. The pull cycle was then repeated in that the solid and liquid metals were reversed and 30 cm of filament added to the original growth. This technique was successfully used to produce the first filaments a few meters long. These filaments were eventually used as seeds for the belt puller and the first filaments continuously grown.

Within a few years, a number of cross sectional shapes had been made and commercialization was to the point where sapphire tubes were being used in street lights in both the U.S. and Europe (on a limited basis) and kilometers of filament were being made.

At this point, knowing that the patent would soon be published, we decided to prepare an article for open publication. The first closed publication on EFG had occurred in 1969 in a government report. While considerable attention had been given to heat flow and the hydrodynamics of the process, Dr. Mlavsky and I felt additional assistance was in order before I gave a presentation to the scientific community. It was at this point that Professor Bruce Chalmers was introduced to the process: Acting as a consultant, he made additional contributions to our understanding of the heat flow, hydrodynamics and the basic stability of the process. I first described the process to the ICCG in Marseille in 1971.

To date, there are clearly hundreds of individuals who have contributed to EFG. I would like to acknowledge those individuals who uniquely contributed to my work as EFG was being invented. Drs. Arthur Rosenberg and Ed Mlavsky at Tyco presented an environment of great enthusiasm in which to work. Dr. Mlavsky in particular who I worked directly for and with for a number of years, afforded me the opportunity to work with minimum direction but with maximum support. He is responsible for the name, EFG. I had in mind to call the process continuous shaped film propagation, but bowed to his more definitive suggestion. Dr. Gunther Wolff invested many evenings with me, explaining a number of the intricacies of crystallography and crystal growth. Mr. John Bailey, who supported my work as a technician, offered ideas of his own, and worked many long hours with me in the early days of the process.

Hydrothermal synthesis of crystals

Robert A. Laudise

AT&T Bell Laboratories

During the past two decades, the needs of the electronics industry for such recalcitrant materials as single crystals has sparked research on methods for preparing those crystals, including even gem materials like emerald, from solvents at near- and supercritical conditions. Chemistry at these conditions, usually called hydrothermal chemistry, is responsible for the formation of many crystalline minerals in nature, and crystal growth under hydrothermal conditions has been used to produce large piezoelectric quartz crystals experimentally and commercially for electronic equipment.

The success of researchers in growing large crystals has encouraged, in turn, the investigation of the physical chemistry of hydrothermal solvents. This article, therefore, focuses on hydrothermal physical chemistry, hydrothermal synthesis, and crystal growth, areas in which the accomplishments of chemists perhaps are most recent and evident. However, chemistry of high-temperature, high-pressure water and other solvents also is of great interest now in geochemical, chromatographic, biological, and separation studies.

In 1839, the German chemist Robert Wilhelm Bunsen contained aqueous solutions in thick-walled glass tubes at temperatures above 200 °C and at pressures above 100 bars. The crystals of barium carbonate and strontium carbonate that he formed under these conditions marked the first use of hydrothermal aqueous or other solvents as reaction media.

Geochemists and mineralogists have studied hydrothermal phase equilibria since the turn of the century. George W. Morey and his coworkers at the Geophysical Laboratory of Carnegie Institution in Washington, D.C., and later Percy W. Bridgman at Harvard University did much to lay the early experimental foundations necessary for the containment of reactive media in the 300 to 700 °C, 1- to 3-kilobar range, where most hydrothermal work is conducted. In particular, they conceived designs for high-pressure autoclaves that are used to this day.

Major recent advances in understanding the physical chemistry of hydrothermal solutions are largely due to the research of a small group of chemists and earth scientists, especially E. Ulrich Franck of the University of Karlsruhe, West Germany. Phase equilibria aspects of hydrothermal chemistry have been particularly emphasized by Rustum Roy and his students at Pennsylvania State University, where many critical oxide-water diagrams have been determined. Rare-earth oxide systems are currently being studied by Stanley Mroczkowski and others at Yale University.

Hydrothermal crystallization typically is from solution. It differs from conventional liquid-solution crystal growth at temperatures near to ambient largely because the viscosity of the liquid is lower. With conventional growth, solubility must be high; otherwise, diffusion problems can be severe. The viscosity of supercritical solutions used in hydrothermal crystal growth can be as much as two orders of magnitude lower than the viscosity of near-ambient solutions, so that slow diffusion is much less of a limit to crystal growth. This results in a high rate of growth at low

Previously published in AACG Newsletter Vol. 21(1) Spring 1991.

temperature. One disadvantage is the need for high pressure. Another is the present relative lack of experimental data, which makes crystal growth under hydrothermal conditions hard to predict.

Because diffusion is significantly faster at hydrothermal conditions, crystals can be grown faster, even from materials of low solubility. And because the diffusion zone close to the growing crystal interface is relatively narrow, constitutional supersaturation, and hence dendritic growth, is less frequent. For example, because of solubility and diffusion limitations, quartz crystals cannot be grown at near-ambient conditions. Under hydrothermal conditions, however, quartz crystals grow at rates as high as 2 mm per day without severe faults or dendritic growth, even though solubility is only a few percent. (Quartz cannot be grown from a melt because the piezoelectric phase necessary for formation of electronic crystals is not stable at the melting point and because silicon dioxide melts are so viscous that they form glasses rather than crystals when they are cooled.)

Mineralizers

The solubility of many inorganic materials can be increased by adding a complexing agent to the solution. These complexing agents often are called mineralizers because they promote the solubilization and recrystallization of naturally occurring minerals. Quartz, for instance, dissolves in water under hydrothermal conditions to form $Si(OH)_4$. But even at temperatures above 400 °C and pressures above 1 kilobar, the solubility of quartz in water is only a fraction of a percent. In 1 M sodium hydroxide solutions, however, the solubility of quartz can be several percent under hydrothermal conditions; the hydroxide ion acts as a complexing agent leading to the formation of such soluble silicate complexates as SiO_3^{2-} and $Si_3O_7^{2-}$.

Many inorganic ions, including OH^-, Cl^-, F^-, S^{2-}, NH_4^+, H^+, and WO_4^{2-}, are effective mineralizers. The complexates formed by these mineralizers should not be so stable that they form a solid that itself precipitates. The most common mineralizers in the formation of natural crystals are Cl^- (from sodium chloride) and CO_3^{2-}.

The supersaturation needed for hydrothermal crystal growth in the lab is produced by imposing a temper-ature gradient within the autoclave in which the crystals are to be grown. Neglecting minor corrections for hydrostatic pressure, pressure is constant throughout the autoclave. For example, in a vessel that is 83% filled and has a temperature of 350 °C at the top and 400 °C at the bottom, the pressure of a 1 M sodium hydroxide solution saturated with quartz will be 1530 bars. Using isothermal data for percent fill and taking density as percent fill divided by 100, a local density of 0.85 g per cc can be calculated for the 350 °C region or of 0.79 g per cc for the region at 400 °C. Procedures of this sort can be used to calculate density gradients and hence the driving force for convective circulation in systems used for hydrothermal crystal growth.

Crystallization of quartz

The material crystallized hydrothermally in greatest commercial volume is quartz, which now is firmly established as a key electronic material, perhaps second in importance in volume only to silicon.

Piezoelectricity—the production of an electric dipole in a crystal when it is deformed—was first discovered in α-quartz by Pierre and Jacques Curie in France in 1880. Piezoelectricity remained a scientific curiosity until World War I, when quartz and Rochelle salt were used independently by Paul Langevin in France and Alexander M. Nicholson at Western Electric Co. in the U.S. for picking up underwater sound in the sea for submarine detection and depth measurement.

During the 1920s, Walter Cady and his students at Wesleyan University discovered the great utility of piezoelectric quartz crystal for controlling oscillators in electric circuits. The frequency of oscillation in the circuit is set by the dimensions and orientation of the quartz crystal, which may be thought of as an electrically driven tuning fork with a resonant frequency which, in today's designs, can be anywhere between about 10 kHz and 4 GHz.

Quartz crystals for Cady's experiments were purchased from mineral supply houses, which imported them mainly from Brazil. They were first used as frequency standards at the National Bureau of Standards and for frequency control in radio transmission at radio station WEAF in New York City. Quartz is particularly useful for frequency control because for certain crystal orientations its frequency-temperature dependence exhibits a very small temperature coefficient. By the outbreak of World War II, quartz had become essential for military radio communication, and the best-quality quartz continued to be imported from Brazil.

German submarine activity in the Atlantic during World War II severely curtailed the availability of Brazilian quartz, and because no other adequate source of natural crystals of suitable quality could be found, quartz was a highly critical material throughout the war. At the end of the war, U.S. intelligence agencies found that Richard Nacken had led a German effort to prepare quartz hydrothermally, based in part on the earlier work by Bunsen and by other 19th century German, French, and Italian chemists, especially the Italian Gregor Spezia, who had prepared very small quartz crystals under hydrothermal conditions.

Any growth process for quartz must be effective below 570 °C, the $\alpha \Rightarrow \beta$ quartz transition, if it is to produce the piezoelectrically useful alpha phase. The melting point of silicon dioxide is above 1700 °C, ruling out melt growth. Solution growth required an effective solvent; because of the high viscosity of silica solutions, molten salt solvents are not attractive. Vapor phase reaction growth might be considered, but because vapor phase processes are slow they also are not attractive for preparing bulk crystals.

Thus hydrothermal growth emerges as the logical method for preparing cultured quartz crystals. Although Nacken's group succeeded in preparing small crystals, wartime exigencies terminated his study. Reports of the German activities, however, stimulated research on the hydrothermal preparation of quartz in the U.S. and Britain, and later in the Soviet Union, Japan, and other countries.

By the early 1950s, Albert Walker, Gerard T. Kohman, and Ernest Buehler at Bell Laboratories in Murray Hill, NJ, had succeeded in synthesizing large crystals of quartz in sodium hydroxide solutions under hydrothermal conditions. Later, Albert Ballman and I carried out a systematic study of the physical chemistry of hydrothermal growth and with Richard Sullivan of Western Electric in North Andover, MA, set up quartz production.

A different process using sodium carbonate solutions, based on the work of Danford Hale and Hans Jaffe at Clevite Corp. in Cleveland is now being used by Sawyer Development Corp., East Lake, OH. A similar process was developed by Cyril Brown and his colleagues at General Electric Co. Ltd. in England.

The Bell Labs process uses aqueous sodium hydroxide as the solvent at pressures in the 1.3- to 2-kilobar range; the Clevite process uses aqueous sodium carbonate in the 0.7- to 1.3-kilobar range. Growth rates in hydroxide solution are about twice those in carbonate solution. The choice of processes is largely a trade-off between the higher cost of high-pressure equipment and the faster growth at high pressures. The reaction

$$SiO_2 + 2H_2O \Leftrightarrow Si(OH)_4$$

is mainly responsible for the solubility of quartz in water. This solubility, however, is only a few tenths of a percent, even at supercritical temperatures.

The addition of a complexing agent or mineralizer, which provides additional ionic species, can raise the solubility of quartz appreciably, however. The complexing agent, though, should not react so strongly that the complexate (solute) becomes a stable solid. Because of the amphoteric nature of SiO_2, OH^- is an excellent mineralizer. It leads to the formation of soluble silicates, so that the solubility of quartz in 1M sodium hydroxide at 350 to 400 °C and 1 to 2 kilobars is several percent at conditions where α-quartz is the stable solid phase.

Solubility studies suggest that dissolving reactions such as

$$2(OH)^- + 3SiO_2 \Leftrightarrow Si_3O_7^{2-} + H_2O$$

are responsible for dissolving quartz in OH^- and

$$2(OH)^- + SiO_2 \Leftrightarrow SiO_3^{2-} + H_2O$$

for dissolving quartz in $(CO_3)^{2-}$. In the latter equation, the OH^- is produced by CO_3^{2-} hydrolysis.

In processes for hydrothermally crystallizing quartz, small (about 0.5 inch) "nutrient" pieces of inexpensive Brazilian quartz feedstock are placed in the bottom, or dissolving, region of a hydrothermal autoclave, suitably oriented seed plates are mounted in the upper, or growth, region, and the autoclave is filled to, say, 82% of its free volume with 1.0 M sodium hydroxide. The dissolving and growth regions are separated by a perforated metal disk or baffle, which restricts convection (which can be very rapid and actually is turbulent under many conditions) and transport. The baffle localizes the temperature difference between the nutrient and growth regions, so that growth of all the seeds in the growth region is at the same rate. External heaters heat the nutrient region to, say, 425 °C and the growth region to 375 °C. Under these conditions, the autoclave fills with a single fluid phase at a pressure of 1.5 kilobars. The solution saturates with quartz in the dissolving zone and moves by convection to the growth zone, where it becomes supersaturated, causing the seeds to grow.

After 15 to 30 days, the grown seeds are harvested, yielding several hundred pounds of quartz crystals from a small commercial autoclave with an internal diameter of 10 inches and a length of 10 feet. (At present, some Japanese autoclaves have dimensions of 26 inches × 25 feet.) The vessel is then recharged for a new growth cycle. New seeds, piezoelectric oscillators, and other electronic devices can be cut from the grown crystals, using silicon carbide saws in an abrasive slurry.

Detailed studies show that the kinetics of quartz crystal growth may be described by a simple rate equation:

$$R_{hkl} = \alpha k_{hkl} \Delta S,$$

where R_{hkl} is the growth rate in a particular crystallographic direction, k_{hkl} is a velocity constant that has an Arrhenius temperature dependence, ΔS is the supersaturation (that is, the actual concentration minus the equilibrium concentration), and α is a dimensional conversion constant. This equation, together with pressure–volume–temperature data, can be used to map the dependence of growth rate on the important engineering parameters temperature and pressure.

Incorporation of impurities is determined by appropriate distribution (partition) constants. The distribution constant is a special case of the equilibrium constant for the reaction that described the incorporation of a particular impurity. Because, at useful growth rates, a diffusion field exists in the boundary layer in front of the growing crystal, the bulk concentration of impurity in the solution must be corrected to obtain the actual concentration near the growing crystal, where impurity incorporation actually occurs.

"Coupled" substitution also is important in determining impurity uptake. For instance, Al^{3+} (a common impurity in natural quartz, such as Brazilian nutrient), which enters the quartz lattice substantially

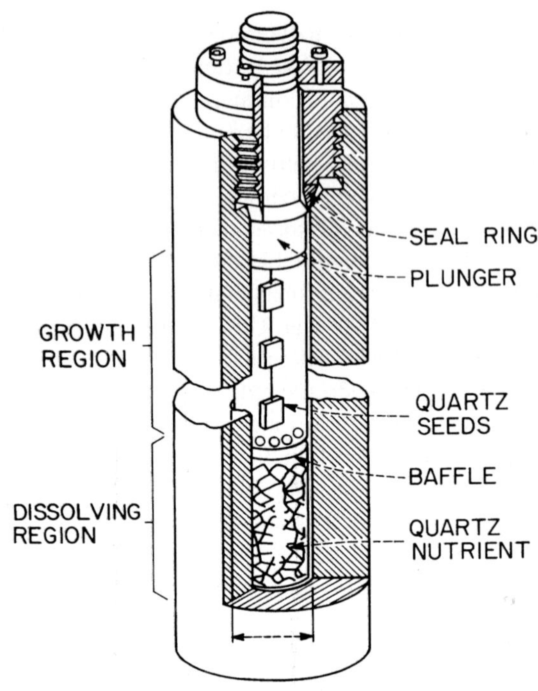

Fig. 1. Schematic of quartz growing autoclave.

at Si^{4+} sites, is charge compensated by an interstitial H^+:

$$Al^{3+}_{in\ solution} + H^+_{in\ solution}$$
$$\Leftrightarrow \left(Al^{3+}_{at\ an\ Si\ site} \bullet H^+_{interstitial}\right)_{in\ the\ solid}.$$

The solid species is written as a complex because charge compensating ions in quartz generally associate. Because growth takes place in an OH^- media and because the solid is an oxide matrix, H^+ enters the lattice as OH^-, and its concentration, as shown by Dorothy Dodd, David Fraser, and Darwin Wood at AT&T Bell Laboratories, is proportional to the optical absorption at 2.86 μm (an OH^- stretch frequency).

Measurements by J.C. King and others at Bell Labs also show that the acoustic loss in quartz at frequencies of importance to piezoelectric devices is dependent on the OH^- content. High acoustic loss results in poor device performance. Conversion between mechanical and electrical energy is inefficient, with energy being dissipated as heat.

These studies provide an understanding of the dependence of acoustic loss on impurity concentration and growth rate. As a result, cultured hydrothermal quartz with acoustic loss lower than that in most natural quartz now is prepared routinely.

In 1974, the Brazilian government embargoed quartz nutrient to encourage production of cultured quartz and electronic devices in Brazil. Natural quartz crystals occur either as vein quartz or pegmatitic quartz. Brazilian quartz is vein quartz. It is deposited in cavities from supercritial hydrothermal fluid in much the same way that quartz crystals are grown in the lab. It often occurs as large, clear crystals, which makes cutting it directly into devices attractive.

High chemical purity, however, is the only requirement for nutrient for recrystallization. Pegmatitic quartz forms by fractional crystallization when a magma freezes, sometimes producing very pure crystals. Such quartz had long been considered unsuitable for cutting into piezoelectric devices, however, because it is poorly formed and milky, rather than clear. But Earle Simpson of AT&T Network Systems and Kurt Nassau and I of Bell Labs have shown that such properties are irrelevant for nutrient quartz. As a result, several vein and pegmatitic sources in North America have proved satisfactory for quartz nutrient once minor process modifications are made. The U.S., consequently, now is independent of overseas sources for electronic quartz.

However, the recent increasing cost competitiveness of Japanese quartz producers now is threatening the domestic U.S. quartz industry, once more raising the possibility that the U.S. will be dependent on foreign quartz supplies. The National Materials Advisory Board recently recommended broader-based support in the U.S. for hydrothermal chemical research on electronic materials to maintain the nation's competitive edge.

Recent research has been aimed at improving the physical quality of quartz. One such improvement is dislocation-free (DLF) quartz, which was first prepared by L.A. Gordienko, L.I. Tsinober, and coworkers at the Institute of Crystallography of the Soviet Academy of Sciences. Studies at Bell Labs show that to produce DLF quartz, DLF seeds must be used and special care taken to avoid inclusions of solid particles, which often are corrosion products formed by reaction with the steel walls of the autoclave. Such inclusions lead to strain and often dislocations when they are covered over by the growing crystal. Present research focuses on further perfecting crystal purity and physical

Fig. 2. Harvesting an early quartz growth run, AT&T Merrimac Valley Factory, Massachusetts.

properties, which affect the performance of advanced piezoelectric devices.

Other hydrothermal crystals

Potassium titanyl phosphate, $KTiOPO_4$, is an interesting optical material with a large nonlinear optical coefficient (a measure of the efficiency by which a material converts light of one wavelength to light of another wavelength) at 1.06 μm, which makes it comparable to lithium niobate, although it is much more resistant to optical damage. Thus, it is an excellent material for converting laser light at 1.06 μm into coherent green light at 0.53 μm for use as a spectroscopic source. The material was first prepared as single crystals by Frederick C. Zumsteg, John D. Bierlein, and Thurman E. Gier at the Du Pont Experimental Station in Wilmington, DE. They grow these crystals by dissolving titanium dioxide in a concentrated aqueous solution of potassium salts contained in a gold tube, pressurized to 3 kilobars, then cooling the solution from 850 to 650 °C for a period of a week.

More recently, Roger Belt and his colleagues at the Airtron division of Litton Industries in Morris Plains, NJ, developed another method for growing $KTiOPO_4$ hydrothermally. They make a nutrient by reacting KH_2PO_4 with titanium dioxide at 1250 °C in platinum containers to produce a hydrothermal solution of 1.5 parts KH_2PO_4 to 1 part TiO_2. Crystals are grown in silver or gold cans in autoclaves at 1.5 to 1.8 kilobars and 520 to 560 °C at rates of between 0.2 and 1.8 mm per week.

Anthony J. Caporaso and I also recently discovered that aqueous K_2HPO_4 is an excellent solvent for $KTiOPO_4$ and that crystals can be grown from such solutions at rates of up to 0.2 to 0.4 mm per day when the temperature is only 375 to 425 °C. At these lower temperatures, ordinary steel autoclaves can be used instead of autoclaves made from special alloys. In a joint project between Airtron and Bell Labs, these conditions have been used to grow large crystals.

Hydrothermal conditions have been used to grow many other crystals for use as electronic materials. For example, Albrecht Rabenau (now at Max Planck Institut für Festkörperforschung in Stuttgart, West Germany) first grew crystals of many elemental metals, including gold, silver, platinum, cobalt, nickel, tellurium, and arsenic, hydrothermally while at the Philips Laboratory in Aachen, West Germany. Because aqueous acid media usually are required for growing these crystals, they are grown in sealed quartz ampoules held in an autoclave. For example, gold crystals are prepared easily by dissolving gold sheet in 10 M hydroiodic acid mineralizer at 480 °C and then recrystallizing it in a hotter region (500 °C) of the tube. Other mineralizers that can be used include other halogen acids, sodium chloride, and potassium chloride. An oxidizing agent, such as chlorine, bromine, iodine, or hydrogen peroxide, must be present for good transport. Hydroiodic acid contains enough free iodine that extra oxidizing agent need not be added deliberately when it is used.

A typical dissolving reaction, which shifts to the left as the temperature is increased, is

$$Au + 3/2 I_2 + I^- \Rightarrow AuI_4^-.$$

Many single crystals of oxides, including Al_2O_3 (corundum), Al_2O_3 doped with chromium (ruby),

ZnO (zincite), and $Y_3Fe_5O_{12}$ (ferrimagnetic yttrium iron garnet), have been grown by me and my colleagues at Bell Laboratories using OH^- mineralizers. We also have grown crystals of chalcogenides, including zinc sulfide, zinc selenide, and zinc telluride, for which, surprisingly, hydrolysis under hydrothermal conditions is not significant.

Hydrothermal researchers at the Institute of Crystallography of the Soviet Academy of Sciences in Moscow, including L.N. Demianets, A.N. Lobachev, and V. Kusnetsov, have systematically studied the phase equilibria and synthesis of rare-earth compounds, especially germanates. They have used non-aqueous solvents, including ammonia, hydrofluoric acid, bromine, sulfur monochloride, and carbon tetrachloride, at near- and supercritical conditions to prepare halides, chalcogenides, and other materials.

Herbert Jacobs and Detlof Schmidt of the Institut für Anorganische Chemie at the Rheinisch Wesfälischen Technischen Hochschüle in Aachen, West Germany, synthesized single crystals of compounds such as $Li_3Na(NH_2)_4$, $BaNH$, and EuN by using near- or supercritical ammonia (ammonothermal synthesis).

Hydrothermal crystallization has been used to prepare single crystals of oxide superconductors by Shinichi Hirano of the University of Nagoya in Japan. Hirano has grown small single crystals of $BaPb_{1-x}Bi_xO_3$ in 4.5 M potassium chloride solution at temperatures of 450 °C. The superconducting transition temperature for these crystals is 11.7 K (which is comparable to the best polycrystalline samples) and the width of the superconducting transition is only 1.8 K (which is better than in polycrystalline or flux-grown samples). This suggests that low-temperature hydrothermal growth has a good potential for preparing superconductors such as $YBa_2Cu_3O_7$ and other perovskites with superconducting transition temperatures above 90 K. Results in my lab indicate, however, that $YBa_2Cu_3O_7$ reacts with water, even at room temperature, with the formation of O_2 and Cu^{2+}. This suggests that non-aqueous hydrothermal solvents will be required.

Hydrothermal synthesis also is of great interest for preparing other materials, although generally in the form of fine polycrystalline aggregates rather than as large single crystals. Examples are ferromagnetic oxides such as δ-Fe_2O_3 and CrO_2 for magnetic recording, zeolites for catalysis ion exchange applications, ultrafine starting materials like zirconium oxide for ceramics, and apatites for the study of teeth and bones.

Hydrothermal chemistry continues to pose research challenges. For one thing, understanding of the physical chemistry of hydrothermal solutions must be further extended so that synthesis and crystal growth can become less of an empirical art. In addition, the range of materials that can be synthesized hydrothermally needs further exploration, focusing especially on compounds from which single crystals of high quality are difficult to grow by alternative methods because of thermodynamic or other reasons.

Quartz crystals already are grown commercially, and hydrothermal growth of crystals of aluminum phosphate, potassium titanyl phosphate, and emerald is likely to be commercially viable. But hydrothermal crystal growth certainly can be further extended, especially because of the method's ability to prepare refractory materials at relatively low temperatures.

Meanwhile, our present considerable understanding of the quartz system can be further built upon so that crystals better than the best obtainable from nature can be uniformly and reproducibly grown on a routine basis.

Hydrothermal methods can be used to synthesize some of the most difficult-to-prepare materials known to modern technology. They will continue to provide a unique experimental milieu, as well as an intriguing intellectual milieu for testing our overall understanding of solution chemistry.

The transformation of graphite into diamond

H. Tracy Hall

Brigham Young University and Smith Megadiamond Provo, Utah 84604

"And the second row shall be an emerald, a sapphire, and a diamond." Exodus 28:18

This earliest known reference to diamond assigns emerald, sapphire, diamond and nine additional precious stones to be set in a breast plate to be worn by Aaron, the high priest. Each stone represented one of the twelve tribes of Israel.

Another Old Testament scripture seems to affirm by its imagery that diamond was known by the ancients to be the hardest of substances. Jeremiah 17:1 states: *"The sin of Judah is written with a pen of iron, and with the point of a diamond: It is graven upon the table of their heart, and upon the horns of your altars."*

It is thought that the earliest diamonds came from India. Centuries later, alchemists seeking to transform ordinary metals into gold also considered transforming common gemstones into more precious types. A scientific discovery presaging the possibility of transforming a common substance into diamond occurred in the year 1792 when Antoine Lavoiser burned diamond in oxygen and obtained carbon dioxide as the only combustion product. He concluded that diamond was comprised of only the element carbon. The common mineral graphite was already known to be carbon.

Thus graphite and diamond were shown to be chemically the same and men began experimenting with ways to transform inexpensive graphite into expensive diamond. If 0.200 g of graphite could be transformed into a one carat (0.200 g) gem quality diamond, a million fold increase in value would be attained.

C. Cagniard de la Tour seems to have been the first to claim success at making diamond. This claim was made in 1823. From that time until December 16, 1954, when I succeeded in transforming graphite into diamond, the "diamond problem" attracted the interest of many people. Those who pursued the problem included rank amateurs, downright charlatans, and some of the world's most honored scientists including Boyle, Bragg, Bridgmen, Crookes, Davey, Despretz, Friedel, Liebig, Ludwig, Moisson, Parsons, Tamman, and Wohler.

British encyclopedias credit J.B. Hannay as the first to make diamond. His diamonds, supposedly made in 1880, are still displayed in the British Museum. Hannay's method employed the use of wrought iron tubes in which lithium metal, bone oil and mineral oil were sealed. The tubes were then heated to redness in a furnace. Some eighty tubes exploded in his experiments. Two survived however, and when cooled and opened were supposedly found to contain three rather large, gem quality diamonds.

Some old school books and encyclopedias credit Henry Moisson as the first to make diamond. He invented the electric arc furnace and used it to synthesize many previously unknown metal carbides and other refractory substances. This success led him to take on the ultimate challenge: the diamond problem.

In the year 1893 Moisson claimed to make diamond by dissolving sugar charcoal in molten iron and

Previously published in AACG Newsletter Vol. 16(1) March 1986.

rapidly cooling the melt by pouring it into water. He thought that a great pressure would develop on cooling and cause diamond to form. After treating the solidified mass with hydrochloric acid, he reported finding a few microscopic diamonds in the undissolved residue.

Sir Charles Parsons, who experimented with diamond making from 1882 to about 1922, repeated Moisson's experiments and the experiments of all previous claimants without success. He also performed many ingenious experiments of his own. In 1922, he concluded that neither he nor anyone else had succeeded in making diamond.

It is worth noting that Parsons was the inventor of the practical steam turbine which rapidly replaced sails as means for ship propulsion in the late 1800's. He amassed a fortune from this enterprise and spent much of it on the diamond problem.

Another noted worker who spent the better part of a lifetime on the problem was Percy W. Bridgman of Harvard University. He started his work in 1905 and concluded it in 1955. Although he never made diamond, he received the Nobel Prize in 1948 for his prodigious work in the general field of high pressure research.

Great secrecy has been companion to most of those who have attacked the diamond problem. In Bridgman's case, David T. Griggs, one of the few graduate students who worked with him, stated in a 1954 article: "It was my privilege to work in Bridgman's laboratory during the period when working pressures were increased from 20,000 to 1,000,000 bars. As each new apparatus was readied for trial, I noticed that Bridgman would become secretive and brusque. During the first run, visitors were not welcome. I subsequently learned that in each case *graphite* was the first substance tried."

Note that 1 bar = 10 million dynes per square cm = 1.02 kg per square cm = 1 Newton per square m = 100,000 Pascals = 0.987 Atmospheres = 750 Torr = 14.5 pounds per square inch. All of these pressure units have been used at one time or another and have made quite a mess of the published literature. The currently decreed unit is the Pascal (Pa). Chemists have traditionally used atmospheres; geologists, bars.

In 1937, a consortium of companies provided very large financial backing for Bridgman's research on diamond. Work on the project ended in 1942. Diamonds were not made. Bridgman never succeeded

in inventing an apparatus that could simultaneously contain a high pressure and a high temperature.

My interest in diamond synthesis began rather early. I had read about the problem as an undergraduate at the University of Utah. Later, while working for a Master's degree, my adviser, G. Victor Beard, encouraged me to conduct experiments concerning the problem, even though my thesis was in an entirely different area. In those days there was no possibility of funding for experimental work on high pressure apparatus, but there was the hope that it might be possible to make diamond without such equipment.

I had been intrigued by a journal article that described the way an ordinary incandescent light bulb had been used to produce sodium metal. The lighted bulb was immersed in a low melting salt solution containing sodium ions. A battery was connected to one terminal of the filament and to an inert electrode in the molten salt. The positive sodium ions passed through the glass and picked up electrons at the surface of the filament to become sodium metal.

I tried, without success, to prepare elemental boron using a borate bath and a boron glass "light bulb." And I pondered how to produce carbon ions and pass them through some kind of barrier onto a heated filament. I hoped, of course, that the carbon would deposit as diamond. Several researchers in the last twenty years or so have produced diamond layers up to 100 atoms thick by decomposing methane and other hydrocarbons on heated filaments.

In a different vein, I tried to selectively oxidize graphite with oxidizing acids, believing that regions in the disturbed graphite structure might coalesce into diamond. I also disturbed the graphite lattice by intercalation with sodium and potassium. Needless to say, I never detected any diamond.

World War II came and I joined the navy. After it was over, I earned a Ph.D. degree with the aid of the GI bill and went to work for General Electric.

In 1951, G.E. Research Laboratory managers called about twenty of their chemists to a meeting and announced that they were going to tackle the diamond problem. I was elated and ready! Volunteers were called for. I was the only one interested and I got the job.

It was revealed at the meeting that personnel from other disciplines were already at work on some aspects of the problem, such as designing high pressure, high temperature apparatus capable of achieving 35,000 atmospheres and 1000 degrees centigrade. It was anticipated that graphite would convert to diamond under these conditions. Others were working on "nonthermodynamic" approaches to making diamond; believing, for instance, that high pressure might not be necessary. These studies were primarily theoretical.

My assignment related to chemistry. Thermodynamics indicated that high pressure and high temperature would be needed to transform graphite to diamond. But nature and theoretical studies did not give any clues as to how high a pressure or how high a temperature might be needed. Indeed, geologists do not yet know how diamonds were formed in nature.

Since the chemistry of diamond formation was not known, several questions presented themselves: Did diamond (in nature) form directly from graphite, or were other reactants required? Were catalysts needed? Did diamond take a million years or more to form? If the latter were true, man might never be able to demonstrate laboratory diamond synthesis. Could there be several different procedures for making diamond?

For a time, I eagerly pursued these questions. Most notable was my determination that the activation volume of carbon in the transition state was of the order of ten cubic centimeters per mole! Graphite's molar volume is 5.34 cubic centimeters and diamond's is 3.42. Thus the pressure that is needed to place graphite in a region where diamond is thermodynamically stable is very detrimental to favorable reaction kinetics. It would take more than a million years to produce diamond this way! If graphite is the starting material, a catalyst is needed. If graphite is not the starting material, several possibilities present themselves. For example, the carbon in carbonates might be replaced by another element such as silicon or sulfur to form silicate or sulfite and diamond. Perhaps copper would, at high temperature, alloy with the tungsten in tungsten carbide and free the carbon as diamond.

By this point it was apparent that the lack of progress in the invention of high pressure, high temperature equipment was the barrier to really getting hold of the diamond problem. I had no assignment in this area but I began to think of non-conventional means for simultaneously generating high pressure and high temperature. My ideas, however, met with resistance as I found myself intruding on the "turf"

of others; a classic problem in industrial R&D. Fortunately, a shop foreman, a machinist, and a manager from another area helped me skirt the roadblocks, and I brought forth the Belt high pressure, high temperature apparatus.

This device advanced into territory far beyond what had been hoped for. It could generate a pressure of 120,000 atmospheres and sustain a temperature of 1800 degrees C simultaneously for periods of several minutes, in a working volume of about one tenth of a cubic centimeter!

Managers and others were reluctant to accept the Belt and its enormous capabilities. Those charged with the responsibility of developing high pressure apparatus continued to work on unworkable ideas. It took several months for them to become believers. But when they finally did, there was a scramble to get a piece of the action. The vice president for research decreed that "this is big enough for all to share." It is worth noting, however, that U.S. Patent 2,941,248, Belt Apparatus, issued June 21, 1960, bears my name only.

At this writing, about 150 tons of man made industrial diamond, valued at about one billion dollars, has been manufactured in the Belt!

An exploded diagram of the Belt is shown in Fig. 1 and a closed diagram is shown in Fig. 2.

The functions of the various parts are as follows (see Fig. 1): Two conical semi-pistons (1) push into each side of a specially shaped cemented tungsten carbide chamber (2). Pressure is transmitted to the sample contained in a metal or graphite tube (3) by wonderstone (pyrophyllite) (4) a special hydrous aluminum silicate mined in South Africa. The pyrophyllite also serves as thermal and electrical insulation. The sample is heated by passage of an electrical current through the heating tube (3). If the sample is a good electrical conductor, it may be necessary to electrically isolate it from (3) in a container of hexagonal boron nitride or some other high temperature electrical insulator.

Current enters tube (3) through a refractory metal disk such as tantalum (6), which touches steel ring (5), which in turn touches the tip of the semi-piston. Under pressure, these various parts (3, 5, and 6) are forced together and make a good electrical contact. The pyrophyllite disks (7) provide thermal insulation.

As the conical pistons advance, a sandwich gasket of pyrophyllite (8) and (10) and steel (9) compresses

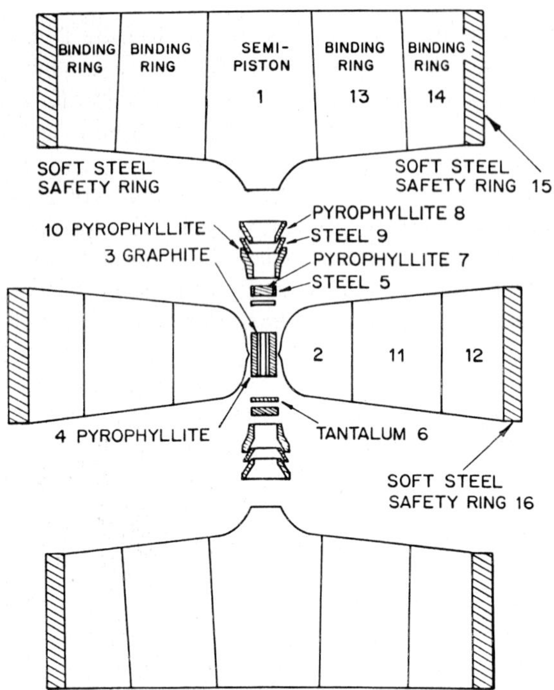

Fig. 1. "Exploded" view of the belt high pressure, high temperature apparatus.

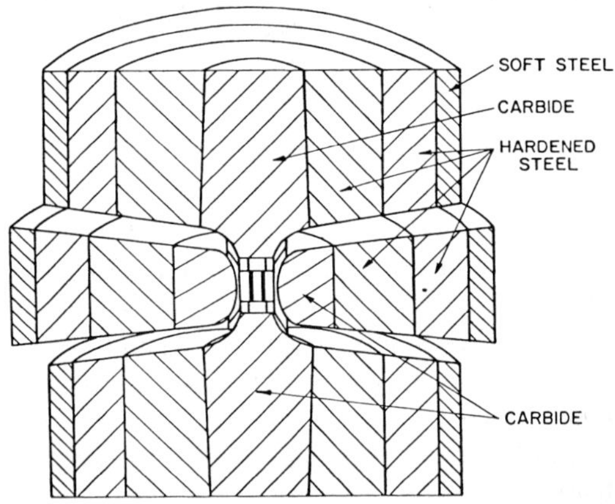

Fig. 2. "Closed" view of the belt high pressure, high temperature apparatus.

that amazingly contains a pressure exceeding a million pounds per square inch!

Interference fit compound binding rings of hardened steel (11) and (12) provide lateral support for the cemented tungsten carbide chamber (2). Lateral support for the cemented tungsten carbide conical semi-pistons is similarly provided by rings (13) and (14).

The low carbon steel, dead soft rings (15) and (16) are safety rings provided to absorb the substantial energy released if the binding rings should fail.

The extreme conditions available in the Belt were thought to be more than sufficient to transform graphite into diamond, but experiment proved otherwise. Since the direct transformation would not occur, I attempted hundreds of indirect approaches along the lines previously mentioned. None were successful and I was becoming discouraged. General Electric was considering abandoning the project.

Then, on the wintry morning of December 16, 1954, I broke open a sample cell after removing it from the Belt. It cleaved near the tantalum disk (6 in Fig. 1). Instantly, my hands began to tremble. My heart beat wildly. My knees weakened and no longer gave support. Indescribable emotion overcame me and I had to find a place to sit down!

My eyes had caught the sparkling light from dozens of tiny octahedral crystals growing out of the tantalum and I knew that diamond had at last been made by man!

It took about twenty minutes for me to regain my composure. Then I examined the crystals under a microscope. The largest was 150 micrometers across and contained triangular etch and growth pits such as those that occur on natural diamonds. The crystals scratched sapphire, burned in oxygen to produce carbon dioxide, and had the density and refractive index of natural diamond. A few days later, an x-ray diffraction pattern positively identified the crystals as diamond.

This first successful experiment contained the mineral troilite (FeS) inside a graphite heating tube. The pressure in the Belt was near 70,000 atmospheres (just a little over 1,000,000 pounds per square inch). The temperature was near 1600 degrees C (2912 degrees F). Troilite is associated with the microscopic diamonds found in the Canyon Diablo meteorite. The meteoritic diamonds were probably formed by the transient pressure and temperature generated on impact with the earth. I thought that the FeS might have been a catalyst for graphite–diamond conversion in the meteorite and consequently tried it in my experiment. I repeated this experiment twenty times in the next two weeks varying pressure and temperature to find the pressure–temperature field in which diamond would form. Diamond was produced in twelve

Fig. 3.

of these runs. Diamond always grew on the tantalum end disks. Since troilite is a non-stoichiometric compound, I wondered whether it was FeS, S, or Fe that was important for the catalytic action. I also wondered what role the tantalum played. Experimentation showed that diamonds grew on the tantalum when either FeS or Fe was in the graphite heating tube. But no diamonds were formed when S alone was in the tube. Under the high temperature, high pressure conditions in the graphite tube, sulfur distills from the FeS and passes through the graphite tube into the pyrophyllite, leaving iron behind to alloy with the tantalum. I concluded, therefore, that an alloy of iron and tantalum acted as the catalyst.

A microphotograph of the first diamonds I saw growing out of the tantalum is shown in Fig. 3. Note the unusual interpenetrating twin in the lower right hand corner. The skeletal morphology of the diamonds resulted from very rapid growth at the high operating temperature. These diamonds grew in just a few seconds, thus evidencing the possibility of economical industrial diamond production. Diamond wheels using this type of diamond proved to be vastly superior to crushed natural diamond grit in the grinding of cemented tungsten carbides. More perfectly formed diamond crystals require growth at a lower temperature and a lower pressure for a longer length of time (a few minutes).

On December 31, 1954, Hugh Woodbury, a company physicist, made diamond under my tutelage using FeS in the graphite heating tube. He thus became the first man to duplicate the diamond synthesis claim of another.

Due to the long history of fraud associated with the diamond problem, company officials carried out "official duplication syntheses" on January 18th and 19th of 1955. I was not allowed to be present. Under the watchful eyes of company officials and attorneys,

Hugh Woodbury and Richard Oriani (a company metallurgist) each made three runs in the Belt according to my procedure, using independent sources of graphite and FeS. They succeeded in making diamonds in all six runs.

Management, thus convinced of the authenticity of my synthesis, sent out an impressive press release on February 15, 1955. Within the next two days, most U.S. newspapers carried front page stories reporting that diamonds had been made at the General Electric Research Laboratory in Schenectady, New York.

The lack of recognition I received for this extraordinary dual achievement, the invention of the Belt (U.S. Patent 2,941,248 issued June 1, 1960) and synthesis of the first diamond (U.S. Patent 2,947,608 issued August 2, 1960) was, simply stated, demeaning.

Saddened and hurt, I left General Electric, a company I had admired and aspired to work for since the age of nine.

In August of 1955, I began a new career as director of research and professor of chemistry at Brigham Young University.

I had anticipated building a Belt to continue high pressure research at my new location. But G.E. officials warned that I could not build a Belt under any circumstances. So, I had to invent another device. I called this invention the Tetrahedral Press. It was the first of a series of "multi-anvil presses" that I was to invent. I succeeded in obtaining a patent on the Tetrahedral Press (U.S. Patent 2,918,699 issued December 29, 1959) before G.E. obtained a patent on my Belt.

Having thus extricated myself from dependence on the Belt, I was free to pursue a 25-year career in high pressure research at Brigham Young University. A photograph of the first Tetrahedral Press is shown in Fig. 4.

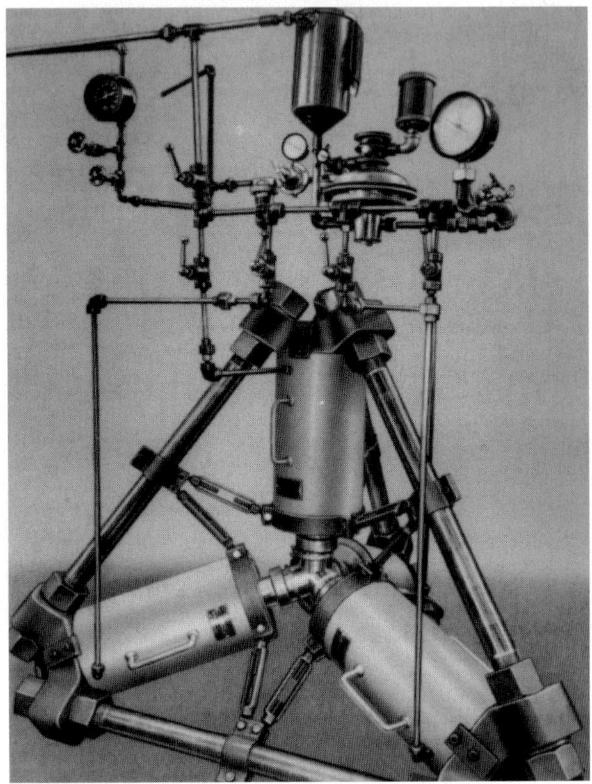

Fig. 4. The first tetrahedral press.

Recollections about the early development of molecular beam epitaxy (MBE)

A.Y. Cho

AT&T Bell Laboratories, Murray Hill, New Jersey 07974

The term molecular beam epitaxy (MBE) was first used in one of our papers in 1970 after five years of extensive studies of atomic and molecular beams interacting with solid surfaces. The request from the American Association for Crystal Growth for me to write about the initial milestones of MBE made me recall some of the excitement of the early years in the development of this technology.

I was recruited to AT&T Bell Laboratories by Jim Goldey and hired by John Galt when he was a director and my acting department head. I started to work with John Arthur in February 1968 because of our mutual interest in surface physics. He was studying Ga and As_4 beams interacting with GaAs surfaces, while my previous work at TRW, California and later my PhD Thesis at the University of Illinois were also concerned with atomic beams interacting with solid surfaces. These interactions were studied by a mass spectrometric pulsed beam technique. John Arthur's study concluded that the sticking coefficient for Ga was unity and the sticking coefficient of As was highly dependent on the Ga atom coverage on the GaAs surface. It takes a Ga atom to trap an As atom; by making the As beam intensity much higher than that of Ga, he was able to grow stoichiometric GaAs.

When I first came to AT&T Bell Laboratories in 1968, the Ga and As beams were produced by heating polycrystalline GaAs in a quartz ampule with tungsten wire wound over the ampule. For equilibrium evaporation, the ampule had a pin hole aperture of less than $1/8''$ in diameter which resembled a Knudsen cell. This configuration was satisfactory for surface physics studies but not desirable as a film growth effusion cell because the deposition rate from the pin hole was limited to less than one atomic layer per minute. The vacuum system we used at that time had a background pressure of about 10^{-8} torr; which meant that the arrival rate of the background atoms was in the same order of magnitude as those of the Ga and As_4 atoms. The GaAs layers deposited under that condition were all semi-insulating. No transport properties or photoluminescence could be performed on these films.

A major advancement occurred in 1969; the author used his previous knowledge of cesium ion engine configuration for ion propulsion to design the effusion cells. The effusion cells with large apertures were heat shielded with layers of corrugated tantalum foil to reduce the heat loss and temperature cross-talk with adjacent cells. All cells were surrounded with liquid nitrogen cooled shrouds to reduce the background pressure. The cell temperatures were also regulated by negative electronic feedback circuits to assure precise effusion fluxes. At the same time, the author introduced a reflection high energy diffraction system in situ with MBE to investigate the initial, successive, and final epitaxially grown films. Then, for the first time, we learned how to clean a GaAs substrate in the vacuum system reproducibly before de-

Previously published in AACG Newsletter Vol. 15(2) July 1985.

Fig. 1. Effusion cells for molecular beam epitaxy (MBE) used a decade ago.

position. Reconstructed surface structures were observed. The author called them Ga-stabilized [$111 - \sqrt{19}$ and $100 - C(8 \times 2)$] and As-stabilized [$111 - 2$ and $100 - C(2 \times 8)$] surface structures because they varied with the As/Ga ratio in the molecular beam and the GaAs substrate temperature. Surface physicists at that time always thought reconstructed surface structures were due to impurity atoms absorbed on the surface. It was not easy to change a school of thought. In 1970, a double oven As$_4$ cracker effusion cell [Solid State Electronics 14, 125 (1971)] and a separate Ga effusion cell were installed to independently vary the As/Ga ratio (rather than being evaporated from polycrystalline GaAs) and to increase the GaAs growth rate. A "surface phase diagram" was then constructed with the Ga- and As-stabilized surface structures. With these epitaxial growth conditions, GaAs layers were grown reproducibly. At this time Bruce Joyce at Mullard, England, also became interested in MBE from the surface physics point of view, and Leo Esaki and Leroy Change at IBM wanted to use MBE to fabricate multilayers of superlattice.

John Arthur continued to contribute to the fundamental understanding of MBE growth with surface physics studies [surface Sci. 43, 449 (1974)], and the author introduced n- and p-type dopants in GaAs for device fabrication. Transport properties and photo-luminescence were measured for the first time in 1970. In 1971, to demonstrate the precise control of MBE, superlattice GaAs/Al$_x$Ga$_{1-x}$As and p–n multiple layers were first reported [Appl. Phys. Lett. 19, 467 (1971)].

Appreciation for MBE increased after the demonstration of microwave devices such as the varactor, mixer diode, IMPATT diode, and field effect transistor in 1974, and room temperature CW laser diodes in 1976. Device demonstration could be a long drawn out process. For instance, the pressure from device development people after the achievement of a pulsed laser pushed us to demonstrate a CW laser, then a low threshold (< 1 kA/cm^2) laser, and finally a reliable (long life) laser. Won Tsang contributed many of the low threshold, long life laser diodes in 1980 [Appl. Phys. Lett. 37, 141 (1980)]. Ray Dingle, Horst Stormer, and Art Gossard introduced dopants in the

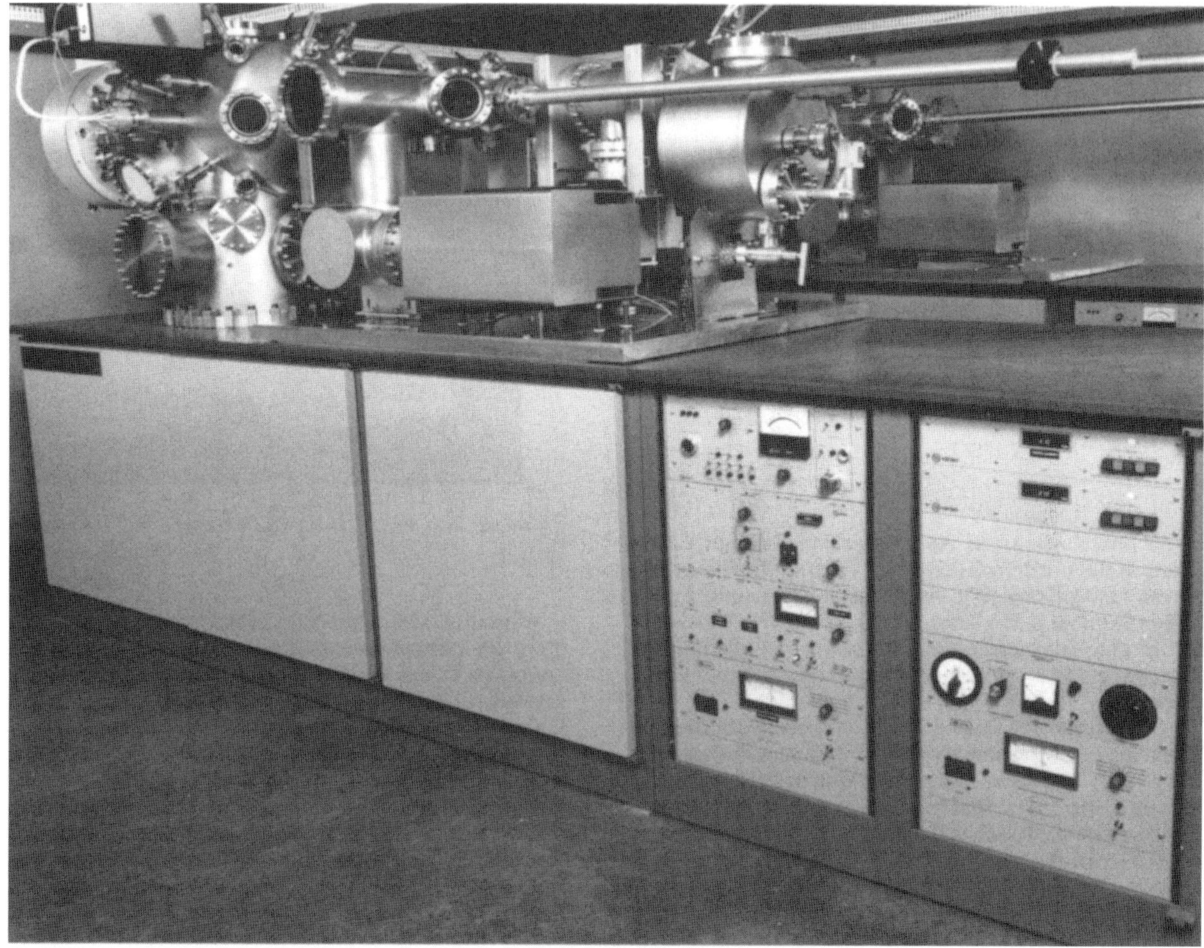

Fig. 2. Modern state-of-the-art MBE system (Varian).

large bandgap region of a heterostructure to increase the carrier mobility which has had far reaching implications to fundamental physics studies and device fabrication [Appl. Phys. Lett. 33, 665 (1978)]. MBE was extended to cover other III–V compounds, Group IV, II–VI compounds, metals and insulators. Newcomers joining the MBE field increase every day and excitement and new ideas are popping up everywhere, here in the United States, in Asia, and in Europe. Federico Capasso has another new idea for novel devices with "band-structure engineering" [Physica 129B, 92 (1985)]. My director, Venky Narayanamurti, greets us every day with "what's new?"

It was over seventeen years ago when I changed from surface physics to crystal growth without any knowledge of crystal growth. Only AT&T Bell Laboratories could provide me this opportunity to develop a technology within this time frame. The emphasis from the management toward good physics and sound engineering, coupled with a close working relationship with the development and manufacturing people, provides us with the atmosphere and the ability to pick the correct subject to study. The opportunity to create, compete, and build provides a unique environment at AT&T Bell Laboratories. It certainly has been an exciting and happy seventeen years for me to remember.

Basic studies of molecular beam epitaxy—past, present and some future directions

Bruce A. Joyce [a,*], Tim B. Joyce [b]

[a] *Department of Physics, The Blackett Laboratory, Imperial College, London SW7 2BZ, UK*
[b] *Department of Materials Science and Engineering, The University of Liverpool, Liverpool L69 3GH, UK*

Abstract

We present a summary of the history, present day activities and some possible future directions of molecular beam epitaxy. We confine our attention to growth-related phenomena and do not attempt to discuss the vast amount of work on device fabrication, although we acknowledge its huge contribution to the subject. We emphasize the extent to which basic studies are now being directed towards an understanding of growth processes at the atomistic level, treating as examples the homoepitaxial growth of GaAs(0 0 1) films and InAs–GaAs quantum dot formation. We also discuss recent work on silicon and on Group III-element nitrides.
© 2004 Elsevier B.V. All rights reserved.

Keywords: A1. Nucleation; A1. Reflection high energy electron diffraction; A3. Molecular beam epitaxy; B1. Nitrides; B2. Semiconducting III–V materials; B2. Semiconducting silicon

1. Historical background

Molecular beam epitaxy (MBE) may be defined as the deposition of epitaxial films onto single crystal substrates using atomic and molecular beams produced from Knudsen cells under ultra-high vacuum (UHV) conditions.

It had its origins in the mid-1960s, when homoepitaxial films of silicon were grown from molecular beams of monosilane (SiH_4) by Joyce and Bradley [1]. Although this work was of interest from the point of view of nucleation behaviour [2,3] and the formation mechanisms of certain crystallographic defects, particularly stacking faults and microtwins [4], the ex-

tremely low growth rates available ($\leqslant 0.01$ monolayers $(ML)\,s^{-1}$), meant that it was not a viable technique for the preparation of devices then in vogue, which required layers up to 10 μm thick. Growth studies on silicon consequently went into abeyance for the next 10 years or so, but the growth of III–V compounds assumed prominence by the end of the 1960s, with an incredible expansion of effort and output. This was triggered by the work of Cho and Arthur at Bell Laboratories. Arthur concentrated on surface kinetics and mechanisms involved in the growth of GaAs from beams of Ga and As_4 [5,6], while Cho predominantly pursued the growth of device-quality material [7,8]. In this paper, we will not directly discuss device related topics, but the interested reader is referred to the compilation by Cho [9], which reviews some of the earlier work in this area. It cannot be emphasized enough,

* Corresponding author.
E-mail address: b.a.joyce@imperial.ac.uk (B.A. Joyce).

however, that without the very successful work on device structures pioneered by Cho and co-workers, the whole subject of MBE would have quietly faded away for lack of funding.

The early work on growth-related phenomena of III–V compounds (predominantly GaAs) was centred on surface reaction kinetics and evaporation behaviour, with some activity on dopant incorporation. The need to understand reaction kinetics arises because although the group III elements could be supplied as atomic beams from Knudsen cells, the group V element beams were either tetramers (As_4, etc.), or dimers (As_2, etc.), so a rather complex chemical reaction is involved in the formation of, say, GaAs. Using an elemental arsenic source in a Knudsen cell produces a flux which is effectively all As_4 (unless a high temperature stage is added to the cell to crack the As_4 to As_2, but this technology was added considerably later). The only species which evaporate from the III–V compounds themselves, however, are dimers. This applies to both Langmuir and Knudsen evaporation, but from an equilibrium cell there will be a mixture of dimers and tetramers depending on the temperature, i.e. the flux composition will be representative of the gas phase equilibrium [10,11].

Investigation of the surface chemistry made use of the fact that the beams were neutral, with thermal velocities and intensities in the range 10^{11}–10^{16} atoms (molecules) $cm^{-2} s^{-1}$. The first collision of the incident species is with the substrate surface and the reaction can be followed by detecting any desorption flux mass-spectrometrically. The problem is to distinguish between background signals and those produced directly by desorbing atoms and molecules. It can be solved by mechanically modulating either the incident beam or the desorbing flux and examining the signal detected in the mass-spectrometer for a correlated response. The simplest method is to determine the response of the desorption signal to a step-function change in the intensity of the incident beam, produced by opening or closing a shutter. It is only suitable for readily condensable species, but Arthur [5] successfully employed the technique to study the desorption of Ga from different orientations of GaAs, and by measuring time constants as a function of temperature was able to determine activation energies of desorption.

Foxon et al. [12] and Foxon and Joyce [13,14] extended this approach by using periodic modulation,

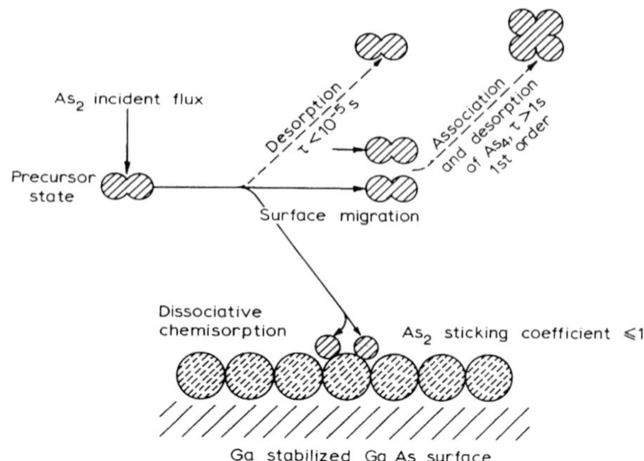

Fig. 1. Model of the growth chemistry of GaAs(0 0 1) from molecular beams of Ga and As_2.

combined with signal averaging to obtain statistically significant data. By Fourier transforming the detected signal and the time dependence of the incident flux, the attenuation and phase shift of the Fourier components of this flux due to all events between beam modulation and detection could be obtained. Information on the particular surface process of interest was extracted by deconvolution. The results for the As_2–Ga interaction on GaAs(0 0 1) surfaces are summarized in Fig. 1. The most important points are (i) that it is a first-order dissociative chemisorption process with respect to As_2, provided there is a Ga adatom population, and (ii) the As_2 exists in a weakly bound precursor state prior to dissociation [14]. A similar, but rather more complex model was derived for arsenic supplied as As_4 [13], although it is known that at the higher temperatures now used for growth ($> 580\,°C$), As_4 first dissociates to As_2 on the surface [15].

The second, and as it turned out, crucial in situ technique introduced in the comparatively early stages of MBE was reflection high-energy electron diffraction (RHEED). Cho [16] initially used it to determine the surface structure of the clean substrate and growing layer. This revealed that in general all surfaces are reconstructed, i.e. have a lower symmetry than the bulk and Cho was the first to propose that the two-fold periodicity observed in the [$\bar{1}$ 1 0] direction on the (0 0 1) surface was the result of dimerization of As atoms on the arsenic terminated surface, which was confirmed many years later by scanning tunnelling microscopy (STM) [17]. RHEED was preferred to the then more

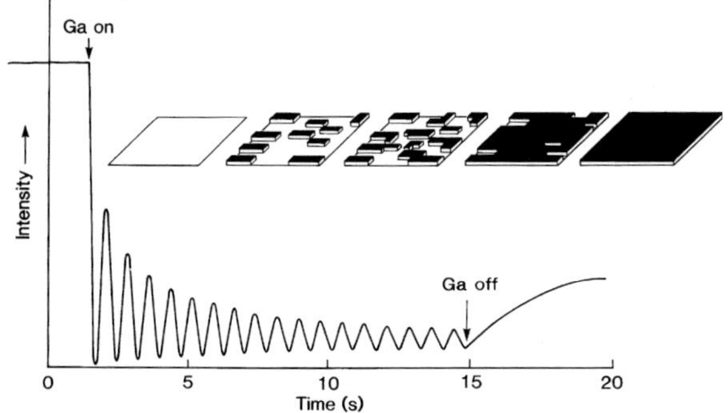

Fig. 2. Intensity oscillations of the RHEED specular beam from a GaAs(0 0 1)-2 × 4 surface viewed in the [1 1 0] azimuth during growth. The period corresponds exactly to the growth of a single molecular (Ga + As) layer.

popular low-energy electron diffraction (LEED) for surface structure determination because it is a forward scattering technique and therefore more compatible with the MBE arrangement of normally incident fluxes than the back-scattering geometry of LEED.

The other application of RHEED was the discovery of the so-called RHEED intensity oscillations [18–20], when it was found that the intensity of any diffraction feature (but usually recorded via the specular beam), oscillated with a period corresponding to the growth of a single ML, i.e. a layer of Ga + As, in the [0 0 1] direction on a (0 0 1) substrate. This was a manifestation of two-dimensional (2-D) layer-by-layer growth, the Frank–van-der-Merwe mode, and a typical result for GaAs is shown in Fig. 2. The general technique was found to be applicable to many other material systems, however, including elemental semiconductors, II–VI compounds, metals, insulators, superconductors and even organic compounds. Its real importance, though, was that it provided for the first time a quantitative means for the in situ measurement of growth dynamics, which could be related to theoretical treatments [21,22]. The origin of the oscillations was controversial at the time of their discovery, and this controversy is still not fully resolved. We will discuss it in more detail in Section 2, when we will also deal with certain other anomalies, in particular, those (many) cases where the oscillation period does not provide a direct measure of the growth rate.

So far, we have considered only homoepitaxy, but the growth of most device structures, especially those for optoelectronic applications such as lasers

and light-emitting diodes (LEDs) required the preparation of heterojunctions [23,24]. Fortunately, the ideal material to provide electron confinement in GaAs is AlAs (or $Al_xGa_{1-x}As$), which not only has the necessary greater band-gap, but also a lattice parameter very close to that of GaAs, so mismatch is negligible ($\approx 0.001\%$) and heterostructures essentially free from extended crystallographic defects can be produced. We will deal with systems in which the mismatch is important in Section 3.

A number of modifications to the basic growth technology of III–V compounds were introduced in the 1980s, and while each have their adherents, none has found sufficient favour to replace conventional systems, except for the growth of nitrides (vide infra). Two related techniques based on gas sources appeared to offer advantages over solid sources; gas source MBE (GSMBE) replaced elemental As and P with arsine and phosphine [25,26], while metalorganic MBE (MOMBE) [27–29], also known as chemical beam epitaxy (CBE), used organometallic sources for the group III elements, i.e. it was effectively a hybridization of MBE with metalorganic vapour phase epitaxy (MOVPE). In principle it combined many of their strengths, such as the abrupt composition and doping profiles, and in situ diagnostics, of MBE, with the external sources, greater throughput and higher growth rates of MOVPE, while avoiding the morphological defects associated with the use of solid source Knudsen cells. In addition, it offered greater precursor flexibility, improved InP quality and lower growth temperatures. There was, however, a price to pay in system complexity, with the need for gas handling and high volume pumping arrangements. It was also found that the group III hydrides were too stable to use without pre-dissociation, which constrained the system geometry, so the potential for high wafer uniformity could not be realized. Furthermore, the standard Al and Ga precursors used in MOVPE (trimethylaluminium and trimethylgallium) produced strongly p-type material when used in MOMBE, due to the incorporation of C as an acceptor, and although triethylgallium proved to be a viable Ga source [28], a suitable source for Al proved more difficult to find, mainly due to oxygen contamination problems. The lack of suitable gaseous dopant sources, particularly for Si, was a further handicap [30], but the deliberate use of C for p-type doping proved a success.

The surface chemistry involved with metalorganic sources is complex and the temperature dependence of surface reactions not only restricted growth conditions, but also had a serious impact on uniformity and reproducibility [31]. A potential advantage of this reaction complexity is in selective area epitaxy, where growth occurs through windows in dielectric masks and MOMBE can be 100% selective, but although a number of device structures have been demonstrated [32], there has been no real demand for this technology.

Both MOVPE and MBE have continued to develop as production techniques, with high uniformity multiwafer design, so MOMBE has never demonstrated a significant advantage over either of its parent technologies.

The second major modification to be introduced to MBE was the use of flux modulation techniques, in particular so-called migration enhanced epitaxy (MEE) [33,34]. In this, both cation and anion fluxes are modulated during growth, which apparently leads to enhanced cation migration distances when the anion surface adatom population is low. In practical terms it certainly appears to improve the quality of material grown at low temperatures, but again it has only really featured in laboratory experiments. RHEED oscillations observed with this technique are undamped and arise simply from surface reconstruction changes which occur directly from the flux modulation [35]. They are not indicative of morphological changes, which is the case for conventional MBE.

The overall impact of the work described very briefly above was to provide the basis of a viable technology able to produce low-dimensional structures for physics and devices, as well as laying the foundations for a much more quantitative evaluation of thin film growth processes than had previously been possible.

2. Present activities

Even if we ignore device aspects, so many developments with regard both to materials and the study of growth processes have occurred during the past few years that it is impossible to refer to them all in an article of this length, so we will select a few where we believe either that real progress has already been made in a particular field, or that the potential exists for major advances.

2.1. Atomistic studies of nucleation in homoepitaxy

The study of nucleation in epitaxial systems has a very long history, but until comparatively recently it was largely limited to macroscopic kinetic measurements, supported by theoretically derived rate equations [36], together with a thermodynamic consideration of growth modes, growth morphology and instabilities and the spatial and size distribution of developing growth centres (islands) [37]. The advent of in situ STM measurements, which in principle provided atomically resolved, real-space images of the growing surface opened a new chapter in the investigation of the atomistics of growth processes. At the same time, the experimental results could be related to more powerful theoretical treatments, such as ab initio calculations of the energetics of growing surfaces, and kinetic Monte Carlo (kMC) simulations which incorporated actual surface structures, as distinct from the solid-on-solid (SOS) simple cubic approach used previously. We will discuss two material systems which between them illustrate the present level of understanding and the limitations of the available methods.

2.1.1. Nucleation and growth on reconstructed GaAs(0 0 1) surfaces from beams of Ga and As$_2$

The GaAs(0 0 1) surface presents a particular challenge because of the presence, under conventional (As-rich) growth conditions, of the As-terminated (2×4) reconstruction, which modifies substantially the three topmost atomic layers of the surface. The equilibrium structure, which is the template on which growth occurs, is universally accepted to be the $\beta 2(2 \times 4)$ reconstruction, illustrated in Fig. 3. This is a highly corrugated surface with effectively two As dimer rows separated by two "missing dimer" trenches, with As dimers also present in the bottom of the trenches, i.e. in the third layer down. Even before STM images were available, Farrell et al. [38] made a reasonable attempt to describe growth atomistics based on the electron counting rule, which requires that under conditions of charge neutrality, all As dangling bond orbitals are filled and all Ga dangling bonds are empty. The limitation is that transient structures, which may not obey the rule, are excluded.

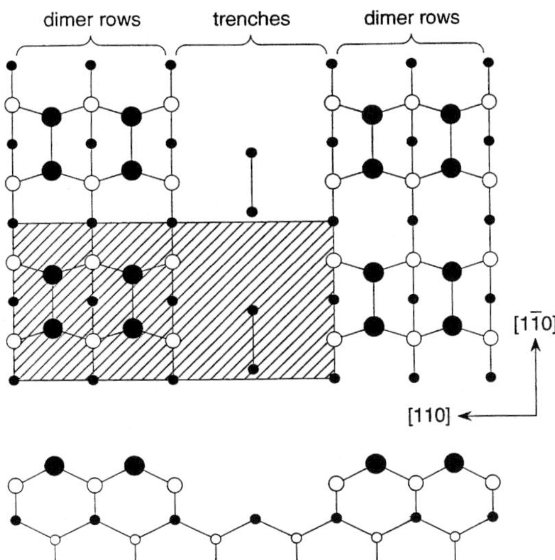

Fig. 3. Model of the ideal GaAs(0 0 1)$\beta2(2 \times 4)$ surface in plan and side view. The black circles represent As atoms and the open circles Ga.

Madhukar and Ghaisas [22] used kinetic MC simulations with explicit inclusion of As reaction kinetics, and although the actual surface structure was not used, they were still able to explain certain trends in the RHEED oscillation behaviour. Itoh et al. [39,40] used kMC simulations which incorporated the $\beta2(2 \times 4)$ surface reconstruction and As$_2$ kinetic effects to reproduce several quantities associated with island kinetics which were observed by STM, including the evolution of island number densities and their size distribution. An important aspect of this work was the necessity to include a loosely bound As$_2$ precursor state in the calculations in order to avoid the need for an unrealistically large As$_2$/Ga flux ratio ($\approx$ 600:1 cf. $\approx$ 6:1 used in practice) to obtain agreement between theory and experiment. This is also fully consistent with the results obtained from macroscopic kinetic studies using modulated beams [14]. It is also in accord with recent ab initio calculations [41]. An arriving As$_2$ molecule is in general not correctly positioned or oriented to bind directly to the GaAs surface bonds, but moves in a very mobile state about 2.5 Å above the undistorted surface. It therefore has a chance to find a favourable binding site before desorption, despite a short surface lifetime. This favoured site, according to the calculations, is via insertion into the broken bonds of two adjacent As surface dimers, with a binding energy of $\approx$ 1.6 eV. Pro-

vided the incoming dimer arrives at the site end-first, there is no barrier to adsorption.

The kinetic processes involved in growth are then (i) random deposition of Ga, (ii) Ga adatom migration, (iii) As$_2$ deposition into the precursor state population, (iv) incorporation of this precursor into the growth front, (v) desorption from the precursor state, and (vi) detachment of incorporated As$_2$ back into the precursor state (essentially local dissociation of GaAs). Deposition rates are fixed by the fluxes used and all other rate processes are assumed to follow the Arrhenius equation ($R = v \exp(-E/k_B T)$, where v is the attempt frequency, k_B is the Boltzmann's constant, T is the substrate temperature and E is the relevant activation energy). In addition to the kMC simulations, a mean field rate equation approach was used to characterize more quantitatively atomistic nucleation, growth and structural transformation kinetics, for comparison with STM-based experiments [40]. The overall agreement between theory and experiment from this work was quite good, in that the simulation uses the correct starting template for the equilibrium starting structure, the importance of the As$_2$ precursor state is identified and the total number density of initial nuclei is accurate, although their individual structure and size distributions may not be valid. The major feature not included is the modified surface structure identified by RHEED as being present during growth [38], but not obviously present in snapshot STM images obtained from quenched surfaces. We have not yet been able to evaluate details of the growing surface structure and the most that can be said at present is that the RHEED patterns are consistent with growth occurring initially in the dimer trenches via Ga adatoms bonding to the As dimers in the trench bottoms. This is also consistent with ab initio calculations of the most favoured Ga adatom sites [42–44], and it is a strongly bound configuration that would influence diffraction features, not just the diffuse background, as would be the case for the mobile As$_2$ precursor. It is not, however, consistent with the STM images from the quenched surfaces, which appear to show that growth commences on the dimer ridges. The problem, of course, is that these, unlike RHEED, are not real time observations and consequently they need to be treated with extreme caution. This work is continuing.

2.1.2. Nucleation and growth during silicon-(0 0 1) homoepitaxy

Although the use of silicon hydrides (SiH_4 and Si_2H_6) in MBE has re-emerged [45] now that much thinner films are required for device fabrication than was originally the case, most of the recent application-based work in this area has used a low pressure, UHV version of chemical vapour deposition (CVD), which we will not discuss here. Rather, we will concentrate on nucleation and initial growth studies from an elemental Si beam using in situ real time electron probe techniques. The major difference from GaAs is that the STM images are obtained directly from the growing surface, not via quenching and post-growth observation [46].

The unit cell of the Si(0 0 1) surface has p(2 × 1) symmetry since the surface atoms form dimers [47] and the total energy is further lowered by their asymmetry, i.e. they are aplanar [48]. The surface also has a domain structure whose boundaries are of ML height ($a_0/4$), so the dimer bonds in adjacent terraces separated by these boundaries are aligned along orthogonal ⟨1 1 0⟩ directions. These features are illustrated in Fig. 4.

In early work RHEED oscillations were used to show that growth could occur either by 2-D nucle-

ation and layer-by-layer growth, or by step flow, where adatoms attached to existing ML high steps. The experiments were supported by kMC simulations of a simple solid-on-solid model [49] and this has also been used to aid the interpretation of direct STM measurements of the transition from island growth to a step-flow mode [50]. The activation barrier to the hopping of surface atoms E is given by $E = E_s + n^{\|} E_N^{\|} + n^{\perp} E_N^{\perp}$, where E_s is the surface contribution, $E_N^{\|}$ and $E_N^{\perp}$ are the nearest neighbour bonding energies parallel and perpendicular to the direction of enhanced bonding and are used to model the influence of surface reconstruction on growth kinetics, and $n^{\|}$ and $n^{\perp}$ are the numbers of these bonds. This model correctly captures the dominating influence of the sticking anisotropy of adatoms on the evolution of surface morphology and does not contain an Ehrlich–Schwoebel barrier to step edge hopping, in agreement with Mo and Lagally [51].

On the Si(0 0 1) surface, each monatomic step separates two perpendicular domains of (2 × 1) reconstruction. The terraces have rows of dimerized atoms either parallel (T_A) or perpendicular (T_B) to the edges of the down-steps (S_A), (S_B) (see Fig. 4). Surface migration of adatoms is much more rapid along dimer rows than across them, which determines island shape. S_B steps are rough and advance faster than S_A steps because adatoms are readily incorporated into B-steps, whereas A-steps grow by propagation of kinks. At lower temperatures islands nucleate on both types of terrace and the coverage of each type remains approximately constant and equal. The transition to step flow growth mode with increasing temperature occurs earlier on T_B terraces, where adatoms migrate much faster in the direction perpendicular to step edges. Higher temperatures are required before island nucleation ceases on T_A terraces.

Finally, it is very interesting to consider these effects of surface reconstruction on growth in the light of simultaneous RHEED and reflectance anisotropy (RA) measurements during growth [52]. RHEED measures long-range order while RA measures only local bonding arrangements, but both produce an oscillatory response during growth, as illustrated in Fig. 5. It is evident that the frequency of the RHEED specular beam signal is double that of the RA response. The RHEED oscillation corresponds to ML growth time, i.e. $a_0/4$, and is morphology related, while the RA cor-

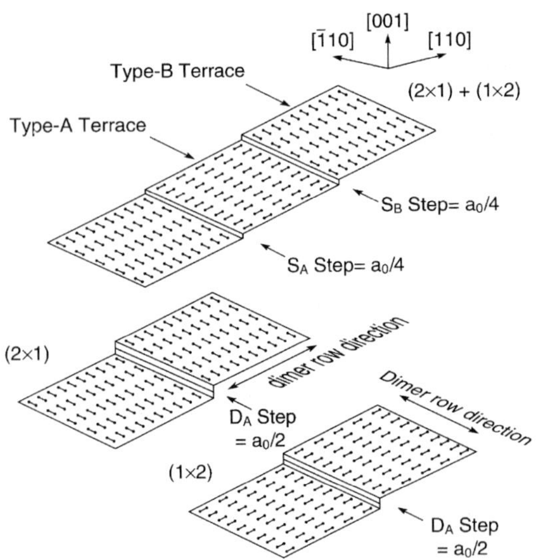

Fig. 4. Schematic illustration of the reconstruction of a Si(0 0 1) surface, showing 1 × 2 and 2 × 1 domains, resulting from Si dimers which are aligned orthogonally on adjacent terraces separated in height by $a_0/4$.

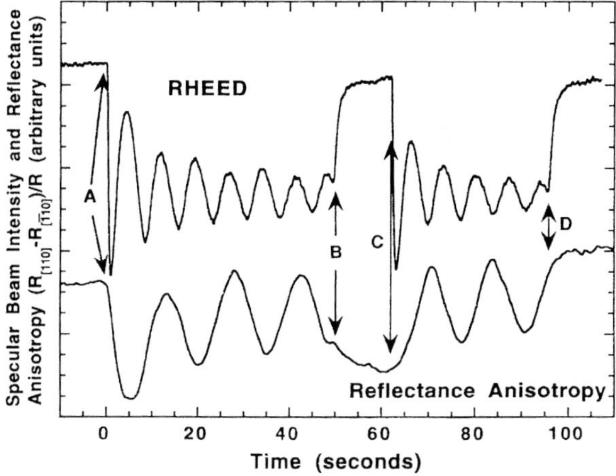

Fig. 5. RHEED and RA oscillations recorded during the growth of Si at 600 °C from a molecular beam of Si_2H_6. The electron beam azimuth is [1 1 0] and RA is measured using light with polarization along [1 1 0] and [$\bar{1}$ 1 1]. The Si_2H_6 flux was initiated at points A and C and terminated at points B and D.

responds precisely to the frequency of domain coverage changes, which repeat every $a_0/2$ and derive from the local bonding arrangement.

2.2. RHEED oscillations—origins and anomalies

In this section, we discuss briefly recent discussions on the origin of RHEED oscillations in terms of surface morphology and electron scattering, and then present a number of apparent anomalies with regard to the oscillation period.

2.2.1. Origin of oscillations

Although this is scarcely a new topic, the origin of the intensity oscillations remains somewhat controversial and several papers have appeared recently dealing specifically with this topic [53–55], but since there is still no universally accepted conclusion, it is worth considering it briefly.

It is generally accepted that the oscillations arise from surface morphological changes, in such a way that their period is an accurate reflection of the ML growth rate as a consequence of a 2-D layer-by-layer growth mode. The point at issue is whether their origin can be based on a two-level interference model [20], dependent on surface coverage, or on step edge scattering [19], which derives from the changing step density during growth of a single ML and between

successive MLs. In essence, the distinction is between diffraction and refraction, and whether or not they are mutually exclusive.

Korte and Maksym [53] claimed from diffraction theory calculations that oscillations could not be accounted for solely by step density fluctuations, but always required a significant component of coverage, and furthermore, an increase in step density at constant coverage would result in an increase in specular beam intensity. Braun et al. [54] stated categorically that step density was not involved, by using a model in which there is an interference effect within the surface reconstructed layer as it forms on the growing surface; i.e. it is coverage dependent, since only the top and bottom surfaces of the reconstructed layer are involved. When the growth morphology is simulated by kMC methods, however, there is excellent agreement between the simulated step density and the RHEED oscillations [56], but in contradistinction to the above models, it requires minimum specular reflectivity for maximum step density. It might reasonably be supposed that the distinction between coverage and step density could be obtained by simply using STM images to measure both separately on the same surface, but such an experiment exposed a basic problem [57]. In earlier work comparing RHEED oscillations and STM images, Sudijono et al. [58] had shown that the oscillation damping was the direct result of the step density reaching a steady state and that the coverage, expressed as the number of MLs over which the growth front was distributed, did not change. They also found a direct relationship between the RHEED specular beam intensity and the step density which was directly contrary to the theoretical predictions, but the same as the kMC simulation results, i.e. minimum intensity corresponds to maximum step density. Very recently, however, a direct comparison of the step density and effective coverage from STM images showed that both were identically periodic and in phase [57]. The corresponding phase of RHEED oscillations varied with diffraction conditions, as expected [59]. In effect, step density and coverage during growth on a singular surface are not separate variables, so it is not possible to resolve directly which dominates the scattering of the incident RHEED beam, but the results did confirm that increasing the step density does not increase the specular beam intensity at the polar angles used (between 0.8° and 1.8°) in this system, again contrary to theoretical predictions. Finally, using a very idealized theoretical model in which each layer is fully completed before the next starts, Mitura et al. [55] concluded that although periodic changes in refraction conditions constitute the major source of RHEED oscillations, diffuse scattering by step edges can have a serious influence. The overall conclusion must be that no single electron scattering or interference process is able to account for all of the observations.

2.2.2. Oscillation anomalies

When there is an excess of the group III element on the surface for any reason, the oscillation period still corresponds to the growth of a ML, but not to the group III element flux. Instead it relates to the flux of the group V element (group V induced oscillations), but only as the product of this flux and the sticking coefficient of the group V element [60]. The period is effectively a measure of the incorporation rate of the group V element and the oscillations continue in this mode until the excess group III element has been consumed. The excess surface concentration can be produced by several different effects; it may be pre-deposited, there may be surface segregation during growth or a very high III–V flux ratio may be used. This type of oscillation has the advantage that it can be applied to measure directly the incorporation kinetics of the group V element [15], since no oscillations occur if it simply condenses on the surface in elemental form.

A similar effect is observed on non-(0 0 1) surfaces [61], because the lifetimes of the group V elements on such surfaces tend to be very short. As a consequence, unless extremely high V:III flux ratios ($\geqslant$ 30:1), or very low substrate temperatures are used, growth naturally occurs under group III-rich conditions and the oscillations again measure the incorporation rate of the group V element. Because there is a tendency for an accumulation of the group III element to occur on the surface, the growth morphology is often rough.

Finally, and as we have already indicated, the undamped oscillations observed during MEE growth are the result of the variation in specular beam intensity between the different surface reconstructions brought about by flux modulation [35].

2.3. The InAs–GaAs system and quantum dot formation

The formation of self-assembled quantum dots (QDs) has become a topic of immense interest due to the potential application of QDs in a wide range of devices, but especially lasers [62] and QD-based architectures for quantum computing. In addition to its technological importance, however, this combination of materials is the archetypal III–V system for the study of growth processes occurring when there is significant lattice mismatch, in this case $\approx$ 7%. We will restrict our discussion here to QD formation, but it is important to emphasize that this growth effect is the exception rather than the rule, and is specific to particular low-index crystal orientations and surface reconstructions, which are not necessarily separate variables.

It is generally accepted that the growth of InAs on GaAs(0 0 1) follows a version of the Stranski–Krastanov (S–K) mode, which implies that following the deposition of 1 or 2 ML in a 2-D pseudomorphic form (sometimes referred to as the wetting layer (WL)), coherent 3-D growth is initiated by a very small increment ($\leqslant$ 0.1 ML) of deposited material, to relax the elastic strain introduced by the lattice mismatch. The QDs rapidly reach a saturation number density having a comparatively narrow size (volume) distribution. We need to examine this apparently simple concept in more detail, however, before we can establish a reasonable growth mechanism. Popular wisdom has distilled from the vast literature a commonly accepted version, but there are many shortcomings within this model and here we briefly summarize the present position. We emphasize, however, that this is a very active research area and here we can do little more than provide an indication of the many problems that remain to be solved.

The first issue concerns the orientation and reconstruction specificity of the growth mode, since if strain relaxation were the only criterion, there should be no differences, because the extent of strain is not dependent on either. In practice, however, we find that on GaAs(1 1 0) and (1 1 1)A growth occurs in a 2-D layer-by-layer mode and strain is efficiently and effectively relaxed by the formation of misfit dislocations. Their geometry and interactions can be observed directly in situ by STM and ex situ by transmission electron microscopy (TEM) [63–67]. A critical point, though, is that on neither orientation is there any alloying, there is no WL formation and even at sub-ML deposition, the coverage corresponds precisely to the integrated InAs flux. Although QDs do form on certain reconstructed (0 0 1) surfaces, the process is by no means universal even on this orientation. While in general Ga-stable reconstructed surfaces cannot be formed in the temperature range available for deposition of InAs ($\leqslant 520\,°C$), it is possible to grow under strongly In-rich conditions. This leads to a 2-D layer-by-layer growth mode and QDs do not form. This has been called virtual surfactant-mediated epitaxy by Tournié et al. [68], but totally conflicting versions of the mechanism have been proposed, based on the surface free energy change of a strained overlayer caused by a surfactant. A reduction should facilitate island formation [69], but only at equilibrium, which is not the case for MBE [70]. In other words, are kinetics or energetics the determining factor, although it is unlikely that free energy considerations alone will provide the answer.

It is only on As-stable GaAs(0 0 1) surfaces that QDs form and even then the situation is not clear cut. The lowest energy structure under these conditions is the $\beta2(2 \times 4)$, as explained previously, but it is very difficult to maintain this structure at the temperatures appropriate for InAs deposition ($\leqslant 520\,°C$). Consequently there are no definitive results to confirm whether or not QDs form on this surface [71]. There is no doubt, however, that on the $c(4 \times 4)$ surface, which is the usual reconstruction during growth at these temperatures, QDs do form, preceded by the formation of an alloy WL. The structure and composition of this WL have been investigated by RHEED and STM and all evidence suggests it is a necessary precursor in the evolution of dots [72], or at least, when there is no WL, there are no coherent dots. Once the WL has formed, at some point the growth mode changes from 2-D to 3-D with a very small incremental deposit (< 0.1 ML). The change can easily be detected from the RHEED pattern, which changes from streaks normal to the shadow edge to transmission spots caused by the incident beam being diffracted as it passes through the newly formed QDs. The dots reach a saturation number density of between $\approx 5 \times 10^9$ and 2×10^{12}, which is both temperature and In flux dependent, with deposition of approximately an additional 0.25 ML. They also have a comparatively narrow size (volume)

distribution and there is very little increase in the size of individual islands prior to coalescence, which starts after ≈ 2 ML total InAs deposition.

The driving force for QD formation is usually attributed to strain relaxation according to the model of Tersoff and LeGoues [73], in which energy gain from the increase in surface area as a result of dot formation more than compensates the increase in interfacial free energy. No other strain relaxation mechanism involving extended defects operates, since both the WL and the QDs are fully coherent, but other factors than strain relaxation alone may be involved in the formation process. If we look more closely we can see that there are several apparent contradictions in addition to those concerning orientation and reconstruction dependencies which mitigate against the simple picture:

(i) The transition thickness has very little In flux dependence over the range 0.01–0.5 ML s^{-1}, but the number density of dots varies strongly with flux, decreasing as it decreases, but the dots are consequently larger at the lower fluxes.

(ii) The amount of material in the QDs (total volume) is considerably greater than the total amount of InAs deposited beyond the 2- to 3-D transition, and at the highest temperature used ($500\,°C$), it is greater than the *total* amount of InAs deposited (Fig. 6). The WL is therefore involved in the formation of the QDs via material transport during growth, and the dots are an (In, Ga)As alloy, not

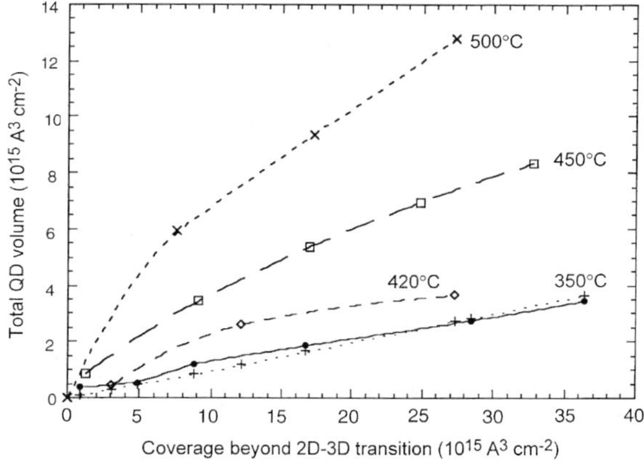

Fig. 6. The total measured volume of QDs as a function of the amount of InAs deposited after the 2- to 3-D growth mode transition. The dashed line is the volume expected when the dots are composed only of the additional amount of InAs deposited.

InAs, even though no GaAs is deposited [74]. A surprising feature, however, is that the lower the In flux, the smaller the amount of Ga incorporated into the QDs [75]. Since the lower rate should allow a closer approach to equilibrium, it might have been expected that the amount of Ga would be greater under these conditions, but that is not the case.

(iii) An unexpected, but still contentious issue is that the size (volume) distribution of the QDs appears to obey the scaling laws derived for homoepitaxial growth, where strain is not a parameter. The effect was first reported by Ebiko et al. [76], but it has recently been challenged by Krzyzewski et al. [77], who claim to show that it only applies when the dots have reached their saturation number density, but not to dots formed close to the transition point. This is actually not different from the original results, which did not consider the evolution of dots, but only the final state. Dot evolution is itself a difficult problem, in that the evidence is based on snapshot STM images [78], so it is impossible to be certain which, if any, of the entities that are first observed eventually turn into the final state QDs. If the scaling laws appropriate to homoepitaxy are obeyed, however, it implies that strain is not involved in determining the QD size distribution, but it does not rule out its possible involvement in their formation.

(iv) The very sudden 2- to 3-D transition implies a very large mobile group III adatom population to be present immediately prior to QD formation and this must be included in any proposed mechanism. It might possibly relate to the scaling behaviour.

There have been numerous attempts to develop theories of QD formation, based variously on the thermodynamics of WL formation, the thermodynamics of the 2- to 3-D transition, equilibrium concepts of 3-D island formation and kinetic models of 3-D island nucleation. They have been reviewed very recently [79], but none is so far able to incorporate all of the following factors:

(a) QDs only form on the $(0\,0\,1)$ orientation and even then only on As-stable reconstructions, so the implication is that strain is a necessary, but not sufficient condition for this growth mode.

(b) Alloying occurs only on the $(0\,0\,1)$ orientation. This is one of several possible strain relief mechanisms and it is dominant on $(0\,0\,1)$ oriented substrates. On other low-index orientations strain relaxation relies exclusively on the formation of misfit dislocations.

(c) The amount of InAs required to form a continuous, homogeneous alloy WL is temperature dependent, so the alloying mechanism is a thermally activated process, perhaps also strain assisted.

(d) The 2- to 3-D transition occurs over < 0.1 ML incremental deposition, so there must be a very large Ga and In adatom population available for island growth once there are stable nuclei, indicating that nucleation is a difficult process. Additionally, the Ga can only be derived by dissociation of the substrate.

(e) It appears that measured size distributions of QDs are consistent with models of homoepitaxy with a critical nucleus of $i^* = 1$, with no discernible influence of strain and no adatom detachment from islands.

At least one of the factors missing from existing theories is any structural component, which is most likely to be involved during WL formation, but perhaps also in the development of coherent 3-D islands. We are presently examining these concepts.

2.4. Group III—element nitrides

An area of MBE growth which has seen an explosion of interest in the past decade is the group III nitrides. This followed the announcement of high brightness blue-emitting InGaN–AlGaN double heterostructure LEDs by Nakamura et al. [80] and the subsequent development of other optoelectronic devices, including laser diodes also emitting in the blue [81]. MOVPE is currently the dominant technology for the growth of these materials for device application and is likely to remain so, with the possible exception of high power microwave transistors [82], but MBE has made a significant contribution to more fundamental studies, largely because of the wide range of in situ diagnostic techniques available. It has also been proved to be a valuable technique for the growth of "dilute nitrides", which are III–V–N materials such as GalnNAs, where the N concentration is $\leqslant 2\%$ [83,84].

The many forms of nitride systems described in the literature are variants of gas source MBE, in which the nitrogen is introduced from a plasma formed using either a radio-frequency (RF) discharge [85] or electron cyclotron resonance (ECR) [86]; alternatively, ammonia may be used, which is dissociated on the growth surface [87]. Initial problems associated with the necessarily high growth temperatures required, inefficient plasma sources and high reactive gas loads have been successfully overcome, and growth rates up to $\approx 1\ \mu m\,h^{-1}$ have been achieved. There are however, several other growth parameters which strongly influence the quality of material produced. For heteroepitaxial growth on the most commonly used substrates, sapphire and silicon carbide, these include substrate cleaning, initial nitridation, the nucleation and coalescence of islands involved in the low temperature growth of a buffer layer which is then annealed at a higher temperature, and the conditions needed for high temperature growth.

Several of these problems can be resolved using "virtual" GaN substrates obtained by growing thick layers on to suitable substrates using MOVPE and then exploiting the advantages of MBE to produce the functional layer on the GaN template. In this way films have been produced with smooth surface morphology and optimum quality in terms of high carrier mobility and/or strong photoluminescent intensity, which has led to Bragg mirrors and InGaN-based multiple quantum well LEDs [85], as well as high electron mobility in bulk and two-dimensional electron gas (2DEG) structures [82].

MBE growth temperatures are lower than those used for MOVPE, so in the growth of InGaN phase separation and In desorption are less problematic [85, 88], but MOVPE nevertheless still produces superior material for light emitters [86], which means that these two factors are not the only important parameters. Local variations in InGaN composition, in some cases sufficiently extreme to form QDs, also play a crucial role [89], and differences in the optical properties of InGaN quantum wells and QDs grown using MBE based on an ammonia nitrogen source have been investigated by Damilano et al. [90].

In the use of active nitrogen from plasma sources, the III/V flux ratio at the substrate during growth is also a critical parameter. GaN layers grown with a low III/V flux ratio (N-stable growth) display a facetted surface morphology and a tilted columnar structure with a high density of stacking faults. Smooth surfaces are only obtained under Ga-rich conditions, where not only is there a dramatic reduction of surface roughness, but significant improvements in structural and electrical properties are also observed. This is, of course, the exact opposite of the growth of most III–V compounds, such as GaAs. In the case of nitrides it is thought that Ga-rich conditions (close to the point where Ga droplets are formed) promote step flow growth, whereas N-stable growth promotes the nucleation of new islands. To compound the problem, growth from NH_3 is smoother under N-rich conditions, but there is no adequate explanation [91].

The generally poor level of understanding of nitride materials is exemplified by recent work on the bandgap of InN, where the widely quoted figure of 2.0 eV has been thrown into question by recent measurements of both MOVPE [93] and MBE [94] grown material, which indicate that it might be as low as 0.7 eV.

The polarity (nature of the outermost layer of atoms) of $\{0\,0\,0\,1\}$ oriented hexagonal structure films also has a crucial influence on material quality, but both N- or Ga-polarity can occur with MBE growth on sapphire substrates [82,92]. Under typical growth conditions with MOVPE, however, Ga-polarity material is exclusively produced. The cubic or zinc-blende polytype of GaN is metastable with respect to the hexagonal phase, but a prospective route to the preparation of cubic GaN films is via low temperature deposition on cubic substrates such as GaAs, which offers higher carrier mobilities and easier cleavage than the hexagonal phase material. In general, therefore, MBE has a role in growth on cubic substrates where a low growth temperature is required, such as Si(1 1 1) [95] and cubic GaN [96]. The low growth temperature also has potential advantages for III–V–N materials, which need a less stable precursor than NH_3 [83,84].

3. Possible future directions

MBE has made some impressive contributions to fundamental studies of the growth of low dimensional structures and to basic concepts of thin film nucleation and growth, as well as to surface studies in general, but it should not be forgotten that it has also become a production technology in its own right. This is an impressive performance for what was for many years regarded merely as a laboratory curiosity. Future advances in this direction are, however, at the mercy of global economic and political factors, which are beyond our ability to predict, a shortcoming we suspect we share with everyone else!

At the fundamental level there is no doubt that the technique will continue to be the vehicle for ever-more detailed in situ investigations of growth processes. At the atomistic level we do not yet have an adequate description of even the simplest III–V system, the homoepitaxy of GaAs, in that most reports are essentially qualitative and the match between theory and experiment is far from complete. A vital advance which is awaited is genuine real-time STM imaging with atomic resolution during growth. As the degree of complexity of the material system increases, for example in QD formation, so the gap to be filled becomes wider, and even growth control and reproducibility are still major issues to be resolved. We are firmly of the opinion that the vast majority of important device applications based on MBE growth have been dependent on our understanding of the fundamental issues. We are equally confident that future device technology will similarly benefit from continued efforts to improve our basic knowledge.

One unfortunate consequence of the almost universal application-oriented funding is that as soon as a device has been demonstrated, if only on a one-off basis, it is often tacitly assumed that no further materials research is needed. The fact that QD lasers can be made is a case in point, where as a result research has been channelled into less than ideal directions by various funding agencies. Fortunately, the potential material demands of quantum computing, which seem certain to be severe, may redress the balance.

It is also clear that MBE will make valuable contributions beyond the more conventional semiconductor field, into areas such as spintronics, spin resonance transistors in relation to quantum computing and magnetic nanotechnology for a variety of applications.

Areas where MBE has made a contribution in the past, but where we suspect its impact will diminish include nitrides, which will increasingly rely on VPE because of its high temperature advantage; SiGe, where UHVCVD technology will dominate, and finally organic films, which in general are much more cheaply and easily prepared by standard polymer-processing techniques.

Acknowledgements

One of us (B.A.J.) is indebted to many friends and colleagues for all the help and encouragement they have provided during the past 40 years, but in particular Jim Neave, Dimitri Vvedensky, Tom Foxon, Don Pashley, Jeff Harris and Jing Zhang. Many research students and post-docs from the Semiconductor Materials IRC have also made valuable contributions, for which grateful thanks are due. T.B.J. would like to thank his colleagues in the Materials Science Department at Liverpool University.

References

[1] B.A. Joyce, R.R. Bradley, Philos. Mag. 14 (1966) 289.
[2] B.A. Joyce, R.R. Bradley, G.R. Booker, Philos. Mag. 15 (1967) 1167.
[3] B.E. Watts, R.R. Bradley, B.A. Joyce, G.R. Booker, Philos. Mag. 17 (1968) 1163.
[4] G.R. Booker, B.A. Joyce, Philos. Mag. 14 (1966) 301.
[5] J.R. Arthur, J. Appl. Phys. 39 (1968) 4032.
[6] J.R. Arthur, Surf. Sci. 43 (1974) 449.
[7] A.Y. Cho, M.B. Panish, I. Hayashi, in: Proceedings of the Symposium on GaAs and Related Compounds, vol. 2, Inst. of Physics, Aachen, Germany, 1970, p. 18.
[8] A.Y. Cho, J. Vac. Sci. Technol. 8 (1971) S31.
[9] A.Y. Cho (Ed.), Molecular Beam Epitaxy, AIP Press, New York, 1994.
[10] J.R. Arthur, J. Phys. Chem. Solids 28 (1967) 2257.
[11] C.T. Foxon, J.A. Harvey, B.A. Joyce, J. Phys. Chem. Solids 34 (1973) 1693.
[12] C.T. Foxon, M.R. Boudry, B.A. Joyce, Surf. Sci. 44 (1974) 69.
[13] C.T. Foxon, B.A. Joyce, Surf. Sci. 50 (1975) 434.
[14] C.T. Foxon, B.A. Joyce, Surf. Sci. 64 (1977) 293.
[15] E.S. Tok, J.H. Neave, J. Zhang, B.A. Joyce, T.S. Jones, Surf. Sci. 374 (1997) 397.
[16] A.Y. Cho, J. Appl. Phys. 41 (1970) 2780.
[17] M.D. Pashley, K.W. Haberern, W. Friday, J.M. Woodall, P.D. Kirchner, Phys. Rev. Lett. 60 (1988) 2176.
[18] J.J. Harris, B.A. Joyce, P.J. Dobson, Surf. Sci. 103 (1981) L90.
[19] J.H. Neave, B.A. Joyce, P.J. Dobson, N. Norton, Appl. Phys. A 31 (1983) 1.
[20] J.M. Van Hove, C.S. Lent, P.R. Pukite, P.I. Cohen, J. Vac. Sci. Technol. B 1 (1983) 741.
[21] S. Clarke, D.D. Vvedensky, Phys. Rev. Lett. 58 (1987) 2235.

[22] A. Madhukar, S.V. Ghaisas, CRC Crit. Rev. Solid State Mater. Sci. 14 (1988) 1.

[23] A.Y. Cho, H.C. Casey Jr., Appl. Phys. Lett. 25 (1974) 288.

[24] A.Y. Cho, R.W. Dixon, H.C. Casey Jr., R.L. Hartman, Appl. Phys. Lett. 28 (1976) 501.

[25] A.R. Calawa, Appl. Phys. Lett. 38 (1981) 701.

[26] M.B. Panish, J. Electrochem. Soc. 127 (1980) 2730.

[27] E. Veuhoff, W. Pletschen, P. Balk, H. Luth, J. Crystal Growth 55 (1981) 30.

[28] N. Putz, E. Veuhoff, H. Heinecke, H. Luth, P. Balk, J. Vac. Sci. Technol. B 3 (1985) 671.

[29] W.T. Tsang, Appl. Phys. Lett. 45 (1984) 1234.

[30] M. Weyers, J. Musolf, D. Marx, A. Kohl, P. Balk, J. Crystal Growth 105 (1990) 383.

[31] B. Marheinecke, M. Popp, H. Heinecke, J. Crystal Growth 164 (1996) 16.

[32] M. Keidler, D. Ritter, H. Baumeister, J. Crystal Growth 188 (1998) 168.

[33] Y. Horikoshi, M. Kawashima, H. Yamaguchi, Jpn. J. Appl. Phys. 25 (1986) L868.

[34] Y. Horikoshi, M. Kawashima, H. Yamaguchi, Jpn. J. Appl. Phys. 27 (1988) 169.

[35] B.A. Joyce, D.D. Vvedensky, C.T. Foxon, in: S. Mahajan (Ed.), Handbook on Semiconductors, vol. 3, Elsevier Science BV, Amsterdam, 1994, p. 275.

[36] J.A. Venables, G.D.T. Spiller, M. Handbüchen, Rep. Prog. Phys. 47 (1984) 399.

[37] J.Y. Tsao, Materials Fundamentals of Molecular Beam Epitaxy, Academic Press, New York, 1992.

[38] H.H. Farrell, J.P. Harbison, L.D. Peterson, J. Vac. Sci. Technol. B 5 (1987) 1482.

[39] M. Itoh, G.R. Bell, A.R. Avery, T.S. Jones, B.A. Joyce, D.D. Vvedensky, Phys. Rev. Lett. 81 (1998) 633.

[40] M. Itoh, G.R. Bell, B.A. Joyce, D.D. Vvedensky, Surf. Sci. 464 (2000) 200.

[41] C.G. Morgan, P. Kratzer, M. Scheffler, Phys. Rev. Lett. 82 (1999) 4886.

[42] A. Kley, P. Ruggerone, M. Scheffler, Phys. Rev. Lett. 79 (1997) 5728.

[43] P. Kratzer, M. Scheffler, Phys. Rev. Lett. 88 (2002) 036102.

[44] K. Shiraishi, T. Ito, Phys. Rev. B 57 (1998) 6301.

[45] S.M. Mokler, W.K. Liu, N. Ohtani, J. Zhang, B.A. Joyce, Surf. Sci. 275 (1992) 16.

[46] B. Voigtlander, M. Kästner, Appl. Phys. A 63 (1996) 577.

[47] R.J. Hamer, R.M. Tromp, J.E. Demuth, Phys. Rev. B 34 (1986) 5343.

[48] P. Krüger, J. Pollmann, Phys. Rev. Lett. 74 (1995) 1155.

[49] S. Clarke, M.R. Wilby, D.D. Vvedensky, Surf. Sci. 225 (1991) 91.

[50] B. Voigtlander, T. Weber, P. Šmilauer, D.E. Wolf, Phys. Rev. Lett. 78 (1997) 2164.

[51] Y.-W. Mo, M.G. Lagally, Surf. Sci. 248 (1991) 313.

[52] J. Zhang, A.G. Taylor, A.K. Lees, J.M. Fernández, B.A. Joyce, D. Raisbeck, N. Shukla, M.E. Pemble, Phys. Rev. B 53 (1996) 10107.

[53] U. Korte, P.A. Maksym, Phys. Rev. Lett. 78 (1997) 2381.

[54] W. Braun, L. Däweritz, K.H. Ploog, Phys. Rev. Lett. 80 (1998) 4935.

[55] Z. Mitura, S.L. Dudarev, L.-M. Peng, G. Gladyszewski, M.J. Whelan, J. Crystal Growth 235 (2002) 79.

[56] T. Shitara, D.D. Vvedensky, M.R. Wilby, J. Zhang, J.H. Neave, B.A. Joyce, Phys. Rev. B 46 (1992) 6815.

[57] G.R. Bell, T.S. Jones, J.H. Neave, B.A. Joyce, Surf. Sci. 458 (2002) 247.

[58] J.L. Sudijono, M.D. Johnson, C.W. Snyder, M.B. Elowitz, B.G. Orr, Phys. Rev. Lett. 69 (1992) 2811.

[59] J. Zhang, J.H. Neave, P.J. Dobson, B.A. Joyce, Appl. Phys. A 42 (1987) 317.

[60] J.H. Neave, B.A. Joyce, P.J. Dobson, Appl. Phys. A 34 (1984) 179.

[61] B.A. Joyce, J.H. Neave, M.R. Fahy, K. Sato, D.M. Holmes, J.G. Belk, J.L. Sudijono, T.S. Jones, J. Mater. Sci. Mater. Electron. 7 (1996) 327.

[62] D. Bimberg, M. Grundmann, F. Heinrichsdorff, N.N. Ledentsov, V.M. Ustinov, A.R. Korsh, M.V. Maximov, Y.M. Shenyakov, B.V. Volovik, A.F. Tsatsalnokov, P.S. Kopiev, Zh.I. Alferov, Thin Solid Films 367 (2000) 235, and references therein.

[63] J.G. Belk, J.L. Sudijono, X.M. Zhang, J.H. Neave, T.S. Jones, D.W. Pashley, B.A. Joyce, Phys. Rev. Lett. 78 (1997) 475.

[64] J.G. Belk, D.W. Pashley, C.F. McConville, J.L. Sudijono, B.A. Joyce, T.S. Jones, Phys. Rev. B 56 (1997) 10289.

[65] J.G. Belk, D.W. Pashley, B.A. Joyce, T.S. Jones, Phys. Rev. B 58 (1998) 16194.

[66] H. Yamaguchi, J.G. Belk, X.M. Zhang, J.L. Sudijono, M.R. Fahy, T.S. Jones, D.W. Pashley, B.A. Joyce, Phys. Rev. B 55 (1997) 1337.

[67] L.A. Zepeda-Ruiz, D. Marondas, W.H. Weinberg, J. Appl. Phys. 85 (1999) 3677.

[68] E. Tournié, O. Brandt, C. Gionni, K.H. Ploog, J. Crystal Growth 127 (1993) 765.

[69] C.W. Snyder, B.G. Orr, Phys. Rev. Lett. 70 (1993) 1030.

[70] N. Grandjean, J. Massies, Phys. Rev. Lett. 70 (1993) 1031.

[71] G.R. Bell, T.J. Krzyzewski, P.B. Joyce, T.S. Jones, Phys. Rev. B 61 (2000) R10551.

[72] J.G. Belk, C.F. McConville, J.L. Sudijono, T.S. Jones, B.A. Joyce, Surf. Sci. 387 (1997) 213.

[73] J. Tersoff, F.K. LeGoues, Phys. Rev. Lett. 72 (1994) 3570.

[74] P.B. Joyce, T.J. Krzyzewski, G.R. Bell, B.A. Joyce, T.S. Jones, Phys. Rev. B 58 (1998) R15981.

[75] P.B. Joyce, T.J. Krzyzewski, G.R. Bell, T.S. Jones, S. Malik, D. Childs, R. Murray, J. Crystal Growth 227–228 (2001) 1000.

[76] Y. Ebiko, S. Muto, D. Suzuki, S. Itoh, K. Shiramine, T. Haga, K. Uno, M. Ikeda, Phys. Rev. B 60 (1999) 8234.

[77] T.J. Krzyzewski, P.B. Joyce, G.R. Bell, T.S. Jones, Phys. Rev. B 66 (2002) 201302(R).

[78] T.J. Krzyzewski, P.B. Joyce, G.R. Bell, T.S. Jones, Phys. Rev. B 66 (2002) 121307(R).

[79] B.A. Joyce, D.D. Vvedensky, in: M. Kotola, N.I. Papanicolaou, D.D. Vvedensky, L.J. Wille (Eds.), Atomic Aspects of Epitaxial Growth, NATO Science Series II, vol. 65, Kluwer Academic Publishers, Dordrecht, 2002, p. 301.

[80] S. Nakamura, T. Mukai, M. Senoh, Appl. Phys. Lett. 64 (1994) 1687.

[81] S. Nakamura, G. Fasol, The Blue Laser Diode, Springer, Berlin, 1997.

[82] H. Morkoç, J. Mater. Sci. Mater. Electron. 12 (2001) 677.

[83] M. Kondow, T. Kitatani, Semicond. Sci. Technol. 17 (2002) 746.

[84] H. Riechert, A. Ramakrishnan, G. Steinle, Semicond. Sci. Technol. 17 (2002) 892.

[85] M. Sanchez-Garcia, J.L. Pau, F. Naranjo, A. Jiminez, S. Fernandez, J. Ristic, F. Calle, E. Calleja, E. Munoz, Mater. Sci. Eng. B 93 (2002) 189.

[86] S.T. Strite, H. Morkoç, J. Vac. Sci. Technol. B 10 (1992) 1237.

[87] N. Grandjean, B. Damilano, J. Massies, J. Phys.: Condens. Matter 13 (2001) 6949.

[88] A.V. Blant, T.S. Cheng, C.T. Foxon, J.C. Bussey, S.V. Novikov, V.V. Tret'yakov, in: Mater. Res. Soc. Symp. Proc. Pittsburgh, PA, 1997, p. 465.

[89] S. Nakamura, Semicond. Sci. Technol. 14 (1999) R27.

[90] B. Damilano, N. Grandjean, S. Vezian, J. Massies, J. Crystal Growth 227–228 (2001) 466.

[91] N. Grandjean, M. Leroux, J. Massies, M. Laügt, Jpn. J. Appl. Phys. 38 (1999) 618.

[92] S. Sonoda, S. Shimizu, S. Kara, H. Okumura, Jpn. J. Appl. Phys. 39 (2000) L202.

[93] T. Matsuoka, H. Okamoto, M. Nakao, H. Harima, E. Kurimoto, Appl. Phys. Lett. 81 (2002) 1246.

[94] Compound Semiconductor.net, November 2002.

[95] M.A. Sanchez-Garcia, E. Calleja, E. Monroy, F.J. Sanchez, F. Calle, E. Munoz, R. Beresford, J. Crystal Growth 183 (1998) 23.

[96] P.R. Chalker, T.B. Joyce, T. Farrell, Diamond Rel. Mater. 8 (1999) 373.

The beginnings of metalorganic chemical vapor deposition (MOCVD) [*]

Harold M. Manasevit

TRW Electronic Systems Group, Electro Optics Research Center, Redondo Beach, CA 90278

The impetus that led to the development of MOCVD at Rockwell International in 1967 was a desire to examine the feasibility of growing GaAs on single crystal insulator substrates. Four years earlier we had demonstrated for the first time that silicon could be grown on sapphire and other insulators such as spinel, BeO and chrysoberyl even though it was not apparent why they should be compatible. In most cases the crystal structures of film and substrate were different, and even though spinel is cubic, its lattice parameter, 8.08 Å, is a far cry from that of Si, 5.43 Å. Even matching atom positions were insufficient to account for many of the orientation relationships found between Si and the insulators. So in an attempt to develop further our understanding of heteroepitaxy, we chose to look at GaAs, close to Si in lattice parameters (5.65 Å) and structure (cubic zinc blende structure) and its compatibility with these insulators. Zanowick of our laboratory initially found that growth of GaAs directly on insulators by a close-spaced HCl vapor transport process was poor. But, by using an intermediate, very thin layer of Ge, he produced heteroepitaxial sandwiches of GaAs/Ge/insulator, and a few orientation relationships were established.

Next, I tried growing GaAs directly on insulators, using the metal alkyls trimethylgallium (TMGa) and triethylgallium (TEGa), with AsH_3 as the source of As. The reason for this choice was my experience in graduate school, where I had compared the Lewis acid (i.e. acceptor) properties of trimethylborane, trimethylaluminum (TMAI) and TMGa with selected silylamines (donors of electrons through the N atom). I wondered if such materials containing the group III and V elements could be used to produce III–V compounds on complete pyrolysis. Could GaAs, for example, be produced by the simple reaction

$$TMGa + AsH_3 \rightarrow GaAs + CH_4?$$

I was not aware at that time that Harrison and Tompkins in 1962 had taken a cursory look at this reaction or that Didchenko, et al. had prepared InP from trimethylindium (TMIn) and PH_3 at 275–300 °C. Actually Harrison probably did not prepare GaAs by this process since he heated only to 200 °C, and later studies by Schlyer and Ring showed only small amounts of GaAs at ~360 °C. I was also not aware of the patent reports of Ruehrwein at Monsanto in the time span 1965–68. He described almost every chemical reaction under the sun, from halides to alkyls to halo-alkyls to hydrides to the elements, and a growth temperature span from 400–1500 °C. Some of the examples given clearly won't work, mainly because the substrates or the films would not survive the indicated growth temperatures. For example, Ruehrwein describes growing AlSb on InP

---

[*] Based on an article in SPIE 233 (1982) 94.

Previously published in AACG Newsletter Vol. 15(1) March 1985.

at 1000 °C, but thermal etching of InP begins at least 300 °C lower. Also growing GaAs on Ge at 1000 °C would be difficult, since Ge melts at 941 °C! His ideas, however, were good considering they were based on what was known in the early 60's.

In the early days, I could not find a commercially available source of TMGa so I chose initially to examine TEGa, which I obtained from Research Organic Chemicals in nearby Sun Valley, California. Meanwhile, I was able to convince Dr. Meloni of Alfa and Ed Lanpher of Orgmet to prepare some TMGa for me. Using a temperature range consistent with GaAs homoepitaxy by other processes, GaAs was produced almost immediately on sapphire, Ge, GaAs and other insulators in a vertical reactor that was similar to that I had used for Si heteroepitaxy. The formation of a continuous film on sapphire rather than balls of Ga, and the differences in color between the film and Ge led credence to the hoped-for chemistry of the proposed reaction. The epitaxial nature of the deposit encouraged further evaluation by reflection electron diffraction and x-ray techniques. These confirmed epitaxy on the substrates. Of course, the low vapor pressure of TEGa did not permit the kind of rapid coverage of sapphire that had been possible with SiH_4 as the source

of Si. The availability of TMGa, with higher vapor pressure, corrected that problem and growth rates up to several microns per minute were examined during the course of GaAs growth on insulating substrates. The growth parameters that we studied were similar to those now used for homoepitaxial growth, for example the effect of AsH_3/TMGa ratio and temperature on film quality, the formation of p-layers at low AsH_3/TMGa ratios and the differences in film properties with different source materials.

Simultaneous growth on (0001) Al_2O_3 and on the (111) A- and (111) B-faces of GaAs produced a film surface on Al_2O_3 that resembled the growth on the GaAs "A" face. This suggested that bonding at the Al_2O_3–GaAs interface may involve an As bridge, perhaps between metal ions at the substrates surface and the succeeding Ga layer in the film. In SOS, bonding seemed to be explained in many cases by a filling-in of Al ion sites and bonding to the oxygens. These differences seem consistent with the fact that a high temperature H_2 treatment of Al_2O_3, which is better for SOS film growth, produces a surface incompatible with epitaxial GaAs growth. Interestingly, a sapphire substrate that was high temperature etched and then immersed in boiling

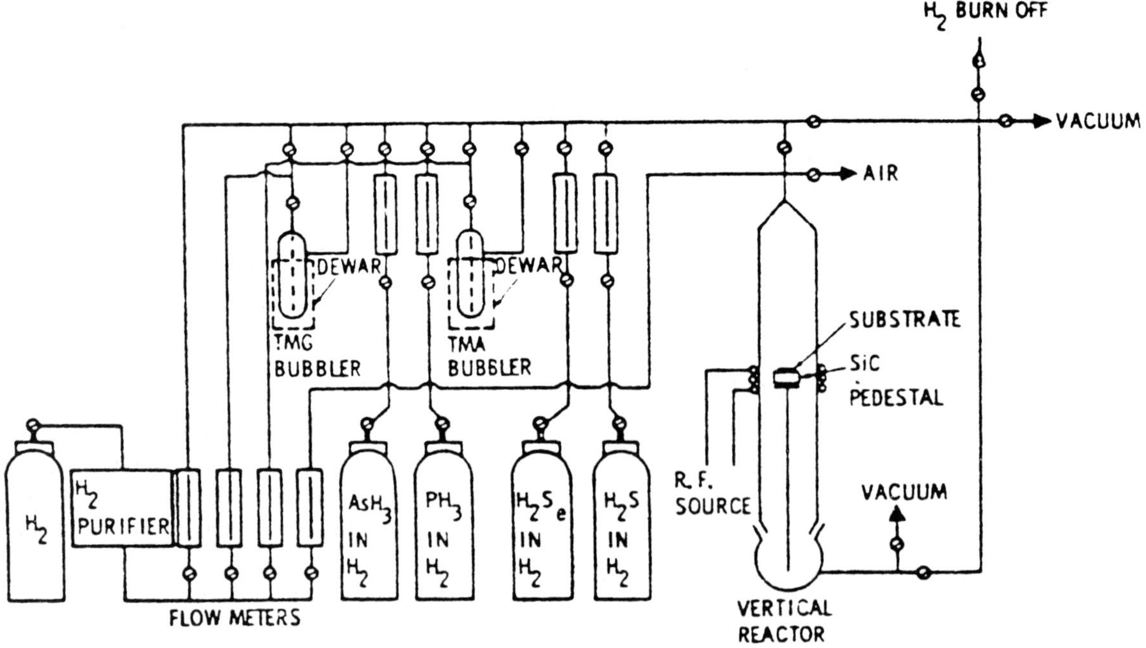

Fig. 1. Schematic of disposition apparatus.

Table 1
III–V compound semiconductors formed on insulators from metalorganics and hydrides

Compound	Insulating substrate	Reactants	Growth temperature (°C)
GaAs	Al_2O_3, $MgAl_2O_4$ BeO, ThO_2	$TMGa$–AsH_3	650–750
GaP	Al_2O_3, $MgAl_2O_4$	$TMGa$–PH_3	700–800
$GaAs_{1-x}P_x$ ($x = 0.1$–0.6)	Al_2O_3, $MgAl_2O_4$	$TMGa$–AsH_3–PH_3	700–725
$GaAs_{1-x}Sb_x$ ($x = 0.1$–0.3)		$TMGa$–AsH_3–$TMSb$	
GaSb	Al_2O_3	$TEGa$–$TMSb$	500–550
AlAs	Al_2O_3	$TMAl$–AsH_3	700
$Ga_{1-x}Al_xAs$	Al_2O_3	$TMGa$–$TMAl$–AsH_3	700
AlN	Al_2O_3, α-SiC	$TMAl$–NH_3	1250
GaN	Al_2O_3, α-SiC	$TMGa$–NH_3	925–975
GaN	Al_2O_3	$TEGa$–NH_3 (unstable)	800
InAs	Al_2O_3	$TEIn$–AsH_3	650–700
InP	Al_2O_3	$TEIn$–PH_3	725
$Ga_{1-x}In_xAs$	Al_2O_3	$TEIn$–$TMGa$–AsH_3	675–725
InSb	Al_2O_3	$TEIn$–$TESb$–AsH_3	460–475
$InAs_{1-x}Sb_x$ ($x = 0.1$–0.7)	Al_2O_3	$TEIn$–$TESb$–AsH_3	460–500

H_2O or acids did encourage epitaxial growth. The results suggested that a hydrated Al_2O_3 surface was preferred to a dehydrated one for III–V epitaxy.

In addition, we obtained (100) GaAs on (110) spinel rather than the parallel orientation we expected. This result indicated that the "parallel rule" was less than adequate, and further studies in these heteroepitaxial combinations seemed worthwhile.

Well, one compound led to another, in a system similar to that shown in the perhaps too familiar schematic of Figure 1. Replacement of the AsH_3 by PH_3 led to GaP, and mixtures of AsH_3 with PH_3 when reacted with TMGa produced $GaAs_xP_{1-x}$. Adding TMAl to TMGa and AsH_3 produced $Ga_xAl_{1-x}As$, and the transparency of the sapphire substrate made it relatively easy to obtain a bandgap for the alloys formed as the reactant ratios changed. Replacing PH_3 with NH_3 led to AlN and GaN formation, except the metalorganics and NH_3 were observed to react at room temperature to form an aerosol which then had to be directed to the hot substrate, and surface coverage was limited by the reactor design. We literally had to paint a surface with AlN by wagging a tube manually over the area to be covered at growth temperatures of about 1250 °C; and even at these high temperatures, in excess NH_3 and H_2 the AlN films were very white and appeared to be free from carbon.

Table 1 lists those III–V semiconductors that have been grown on insulating substrates prior to 1981.

Anyone interested in further details of the development of MOCVD may refer to the SPIE article or to J. Crystal Growth 55 (1981) 1, where references to the work quoted here may be found.

Development and current status of organometallic vapor phase epitaxy

G.B. Stringfellow *

Department of Materials Science and Engineering, College of Engineering, University of Utah, Rm 214 KRC, 1495 E 100 S, Salt Lake City, UT 84112-1114, USA

Abstract

The first success with the growth of III/V semiconductor materials by OMVPE dates back to the mid-1950s. Today, it is the largest volume technique for the production of III/V photonic and electronic devices with commercial reactors yielding 2000 cm^2/run. This paper will briefly trace the history and the development of key concepts in our understanding of this complex growth process, including brief discussions of the precursors and thermodynamics and kinetics of growth. Special attention will be paid to surface processes and the use of surfactants to control the properties of the resulting materials. Our understanding of this topic is still under rapid development. The discussion will extend to the control of surface processes for the growth of low dimensional structures such as superlattices, and quantum-wells, -wires, and -dots. The emphasis will be mainly semiconductor materials, including novel alloys, but the rapidly developing area of oxides for dielectrics and ferroelectrics for integrated circuits will be discussed briefly.
© 2004 Elsevier B.V. All rights reserved.

1. Brief history of the development of OMVPE

Organometallic vapor phase epitaxy (OMVPE) is a vapor phase epitaxial growth technique where the layer constituents are transported to the growing surface using organometallic and/or hydride precursor molecules. Thus, for example, an epitaxial layer of GaAs is produced by heating a GaAs substrate and using trimethylgallium (TMGa), an organometallic Ga compound, and either arsine, As hydride, or tertiary-butylarsine (TBAs), an organometallic As compound, to supply the Ga and As nutrients to the growing film. This process is sometimes referred to as metalorganic

VPE (MOVPE). A very similar technique is metalorganic chemical vapor deposition (MOCVD). As the name implies, this is slightly more general in that the layer produced may not be epitaxial. These "cold wall" growth techniques involve precursor molecules that are stable at room temperature, but decompose at the elevated temperature of the heated substrate to release the constituent elements of the film being grown. Thus, they are closely related to the silane growth of Si. In fact, the early researchers used apparatus and approaches similar to those developed earlier for Si epitaxial growth.

The OMVPE technique apparently began with the work of Scott et al. in the UK [1]. Their work, involving the use of triethylindium and stibine to form InSb layers, appears in a patent originally filed in 1953, very early in the history of III/V semiconductor ma-

---

* Tel.: +1-801-581-6911; fax: +1-801-581-8692.
E-mail address: stringfellow@coe.utah.edu (G.B. Stringfellow).

0022-0248/$ – see front matter © 2004 Elsevier B.V. All rights reserved.
doi:10.1016/j.jcrysgro.2003.12.037

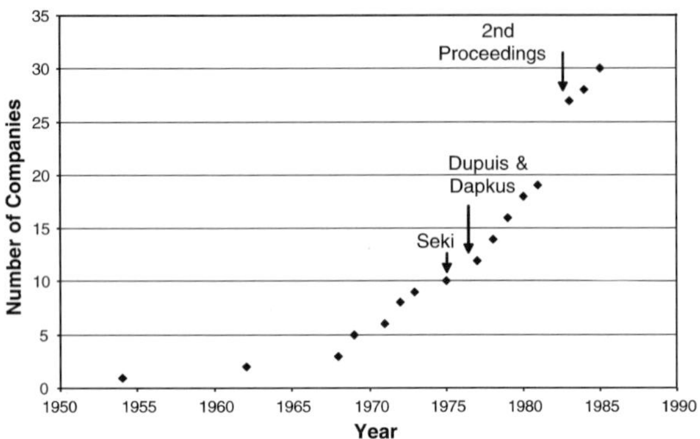

Fig. 1. Time evolution of the number of companies involved in OMVPE research.

terials. Later, Miederer et al. [2] filed a patent describing the OMVPE growth of GaAs. However, neither of these patents was widely known in the epitaxial growth community until fairly recently. In 1968 Manasevit and Simpson [3] published the first of a long series of papers describing the growth of a wide range of semiconductors using what they termed MOCVD. This started the rapid development of OMVPE, primarily for III/V and II/VI semiconductor materials, as indicated by the plot of the number of companies around the world working on this technique versus year, shown in Fig. 1.

Two particularly noteworthy milestones in the development of OMVPE are indicated in Fig. 1. The first, labeled "Seki" represents the first OMVPE growth of truly device quality GaAs, i.e., material having the very high electron mobilities indicative of high purity, highly perfect epitaxial layers [4]. This provided a major stimulus to the field in 1975. The second, labeled "Dupuis and Dapkus", was the initial demonstration, in 1977, of the OMVPE growth of AlGaAs, a material very difficult to grow due to oxygen and carbon contamination problems, having the excellent minority carrier properties necessary for the fabrication of light emitting devices such as injection lasers [5]. As seen in Fig. 1, by 1985 30 companies were involved in research and development activities involving the OMVPE growth of semiconductor materials.

At this time, both GaAs and AlGaAs could be grown by other epitaxial growth techniques, such as liquid phase epitaxy (LPE) and vapor phase epitaxy (VPE), with excellent properties for majority and mi-

nority carrier devices [6]. A major factor propelling OMVPE to the forefront of the epitaxial growth techniques for III/V semiconductor materials was the discovery that OMVPE could be used to grow important alloys that could not be grown by either LPE or chloride or hydride VPE. In 1975, Stringfellow proposed, based on the results of thermodynamic calculations, that OMVPE could be used for the growth of AlInP and AlGaInP alloys that were virtually impossible to grow by LPE and hydride and halide VPE techniques [7]. Subsequent experimental investigations proved this to be correct [8]. Today, these particular alloys are used in the large-scale production of high brightness red, orange, and amber light emitting diodes (LEDs) [9] and the highest efficiency tandem solar cells [10] using OMVPE as the exclusive growth technique.

Much later, the OMVPE technique received another boost by the discovery that OMVPE is by far the best technique for the growth of AlGaInN alloys for the commercial production of blue and green LEDs and blue lasers [11,12].

Today, OMVPE is the favored growth technique for the commercial production of LEDs covering the entire visible spectrum for indicator lamps, large, full-color displays, and, increasingly, lighting. It is estimated that, due to their high efficiencies and long operating lifetimes, white LEDs will eventually replace the last of the vacuum tubes, those used for illumination. OMVPE is also used for the production of many other electronic and photonic devices, including both IR and visible laser diodes, high electron mobility transistor (HEMT) and heterojunction bipolar transistor (HBT) devices and integrated circuits, as well as many others. Today, there are major "foundry" operations that supply wafers to industry for fabrication of diverse devices. It is estimated that 25 million sq in of epitaxial III/V materials are produced each year, generating revenues exceeding \$10 billion [13]. In addition, there are several commercial manufacturers of both research and "turn-key" production scale OMVPE reactors, capable of producing 2000 cm^2 of epitaxial material per run.

One reason that OMVPE is so widely used in commercial operations today is that it is the most versatile technique for the growth of materials and structures for a wide range of devices. Essentially all III/V and II/VI semiconductors can be grown by

OMVPE, including metastable alloys. Many exotic alloys with large miscibility gaps have been grown by OMVPE [8]. Perhaps the most impressive is Ga(In)As:N. The solid solubility of N in GaAs at normal growth temperatures is miniscule [14] and yet alloys have been grown containing more than 5% N [15–18]. A number of Sb alloys with large miscibility gaps have also been produced [8]. In addition to the wide range of materials, OMVPE can also be used to grow the structured materials required for the most advanced devices, such as quantum wells, strain layer superlattices, quantum wires, and quantum dots [19].

In addition to these semiconductor materials and structures, OMVPE has also been used to produce other useful materials, including magnetic semiconductors [20]. The need for controlled growth of very thin, single-crystalline oxide materials has also led to the OMVPE growth of a wide range of dielectrics [21], ferroelectrics [22], electrodes [23], and superconductors [24].

2. Basic aspects of OMVPE

A schematic diagram of a typical OMVPE system for the growth of GaAs is shown in Fig. 2. Here the TMGa and AsH_3 precursors are carried to the growing surface in pure hydrogen in a horizontal reactor geometry. The basic elements of the growth process include precursor chemistry, thermodynamics, mass transport, hydrodynamics, and the nature and rates of gas phase and surface reactions. These topics are discussed in some detail in Refs. [8,25–27] so will be reviewed only very briefly here. However, the understanding and control of surface processes is advancing so rapidly that a more detailed review is included here to supplement these references.

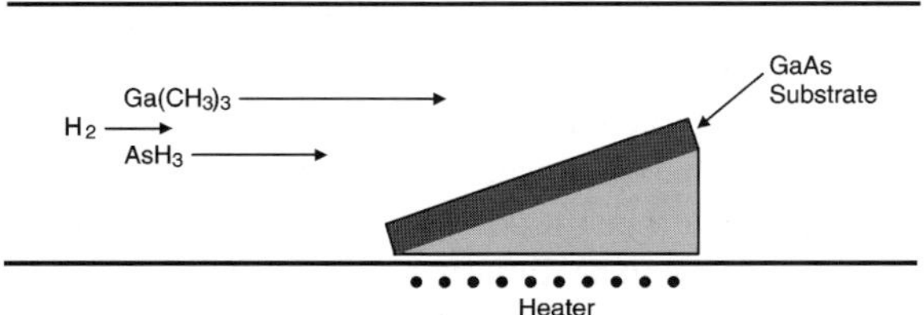

Fig. 2. Schematic diagram of OMVPE growth process for the production of GaAs from TMGa and arsine.

A very important and active research area has been the development of new precursor molecules especially designed for OMVPE. The early OMVPE experiments used Ga, In, As, and Sb compounds developed for other applications. While these precursors were satisfactory in many cases (we still use predominately TMGa, AsH_3 and PH_3 for OMVPE) the development of novel Al, In, As, P, N, and Sb precursors has been an area of important research [8]. As an example, the search for less hazardous As and P precursors led to the development of tertiarybutyl-As and -P compounds specifically for OMVPE. Novel Al precursors have also been developed in order to avoid troublesome C contamination problems. New Sb and N precursors have also played a role in the growth of, respectively, small and large bandgap semiconductors.

An understanding of the thermodynamic aspects of OMVPE has been an important guide at each step of the development of the technique. Of course, thermodynamics provides the basic driving force for all epitaxial growth processes [8]. In addition, solution thermodynamics is a powerful tool for the prediction and control of solid composition [28] and dopant incorporation [29]. An unexpectedly important application has been the use of surface thermodynamics to explain the microscopic arrangement of atoms in solid alloys. Bulk thermodynamics indicates that the solid microstructure should include clustering and phase separation for most III/V alloys; however, the experimental observations show ordering, which is never thermodynamically stable in the bulk [8,30]. Ordering is stabilized in the very top surface layers due to reconstruction of the surface during growth [8,30]. After growth, the thermodynamically unstable phase (in the bulk) is frozen in due to slow solid state diffusion coefficients.

As we have only recently realized, for these reasons, the chemical and physical structure of the surface during growth is an important determinant of the microstructure and the resulting properties of the solid. To date, this has been demonstrated for ordering, clustering, and phase separation, as will be discussed below. The surface structure has also been demonstrated to have a marked effect on solid composition and dopant incorporation [31–33]. In addition, it is anticipated, but largely unproven, that the surface structure will also have a role in determining such important properties of the solid stoichiometry, native defect concentration, and interface states at heterojunctions.

The kinetics of OMVPE reactions are extremely complex; thus, even today, our understanding is incomplete. Gas phase reactions include the pyrolysis reactions yielding the components of the epitaxial layer, as well as complex reactions involving adduct formation in the vapor, due to the Lewis acid and Lewis base natures of many of the respective group III and group V precursor molecules. As a further complication, the gas phase pyrolysis reactions are seldom complete, so heterogeneous pyrolysis reactions occurring on the growing surface often play a key role in the pyrolysis and growth reactions [8,25,27].

The reaction kinetics are closely linked to the hydrodynamic and mass transport aspects of the OMVPE growth process, which further complicates the analysis and understanding of these processes. First principles calculations are frequently used to help sort out these complex problems. This topic is treated in some detail in the literature [25] so will not be treated further here. Such calculations are often used as an aid in reactor design and are expected to become even more useful as we unravel the complexities of the homogeneous and heterogeneous chemical reactions occurring during deposition.

Since heterogeneous pyrolysis reactions are often an important part of the overall OMVPE growth process, it is expected that the chemical and physical state of the surface will have an important role. This is a topic that is somewhat neglected. Nevertheless, it is clear that surface reconstruction, as controlled by the temperature and gas phase composition as well as the presence of surfactants, as described below, will play an essential role in the overall growth process.

Clearly, the surface structure plays such an important role in the OMVPE growth process and the properties of the resulting epitaxial layers. Since this topic is perhaps the least understood and most rapidly advancing fundamental aspect of OMVPE, it will be reviewed in more detail in what follows.

The unreconstructed (0 0 1) surface of a diamond cubic or zincblende semiconductor has two dangling bonds per atom. This suggests that a reconstruction of the bonding at the surface would significantly lower the free energy. The tetragonal geometry of covalent sp^3 bonds on a group V rich surface, combined with the propensity of these atoms to form dimers and tetramers in the vapor, suggests the formation of dimer bonds on the surface. Generally reliable estimates of the surface bonding and reconstruction come from the so-called "electron counting" rule [34]. This has led to several proposed reconstructions having the (2×4) symmetry observed by electron diffraction during MBE growth [35]. The development of in situ tools for observing the surface during OMVPE growth has been much slower because a blanket of hydrogen or nitrogen is typically present over the growing surface which absorbs the electron beam.

The development of optical techniques such as reflection difference spectroscopy (RDS) [36] and surface photo absorption (SPA) [37] has allowed the clarification of the surface during OMVPE growth. The results of studies using these optical techniques indicate that the surface reconstruction during OMVPE growth is nearly the same as that determined during MBE growth [8,30]. This is reassuring, since it indicates that thermodynamics, i.e., the surface phase diagram, determines the surface structure during growth. The phase diagram specifies the equilibrium surface reconstruction as a function of extensive thermodynamic parameters, typically temperature and the group V partial pressure. This is extremely important since it allows us to use extensive literature exploring the surface structure obtained in UHV systems in our efforts to understand the surface processes occurring during OMVPE growth.

A dramatic effect of the surface reconstruction observed for III/V semiconductors grown by OMVPE relates to the microstructure of alloys, as indicated above. The DLP model predicts that the enthalpy of mixing of III/V alloys is always positive. This means that we expect the alloys to evidence clustering and phase separation and that ordering should not be observed [8,30]. However, TEM investigations of many III/V alloys indicate that ordered structures are formed spontaneously during OMVPE growth [8,30,38]. In particular, the CuPt structure, with ordering on the {1 1 1} planes, is observed in most III/V alloys, including GaInP. The formation of this ordered structure is extremely significant, because it has a direct effect on the bandgap energy. Bandgap differences as large as 160 meV between partially ordered and disordered materials have been reported for GaInP [39]. The order parameter can be directly linked to the surface SPA spectrum measured in situ during growth. The change in order parameter induced by changes in the temperature and the partial pressure of the P precursor during

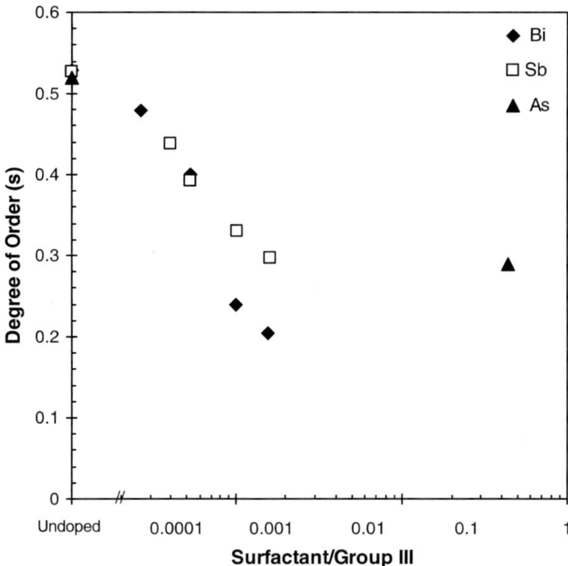

Fig. 3. Degree of order for GaInP layers grown by OMVPE plotted versus the surfactant/III ratio in the vapor. Data are for Bi (◆), Sb (□), and As (▲) (after Stringfellow et al. [41]).

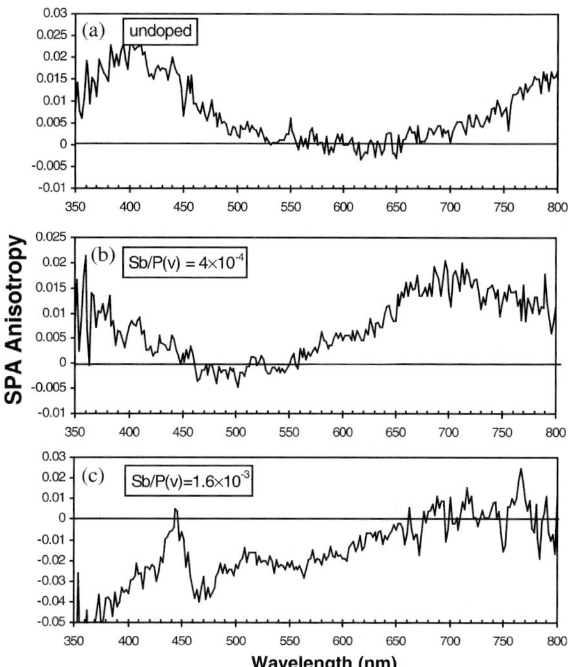

Fig. 4. SPA spectra for undoped GaInP and for samples grown with the addition of TESb with values of Sb/P in the vapor indicated. The SPA spectra were obtained in situ at 620 °C, the growth temperature (after Stringfellow et al. [41]).

growth is linearly related to the magnitude of the SPA signal at 405 nm due to the P dimers characteristic of the (2×4) surface [8,30,38].

A powerful tool for controlling the surface bonding and structure during OMVPE growth is the use of surfactants. Surfactants, in this context, are elements that accumulate at the surface during growth. For example, adding a small amount of an Sb precursor, such as TESb, during the OMVPE growth of GaInP results in the displacement of some surface P dimers by larger Sb dimers. This is indicated directly by the SPA spectra [38] supported by the results of first principles calculations [40]. The Sb is rejected from the solid due to its' large size (relative to P) and does not leave the surface rapidly by evaporation due to its' relatively low volatility.

The effect of a small concentration of the Sb precursor, TESb, on the degree of order of GaInP lattice matched to GaAs is shown in Fig. 3 [41]. The TESb partial pressure is normalized by the total group III precursor partial pressure, since both Sb and the group III elements are relatively non-volatile. The degree of CuPt order is clearly decreased as Sb is added to the surface. This is not a bulk effect, since the mole fraction of Sb incorporated into the solid, determined from SIMS analysis, is only approximately 5×10^{-5}

(or 10^{18} cm^{-3}) for an Sb/III ratio in the vapor of 2×10^{-2}.

The SPA anisotropy spectra are plotted for several Sb/P concentrations in the vapor in Fig. 4 [41]. Fig. 4(a) is for a sample grown without Sb and Fig. 4(b) is for a sample grown with an Sb/III ratio of 1.6×10^{-2}, which results in the growth of nearly disordered material. Correlation of the decrease in order parameter with the decrease in the magnitude of the SPA signal at 405 nm due to [1 1 0] P dimers [30, 38,42] indicates that the reduction in order parameter occurs due to the elimination of the P dimers, which are predicted to provide the driving force for CuPt ordering [8,30,38].

The SPA spectrum in Fig. 4(b) shows that as the intensity of the signal at 405 nm due to P dimers decreases, a peak grows at approximately 650 nm. This is most likely due to [$\bar{1}$ 1 0] Sb dimers [43]. Sb apparently accumulates at the surface, forming a $(2 \times n)$ type of reconstruction. The reduction in degree of order produced by Sb addition to the system is apparently due to the replacement of [$\bar{1}$ 1 0] P dimers by Sb dimers having the same orientation. The larger spacing of the Sb dimers gives a smaller amount of strain in the subsurface layers, resulting in a reduced thermodynamic driving force for CuPt ordering. This interpretation is verified by recent first principle calculations Refs. [40, 44]. For small Sb coverage of the surface, the lowest energy configuration is for Sb to substitute directly for P in the $\beta2(2 \times 4)$ structure.

The SPA spectrum at an even larger Sb/III ratio in the vapor of 6.4×10^{-2} is seen in Fig. 4(c) to be distinctly different, indicative of formation of a non-(2×4)-like structure [45]. TED patterns of the material produced using this TESb concentration indicate that the A variants of a triple-period ordered (TPO) structure are formed [45]. The first principles calculations of Wixom et al. [40] indicate that (4×3) or (2×3) reconstructions will form at higher Sb surface concentrations. This would stabilize the A variants of the TPO structure. This was the first report of the use of a surfactant to change the ordered structure by changing the surface reconstruction [46].

Even higher TESb concentrations added during the OMVPE growth of GaInP lead to a decrease in the low temperature PL peak energy [45,46]. In addition, the low-temperature PL emission is highly polarized [47]. TEM images indicate that the high Sb concentration used in the growth of these layers leads to a composition modulation in the [1 1 0] direction. This phenomenon is driven by the large enthalpy of mixing in the alloy system [8] and is found to occur only when the surface diffusion coefficient is sufficiently high to kinetically allow the early stages of spinodal decomposition to occur at the surface during growth [48]. It appears that high concentrations of surfactant Sb increase the group III adatom surface diffusion coefficients, leading to the compositional modulation that results in a decrease in the PL peak energy and the highly polarized PL [48].

The surfactant effect of Sb can be used to modulate the bandgap energy during growth by varying the TESb flow rate to produce heterostructures. An example is the growth of an undoped layer followed by a layer grown with the addition of a small concentration of TESb to reduce the degree of CuPt order. The TEM dark field cross-sectional images indicate an abrupt change in the order parameter when a 3-minute interruption at the interface is used to accumulate Sb on the surface [49]. The corresponding

20-K PL data clearly show the bandedge PL from both layers. The difference in bandgap energy is 135 meV. This technique has also been used to produce double heterostructures and quantum wells with well layers as thin as 6.7 nm [49,50].

From these results it is clear that a small concentration of TESb, added during OMVPE growth, can be used to modify the surface reconstruction. This leads to a marked change in the microstructure and, hence, the semiconducting properties of the solid. Other group V surfactants, isoelectronic with P, have similar effects. For As (from the pyrolysis of TEAs) rejection from the solid is much less than for Sb due to the decreased size difference relative to the host P [8, 51]. It is also more volatile than Sb. Thus, it is expected to have less of a surfactant effect. Indeed, at low ratios of TEAs to phosphine in the vapor, both PL and TEM analysis indicate that the layers are highly ordered. However, TEM results show that $(As/III)_v = 0.45$ produces a significant reduction in the order parameter, as shown in Fig. 3. SPA spectra show a clear decrease in intensity at 405 nm [46] indicating that, as for Sb, the decrease in CuPt ordering is due to displacement of the $[\bar{1}10]$ P dimers that drive the CuPt ordering process.

Bi is the largest of the surfactants isoelectronic with P studied and is, thus, much more difficult to incorporate into the solid [8,51]. It is also the least volatile of the group V surfactants studied. The order parameter deduced from the 20 K PL peak energy for GaInP layers lattice matched to GaAs grown with several ratios of Bi/III in the vapor are shown in Fig. 3. The addition of Bi results in a decrease in the order parameter similar to that seen for Sb [52]. This is supported by TEM results. The SPA spectrum is changed markedly when sufficient Bi is added to the system to cause disordering [52].

These results confirm that the group V elements larger than P (As, Sb, and Bi) all give reduced strain in the subsurface GaInP layers, leading to a reduction in the thermodynamic driving force for CuPt ordering. Another group V surfactant, N, is *smaller* than P and so has the potential to *increase* the subsurface strain, if, indeed, $[\bar{1}10]$ N dimers are formed on the surface. Sb and Bi are obvious choices as surfactants, since they are rejected from the solid and have low vapor pressures, so are expected to accumulate at the surface. N will also be rejected from the solid, as known from the results of previous thermodynamic calculations [14,51], but it is much more volatile than P. However, the As results indicate that even relatively volatile group V elements can be effective surfactants. For N, high partial pressures of a relatively labile precursor are required to obtain a significant N coverage of the surface. In fact, a change in surface reconstruction using N during MBE growth has been reported for GaAs [53]. This leads one to expect significant N surface coverages during the OMVPE growth of GaInP under suitable conditions, i.e., low temperatures and high N/P ratios in the vapor.

The experimental results obtained using DMHy as the N precursor at 620 °C on singular GaAs substrates with DMHy/TBP ratios as high as 0.8 indicate a clear decrease in order parameter [54]. In situ SPA results indicate a decrease in the 405 nm peak due to P dimers. The results were interpreted as indicating that N does, indeed, replace P on the surface. However, the decrease in order parameter may indicate that N dimers do not form. This may be due to the large strain energy required to form N dimers on the GaInP surface and is consistent with previous work of N on GaN surfaces, where N dimers are not formed [55] even though the GaN lattice constant is much smaller than that of GaInP.

Another striking effect of surfactants added during OMVPE growth is the change in incorporation coefficients of dopants and alloying elements. Surfactants isoelectronic with As have been demonstrated to significantly affect dopant incorporation in GaAs. Consider first the effect of surfactant Sb on the incorporation of Zn and In. Three layer Zn and In doped structures grown at a temperature of 620 °C with TESb added only in the middle layers show that addition of Sb leads to an increase in both the Zn and In concentrations, as indicated by the SIMS depth profiles shown in Fig. 5 [32]. For a small amount of TESb in the vapor (Sb/III = 0.012) the Zn concentration in the layer increases sharply to 8.5×10^{18} atoms/cm^3, a 60% increase. As can be seen, the Sb concentration in the layers is very small (2–3×10^{17} atoms/cm^3). Note that after the TESb is removed from the vapor, as indicated by a decrease in the Sb concentration in the epilayer, the Zn concentration decreases as well. The correlation between the change in the Zn and Sb concentrations in the layer clearly indicates that sur-

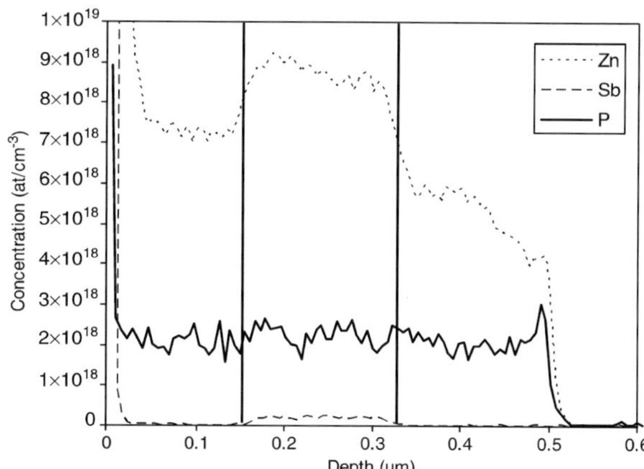

Fig. 5. SIMS depth profile of Zn doped GaAs epilayer grown with a Zn/III ratio in the vapor of 0.08. After 12 min of growth, a small amount of TESb was added to the system. After 12 min, the Sb was removed and GaAs:Zn was grown for an additional 12 min (after Shurtleff et al. [32]).

face Sb increases the incorporation of Zn in GaAs. The SIMS depth profile of a GaAs epilayer that was inadvertently doped with In shows a similar correlation between an increase in the In and the presence of Sb during growth [32]. As shown in Fig. 5, the concentration of P inadvertently present in the GaAs epilayers was also measured. Apparently, Sb has little affect on the concentration of P, which is incorporated on group V sites. The In and the P in the system came from memory effects associated with the growth of GaInP in the same system. The results were interpreted in terms of either an Sb-induced increased group III adatom surface diffusion coefficient or an increase in the group III sticking coefficient at the step edge induced by Sb. Either would cause an increase in In and Zn incorporation into the GaAs, but would have no affect on P incorporation [32].

Materials with bandgap energies of less than 1.4 eV grown lattice matched on GaAs are of great interest for devices [56,57]. The major problem with these alloys is the small equilibrium solubility of N in GaAs: the calculated thermodynamic solubility is only approximately 10^{14} atoms per cm^3 at typical growth temperatures [14]. Still crystals with up to 5% N in GaAs have been grown by OMVPE [58], molecular beam epitaxy [59], metalorganic-MBE [60], and chemical beam epitaxy [61]. Low growth temperatures, below 600 °C, and small V/III ratios were used in the case of OMVPE growth to kinetically limit phase separation.

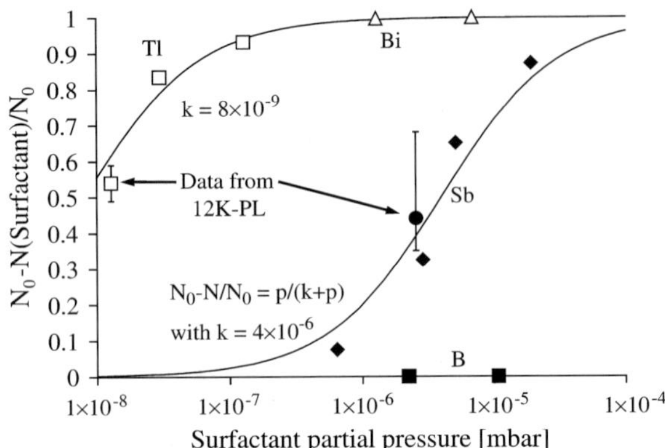

Fig. 6. Influence of the Tl, Bi, Sb, and B precursor partial pressure on the N content of GaAs:N layers grown using similar conditions. The data are plotted as the N concentration without surfactant (No) minus that with surfactant (N), normalized by the N content without surfactant. Data from SIMS analysis and 12 K PL are shown. The curves correspond to the best fit of the data to the Langmuir competitive adsorption model (after Dimroth et al. [33]).

Under these conditions, kinetic effects at the growth surface play a dominant role. An important question is the influence of surfactants like Sb or Bi on phase separation in metastable alloys, in this case, the N content of GaAs:N.

Dimroth et al. [33] used SIMS and PL to study the effects of Sb, Tl, Bi, and B on N incorporation in GaAs grown by OMVPE. Fig. 6 shows the influence of TESb, TMBi, $(C_5H_5)_2$Tl, and B_2H_6 partial pressure on N incorporation into GaAs:N, for growth under otherwise identical conditions. The data are normalized to the N content of GaAs:N grown without surfactant addition. Tl, Bi, and Sb were all found to decrease the incorporation of N in GaAs. B has no observable influence. This behavior was explained using the Langmuir model where the reduced N incorporation results from a competition of N and the surfactant atom for the same surface sites.

3. Future directions

Even though OMVPE is now the dominant commercial technique for the production of III/V materials for electronic and photonic devices, our understanding of the technique remains far from complete. One example is our incomplete understanding of the surface processes occurring during OMVPE growth.

Fundamental research is still required to more fully understand the technique so that the process can be improved, with an improved degree of control. Unfortunately, with the commercial success of the OMVPE technique has come an emphasis on development activities related to specific products. Long-range, fundamental research has nearly ceased in industry, due to the need to produce immediate profits. Furthermore, governmentally supported research was decreased because such work is deemed the realm of industry because of the extensive and successful commercial activities. This leads to a dearth of the basic studies that will certainly payoff in the long term.

A few areas of basic research anticipated to lead to important new applications include: growth on dissimilar substrates and the growth of novel (for OMVPE) materials such as metastable materials, magnetic semiconductors, and oxides. Each area will be discussed very briefly below.

One of the important constraints to the production of new and useful semiconductor materials is the need for substrates having the same crystal structure, lattice constant, and thermal expansion coefficient as the epitaxial layers without which highly defected layers are produced. A good example is the epitaxial growth of GaN and related alloys. The lack of bulk GaN substrates (due to the extreme difficulty in producing this material) has led to the use of sapphire substrates. This presented major difficulties in epitaxial growth, including the inclusion of very high (10^{10} cm^{-3}) dislocation densities, stacking faults, twins, and other defects in the layers. In spite of these defects, high performance blue and green LEDs have been produced [11,12]. However, the defects resulted in poor performance for injection laser devices [12]. A reduction in defect density was required. This led to the adoption of lateral overgrowth schemes such as ELO [62] and pendeoepitaxy [63], following earlier work on the use of similar techniques for the growth of GaAs on Si substrates [64]. Clearly, new approaches to this problem are required. One example of a recent approach was the use of so-called "compliant" substrates [65]. The ability to grow high-quality semiconductor materials on a wide range of substrates would clearly enable the fabrication of a number of improved devices.

The growth of SiO$_2$ layers has been a key to the success of the integrated circuit industry since the thermal growth of oxides was out-dated. However, as feature sizes diminish, processes for the epitaxial growth of SiO_2 are coming under intense pressure since thinner oxides, down to thicknesses of a few atomic layers, are required for the laboratory scale circuits in development today. The growth of uniform SiO_2 with these dimensions is a tall order. One way of alleviating this problem is to develop higher dielectric constant epitaxial oxides. The requirements on techniques for deposition of these oxides include that ability to deposit on large areas with good compositional and thickness control combined with the ability to give excellent conformal step coverage on complex surfaces. It is also likely that ferroelectric oxides will be used for the fabrication of non-volatile memories. Other applications of epitaxial oxides include piezoelectric, ferromagnetic, and non-linear optical materials. In addition, it seems likely that, similar to advanced semiconductors for devices, nanostructured oxides will give enhanced performance. Early studies have used MBE [66]. However, MOCVD is also an attractive process for many applications [67], as described above.

4. Conclusions

OMVPE has moved from the early patents and the first papers of Manasevit and co-workers to become the dominant technique for the growth of compound semiconductor materials in the relatively short time span of 35 years. Today, commercial reactors and foundary services are available for the production of materials and structures for photonic devices such as LEDs, injection lasers, detectors, and solar cells, and the electronic switching devices and integrated circuits. The success of this growth technique is mainly a result of its extreme flexibility. OMVPE can be used for the growth of essentially all III/V compounds and alloys, including metastable alloys. It can also be used for fabrication of the most advanced structures, such as quantum wells, wires, and dots.

Today, one of the areas on the forefront of OMVPE research is the use of surfactants to control surface thermodynamics and kinetics during growth. Striking effects have been demonstrated. CuPt ordering occurring in GaInP alloys is markedly reduced when a small amount of an isoelectronic, group V surfactant is added to the system, resulting in an increase in the bandgap energy by as much as 135 meV. N, As, Sb, and Bi are all found to result in the growth of nearly disordered materials for conditions producing highly ordered materials in the absence of surfactants. Surfactants have also been reported to change the ordered structure formed during growth and to affect phase separation in alloys, via conductivity modulation. Surfactants are also observed to alter the incorporation of elements at dopant concentrations into GaAs during OMVPE growth. Sb is found to *increase* the incorporation of Zn and In. The surfactants Sb, Bi, and Tl were all found to *reduce* the incorporation of N into GaAs. The ability to control major semiconductor properties, such as conductivity type and solid composition by simply adding a small amount of surfactant during OMVPE growth, may profoundly affect the manufacture of many important semiconductor devices.

Other changes in the growth process and properties will undoubtedly be discovered as research on this topic continues. It is expected that this will lead to a new mechanism for controlling the OMVPE growth process for the production of semiconductor materials and structures for advanced devices.

Future basic OMVPE research should be devoted not only to the use of surfactants, but also to the growth of metastable semiconductor alloys, magnetic semiconductors, and thin oxide films. Efforts to develop techniques for growth on dissimilar substrates is also expected to yield major technical advances.

Acknowledgements

The author wishes to thank the Department of Energy for support of the research presented in this paper.

References

[1] T.R. Scott, G. King, J.M. Wilson, UK Patent 778, 383.8.
[2] W. Miederer, G. Ziegler, R. Dotzer, US Patent 3,226,270;
   W. Miederer, G. Ziegler, R. Dotzer, German Patent 1,176,102.
[3] H.M. Manasevit, Appl. Phys. Lett. 116 (1969) 1725;
   H.M. Manasevit, W.I. Simpson, J. Electrochem. Soc. 12 (1968) 156;
   H.M. Manasevit, J. Crystal Growth 13/14 (1972) 306.
[4] Y. Seki, K. Tanno, K. Iida, E. Ichiki, J. Electrochem. Soc. 122 (1975) 1108.

[5] R.D. Dupuis, P.D. Dapkus, Appl. Phys. Lett. 32 (1978) 406.

[6] G.B. Stringfellow, Rep. Prog. Phys. 45 (1982) 469;
G. Beuchet, in: W.T. Tsang (Ed.), Semiconductors and Semi-metals, vol. 22A, Academic Press, Orlando, 1985, pp. 261–298.

[7] G.B. Stringfellow, Annu. Rev. Mater. Sci. 8 (1978) 73.

[8] G.B. Stringfellow, Organometallic Vapor Phase Epitaxy: Theory and Practice, second ed., Academic Press, Boston, 1999 (Section 8.5.3).

[9] G.B. Stringfellow, M.G. Craford, High Brightness Light Emitting Diodes, Academic Press, Boston, 1997 (Chapters 1, 2, 4, 5).

[10] T. Takamoto, E. Ikeda, H. Kurita, M. Ohmori, Appl. Phys. Lett. 70 (1997) 381.

[11] G.B. Stringfellow, M.G. Craford, High Brightness Light Emitting Diodes, Academic Press, Boston, 1997 (Chapters 7, 8).

[12] S. Nakamura, G. Fasol, The Blue Laser Diode, Springer, Berlin, 1997.

[13] R.L. Moon, V. Robbins, AACG Newsletter 29 (2001) 7.

[14] I.H. Ho, G.B. Stringfellow, Appl. Phys. Lett. 69 (1996) 2701.

[15] D.J. Friedman, A.G. Norman, J.F. Geisz, S. Kurtz, J. Crystal Growth 208 (2000) 11.

[16] J.C. Harmand, G. Ungaro, L. Largeau, G.L. Roux, Appl. Phys. Lett. 77 (2000) 2482.

[17] C. Jin, Y. Qiu, S.A. Nikishin, H. Temkin, Appl. Phys. Lett. 74 (1999) 3516.

[18] T. Miyamoto, T. Kageyama, S. Makino, D. Schlenker, F. Koyama, K. Iga, J. Crystal Growth 209 (2000) 339.

[19] W. Seifert, N. Carlsson, J. Johansson, M.E. Pistol, L. Samuelson, J. Crystal Growth 170 (1997) 39.

[20] A.J. Blattner, J. Lensch, B.W. Wessels, J. Electron. Mater. 30 (2001) 1408.

[21] Y. Ohshita, A. Ogura, A. Hoshino, T. Suzuki, S. Hiiro, H. Machida, J. Crystal Growth 235 (2002) 365;
Y. Ohshita, A. Ogura, A. Hoshino, S. Hiiro, H. Machida, J. Crystal Growth 233 (2001) 292.

[22] T. Watanabe, K. Saito, H. Funakubo, J. Crystal Growth 235 (2002) 389;
A. Li, D. Wu, H. Ling, M. Wang, Z. Liu, N. Ming, J. Crystal Growth 235 (2002) 394.

[23] K. Froehlich, D. Machajdik, V. Cambel, I. Kostic, S. Pignard, J. Crystal Growth 235 (2002) 377.

[24] Y. Ma, K. Watanabe, S. Awaji, M. Motokawa, J. Crystal Growth 233 (2001) 483.

[25] K.F. Jensen, in: D.T.J. Hurle (Ed.), Handbook of Crystal Growth, vol. 3b, Elsevier, Amsterdam, 1994 (Chapter 13).

[26] D.W. Kisker, T.F. Kuech, in: D.T.J. Hurle (Ed.), Handbook of Crystal Growth, vol. 3b, Elsevier, Amsterdam, 1994 (Chapter 3).

[27] G.B. Stringfellow, in: D.T.J. Hurle (Ed.), Handbook of Crystal Growth, vol. 3b, Elsevier, Amsterdam, 1994 (Chapter 12).

[28] G.B. Stringfellow, J. Crystal Growth 62 (1983) 225.

[29] G.B. Stringfellow, J. Crystal Growth 75 (1986) 91.

[30] G.B. Stringfellow, in: M. Santos, W.K. Liu (Eds.), Thin Films: Heteroepitaxial Systems, World Scientific, Singapore, 1998, pp. 64–116.

[31] M. Kondo, T. Tanahashi, J. Crystal Growth 145 (1994) 390.

[32] J.K. Shurtleff, S.W. Jun, G.B. Stringfellow, Appl. Phys. Lett. 78 (2001) 3038.

[33] F. Dimroth, A. Howard, J.K. Shurtleff, G.B. Stringfellow, J. Appl. Phys. 91 (2002) 3687.

[34] M.D. Pashley, Phys. Rev. B 40 (1989) 10481.

[35] H.H. Parrel, C.J. Palmstrom, J. Vac. Sci. Technol. B 8 (1990) 903.

[36] D.E. Aspnes, et al., J. Crystal Growth 120 (1992) 71.

[37] N. Kobayashi, Y. Kobayashi, K. Uwai, J. Crystal Growth 174 (1997) 544.

[38] G.B. Stringfellow, in: A. Mascarenhas (Ed.), Spontaneous Ordering in Semiconductor Alloys, Kluwer Academic/Plenum, New York, 2002 (Chapter 3).

[39] L.C. Su, I.H. Ho, N. Kobayashi, G.B. Stringfellow, J. Crystal Growth 145 (1994) 140.

[40] R.R. Wixom, G.B. Stringfellow, N.A. Modine, Phys. Rev. B 64 (2001) 201322.

[41] G.B. Stringfellow, J.K. Shurtleff, R.T. Lee, C.M. Fetzer, S.W. Jun, J. Crystal Growth 221 (2000) 1.

[42] H. Murata, I.H. Ho, L.C. Su, Y. Hosokawa, G.B. Stringfellow, J. Appl. Phys. 79 (1996) 6895.

[43] W.A. Harrison, Electronic Structure and the Properties of Solids: Physics of the Chemical Bond, Freeman, New York, 1980.

[44] R.R. Wixom, G.B. Stringfellow, N.A. Modine, Phys. Rev. B 67 (2003) 115309.

[45] C.M. Fetzer, R.T. Lee, J.K. Shurtleff, G.B. Stringfellow, S.M. Lee, T.Y. Seong, Appl. Phys. Lett. 76 (2000) 1440.

[46] G.B. Stringfellow, C.M. Fetzer, R.T. Lee, S.W. Jun, J.K. Shurtleff, in: 583 MRS Symposium Proceedings, 2000, p. 261.

[47] C.M. Fetzer, R.T. Lee, S.W. Jun, G.B. Stringfellow, S.M. Lee, T.Y. Seong, J. Appl. Phys. 78 (2001) 1376.

[48] C.M. Fetzer, R.T. Lee, D.C. Chapman, G.B. Stringfellow, J. Appl. Phys. 90 (2001) 1040.

[49] R.T. Lee, J.K. Shurtleff, C.M. Fetzer, G.B. Stringfellow, S. Lee, T.Y. Seong, J. Crystal Growth 234 (2002) 327.

[50] J.K. Shurtleff, R.T. Lee, C.M. Fetzer, G.B. Stringfellow, Appl. Phys. Lett. 75 (1999) 1914;
G.B. Stringfellow, R.T. Lee, C.M. Fetzer, J.K. Shurtleff, Yu. Hsu, S.W. Jun, S. Lee, T.Y. Seong, J. Electron. Mater. 29 (2000) 134.

[51] G.B. Stringfellow, J. Crystal Growth 27 (1974) 21.

[52] S.W. Jun, R.T. Lee, C.M. Fetzer, J.K. Shurtleff, G.B. Stringfellow, C.S. Choi, T.Y. Seong, J. Appl. Phys. 88 (2000) 4429.

[53] J. Lu, D.I. Westwood, L. Haworth, P. Hill, J.E. Macdonald, Thin Solid Films 343–344 (1999) 567.

[54] D. Chapman, G.B. Stringfellow, unpublished results.

[55] A.D. Bykhovski, M.S. Shur, Appl. Phys. Lett. 69 (1996) 2397.

[56] D.J. Friedman, J.F. Geisz, S.R. Kurtz, J.M. Olson, J. Crystal Growth 195 (1998) 409;
S.R. Kurtz, A.A. Allerman, E.D. Jones, J.M. Gee, J.J. Banas, B.E. Hammons, Appl. Phys. Lett. 74 (1999) 729.

[57] X. Yang, J.B. Heroux, M.J. Jurkovic, W.I. Wang, Appl. Phys. Lett. 76 (2000) 795;
S. Sato, S. Satoh, J. Crystal Growth 192 (1998) 381.

[58] D.J. Friedman, A.G. Norman, J.F. Geisz, S. Kurtz, J. Crystal Growth 208 (2000) 11.

[59] J.C. Harmand, G. Ungaro, L. Largeau, G.L. Roux, Appl. Phys. Lett. 77 (2000) 2482.

[60] C. Jin, Y. Qiu, S.A. Nikishin, H. Temkin, Appl. Phys. Lett. 74 (1999) 3516.

[61] T. Miyamoto, T. Kageyama, S. Makino, D. Schlenker, F. Koyama, K. Iga, J. Crystal Growth 209 (2000) 339.

[62] S. Nakamura, et al., J. Crystal Growth 189/190 (1998) 820.

[63] K.J. Linthincum, T. Gehrke, D.B. Thompson, E.P. Carlson, P. Rajagopal, T.P. Smith, A.D. Batchelor, R.F. Davis, Appl. Phys. Lett. 75 (1999) 196.

[64] T. Nishinaga, T. Nakano, S. Zhan, Jpn. J. Appl. Phys. 27 (1988) L964.

[65] F.E. Ejeckam, Y.H. Lo, S. Subramanian, H.Q. Hou, B.E. Hammons, Appl. Phys. Lett. 70 (1997) 1685.

[66] D.G. Schlom, J.H. Haeni, J. Lettieri, C.D. Theis, W. Tian, J.C. Jiang, X.Q. Pan, Mater. Sci. Eng. B 87 (2002) 282.

[67] P.A. Williams, A.C. Jones, P.J. Wright, M.J. Crosbie, J.F. Bickley, A. Steiner, H.O. Davies, T.J. Leedham, Chem. Vapor Deposition 8 (2002) 110.

Subject index

Author index

Key to illustrations on back cover[*]

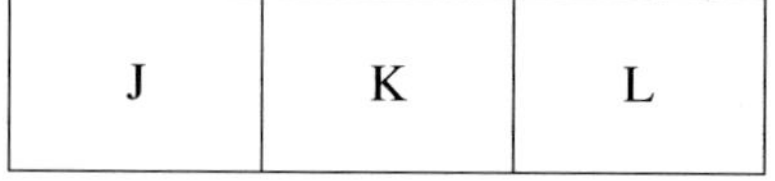

| J | K | L |

Plate J: A computer simulation of flows in a KDP aqueous growth system. The dark blue pyramid in the center represents the growing crystal (along the vertical axis). The cross bars are part of the seed rod assembly and the black traces and arrows indicate the flow patterns. *(Courtesy Jeff Derby and Andrew Yeckel, University of Minnesota.)*

Plate K: Multifaceted quartz crystal grown by the hydrothermal method. *(Courtesy Sawyer Research Products.)*

Plate L: A view of a GaAs crystal being grown by the Liquid Encapsulated Czochralski (LEC) method at the Royal Signals and Radar Establishment. The dark conical section of the boule attached to the seed rod is the part of the crystal emerging from the surface of the molten boron oxide encapsulant. The lighter dumbbell shaped region just below the conical section is the crystal within the transparent layer and whose bottom is touching the surface of the GaAs melt. *(Courtesy Brian Mullin.)*

---

[*] See p. ii in preliminary pages for key to illustrations on front cover.